Advances in Surface Engineering and Biocompatible Coatings for Biomedical Applications, 2nd Edition

Advances in Surface Engineering and Biocompatible Coatings for Biomedical Applications, 2nd Edition

Guest Editors

Egemen Avcu
Mert Guney
Yasemin Yıldıran Avcu

MDPI

Basel • Beijing • Wuhan • Barcelona • Belgrade • Novi Sad • Cluj • Manchester

Guest Editors

Egemen Avcu
Department of Mechanical Engineering
Kocaeli University
Kocaeli
Turkey

Mert Guney
Department of Civil and Environmental Engineering
Nazarbayev University
Astana
Kazakhstan

Yasemin Yıldıran Avcu
Department of Mechanical Engineering
Kocaeli University
Kocaeli
Turkey

Editorial Office

MDPI AG
Grosspeteranlage 5
4052 Basel, Switzerland

This is a reprint of the Special Issue, published open access by the journal *Coatings* (ISSN 2079-6412), freely accessible at: https://www.mdpi.com/journal/coatings/special_issues/Z9777155JV.

For citation purposes, cite each article independently as indicated on the article page online and as indicated below:

Lastname, A.A.; Lastname, B.B. Article Title. *Journal Name* **Year**, *Volume Number*, Page Range.

ISBN 978-3-7258-8287-8 (Hbk)
ISBN 978-3-7258-8288-5 (PDF)
https://doi.org/10.3390/books978-3-7258-8288-5

Contents

About the Editors

Egemen Avcu

Egemen Avcu is affiliated with the Department of Mechanical Engineering at Kocaeli University, Turkey, and is currently a visiting researcher at Chemnitz University of Technology, Germany. His research focuses on surface engineering, mechanical surface treatments, metallic biomaterials, metal matrix composites, tribology, corrosion, and the mechanical behavior of advanced engineering materials. His work particularly addresses the relationships between processing, microstructure, surface integrity, and functional performance in biomedical and structural materials.

Mert Guney

Mert Guney is affiliated with the Department of Civil and Environmental Engineering and the Environment and Resource Efficiency Cluster at Nazarbayev University, Kazakhstan. His research focuses on surface engineering, materials characterization, bioavailability, risk assessment, sustainability, and the environmental performance of engineered materials. His work particularly addresses the interactions between materials, environmental exposure, human health, and resource-efficient engineering applications.

Yasemin Yıldıran Avcu

Yasemin Yıldıran Avcu is affiliated with the Department of Mechanical Engineering at Kocaeli University, Turkey, and is currently a visiting researcher at Chemnitz University of Technology, Germany. Her research focuses on mechanical surface treatments, tribology, biomaterials, coatings, and the surface modification of engineering materials. Her work particularly addresses the relationships between surface integrity, microstructure, mechanical behavior, tribological performance, and biomedical functionality.

Preface

This reprint brings together recent advances in surface engineering and biocompatible coatings for biomedical applications, with a particular focus on the modification of material surfaces to improve biological, mechanical, tribological, and corrosion-related performance. The contributions included in this reprint address a broad range of approaches, including mechanical and physical surface treatments, chemical and electrochemical modification strategies, biocompatible coating systems, and surface-related degradation mechanisms of biomaterials.

This reprint arises from the increasing importance of surface functionality in the design and long-term performance of biomedical materials. Since the surface is the first region to interact with the biological environment, careful control of the surface chemistry, topography, wettability, mechanical response, and degradation behavior is essential for improving implant performance, reducing failure risks, and supporting the development of more reliable biomedical devices.

This reprint is intended for researchers, graduate students, engineers, and clinicians working in the fields of surface engineering, biomaterials, biomedical coatings, tribology, corrosion, and materials characterization. By collating this Special Issue, we aim to provide a useful reference for readers interested in current strategies and future directions in the surface modification of biomedical materials.

We would also like to take this opportunity to acknowledge Dr. Mustafa Armağan, Dr. Eray Abakay, and Dr. Berzah Yavuzyeğit for their valuable support during the preparation of this Special Issue and the process of inviting submissions.

Egemen Avcu, Mert Guney, and Yasemin Yıldıran Avcu
Guest Editors

coatings

MDPI

Editorial

Advances in Surface Engineering and Biocompatible Coatings for Biomedical Applications

Egemen Avcu [1,2,3,*], Yasemin Yıldıran Avcu [1,3], Berzah Yavuzyegit [4,5] and Mert Guney [6,*]

1 Department of Mechanical Engineering, Kocaeli University, Kocaeli 41001, Türkiye; yasemin.yildiran@kocaeli.edu.tr
2 Ford Otosan Ihsaniye Automotive Vocational School, Kocaeli University, Kocaeli 41650, Türkiye
3 Institute of Materials Science and Engineering, Chemnitz University of Technology, Erfenschlager Str. 73, 09125 Chemnitz, Germany
4 Department of Mechanical Engineering, Recep Tayyip Erdogan University, Rize 53100, Türkiye; berzah.yavuzyegit@erdogan.edu.tr or berzah.yavuzyegit@port.ac.uk
5 Faculty of Science and Health, Pharmacy and Biomedical Sciences, School of Medicine, University of Portsmouth, St. Michael's Building, White Swan Road, Portsmouth PO1 2DT, UK
6 Department of Civil and Environmental Engineering, School of Engineering and Digital Sciences, The Environment & Resource Efficiency Cluster (EREC), Nazarbayev University, Astana 010000, Kazakhstan
* Correspondence: egemen.avcu@kocaeli.edu.tr (E.A.); mert.guney@nu.edu.kz (M.G.)

1. Introduction

The two-volume Special Issue *"Advances in Surface Engineering and Biocompatible Coatings for Biomedical Applications"* has brought together a comprehensive collection of 26 peer-reviewed articles, reflecting the rapid scientific progress and sustained global interest in advanced surface modification strategies for biomaterials. Across both volumes, the contributions have accumulated over 73,000 views and over 170 citations in less than two years, a clear indication of the strong and growing impact of this collection within the biomaterials and surface engineering community.

Volume I established the scientific foundation of the collection by presenting advances on innovative surface treatments, antibacterial and bioactive coatings, and tribological and corrosion analyses, including two highly cited review articles that helped shape the broader landscape of biomaterial surface modification. Building on this momentum, Volume II expanded the scope with new contributions including on micro-arc oxidation systems, hybrid ceramic–polymeric coating architectures, multifunctional antibacterial surfaces, corrosion and wear mechanisms in biomedical alloys, and advanced in vitro and in vivo assessment models. Together, these two volumes reveal the evolution of the field from more traditional single-purpose coatings toward more innovative multifunctional and mechanistically engineered surface systems designed to address complex clinical challenges such as osseointegration, infection control, biodegradation, and long-term implant reliability.

The overarching premise of this Special Issue is that while the bulk material provides structural support, it is ultimately the surface and its chemistry, topography, charge, energy, and biological interactions that govern the clinical performance of implants. By presenting advances in physical, mechanical, chemical, electrochemical, and bioactive surface engineering, this two-volume collection offers an integrated, forward-looking perspective on emerging concepts, methodologies, and design principles in biomaterial surface science.

To provide a coherent and structured overview of these contributions, the articles published across both volumes have been thematically classified according to their underlying surface engineering strategies and targeted biological or functional outcomes.

 https://doi.org/10.3390/coatings15121475

The subsequent sections of this editorial highlight five major scientific themes that have emerged from this collection: (i) mechanical and physical surface treatments; (ii) bioactive and ceramic-based coatings; (iii) antibacterial and multifunctional hybrid surfaces; (iv) biological response, biomechanics, and implant–tissue interface studies; and (v) high-impact review papers defining the state of the art. The distribution of the 23 original research articles across the four scientific categories is visually summarized in Figure 1, while Table 1 provides a detailed mapping of only the research articles, categorized by material system, surface modification technique, and biomedical focus. Review papers are discussed separately in Section 2.5 and are not included in Figure 1 and Table 1. This structured synthesis sets the stage for a critical discussion of the key advances, mechanistic insights, and emerging research directions identified throughout the Special Issue.

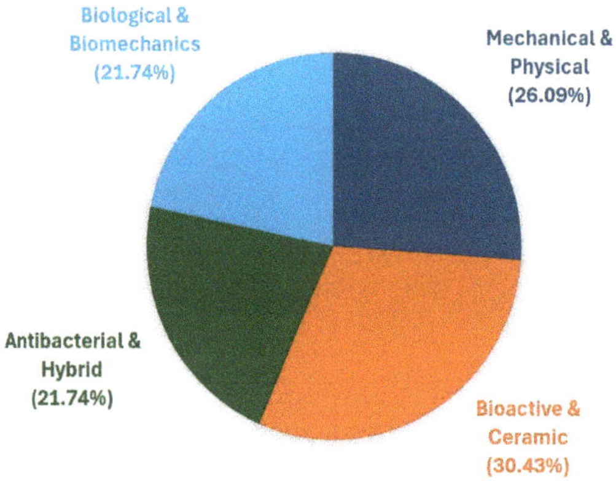

Figure 1. Distribution of the 23 original research articles published across the two-volume Special Issue according to four scientific categories: (i) mechanical and physical surface treatments (n = 6); (ii) bioactive and ceramic-based coatings (n = 7); (iii) antibacterial and multifunctional hybrid surfaces (n = 5); and (iv) biological response, biomechanics, and implant–tissue interface studies (n = 5).

Table 1. Research articles included in this Special Issue (Volume I and II) categorized by material system, surface modification technique, and biomedical function.

Ref.	Full Title	Authors	Material/Substrate	Technique	Biomedical Focus
i. Mechanical and Physical Surface Treatments					
[1]	Impact of Sterilization on the Adhesion Properties of a Polyamide 11 Coating on Textured Metal Substrates	Badji et al.	PA11 on textured metal	Laser texturing + steam sterilization	Adhesion, coating durability
[2]	Corrosion Behavior of Shot-Peened Ti6Al4V Alloy Produced via Pressure-Assisted Sintering	Avcu et al.	Ti6Al4V (PM)	Shot peening	Corrosion, microstructure
[3]	The Surface Modification of ZrO2 Film by Zr/Nb Ion Implantation and First-Principles Calculation	Gao et al.	ZrO_2	Ion implantation	Mechanical strength, wear
[4]	Surface and Tribological Properties of Powder Metallurgical Cp-Ti Titanium Alloy Modified by Shot Peening	Yıldıran Avcu et al.	Cp-Ti (PM)	Shot peening	Hardness, wear, deformation
[5]	Comparative Effects of Fine and Conventional Shot Peening on Surface Morphology, Topography, Wettability, and Antibacterial Activity of Biomedical Ti6Al4V Alloy	Avcu et al.	Ti6Al4V	Fine SP vs. conventional SP	Wettability, antibacterial
[6]	Enhancing Passivation Behaviors and Wear Resistance of Biomedical Ti-15Mo Alloy via {332} Twinning Pre-Tension and Aging	Zhang et al.	Ti–15Mo	Pre-tension + aging	Wear + corrosion resistance

Table 1. *Cont.*

Ref.	Full Title	Authors	Material/Substrate	Technique	Biomedical Focus
ii. Bioactive and Ceramic-Based Coatings					
[7]	Calcium Hydroxyapatite Coatings: Low-Temperature Synthesis and Investigation of Antibacterial Properties	Lukaviciute et al.	Ti	Sol–gel + $CaCO_3 \rightarrow$ HA	Antibacterial HA
[8]	Electrophoretic Deposition of ZnO-Containing Bioactive Glass Coatings on AISI 316L Stainless Steel for Biomedical Applications	Heidari Laybidi et al.	Stainless steel	EPD (BG + ZnO)	Bioactivity, corrosion
[9]	Antibacterial Calcium Phosphate Coatings with Zinc Oxide Nanoparticles	Chebodaeva et al.	Ti	MAO + CaP + ZnO	Antibacterial CaP
[10]	Designing Gallium-Containing Hydroxyapatite Coatings on Low Modulus Beta Ti–45Nb Alloy	Vishnu et al.	Ti–45Nb	Ga-doped HA	Bioactivity + antibacterial
[11]	Preparation of Superhydrophobic Hydroxyapatite Coating on AZ31 Mg Alloy by Combining Micro-Arc Oxidation and Liquid-Phase Deposition	Hu et al.	AZ31 Mg	MAO + LPD	Corrosion, osseointegration
[12]	Corrosion Resistance and In Vitro Biological Properties of TiO_2 on MAO-Coated AZ31 Magnesium Alloy via ALD	Han et al.	AZ31 Mg	MAO + ALD TiO_2	Corrosion + biocompatibility
[13]	Influence of TiO_2 Nanoparticle Concentration on Micro-Arc Oxidized Calcium–Phosphate Coatings: Corrosion Resistance and Biological Response	Zhassulan et al.	Ti	MAO + TiO_2	Corrosion + cell response
iii. Antibacterial and Multifunctional Hybrid Surfaces					
[14]	Effect of Carrier Materials for Active Silver in Antibacterial Powder Coatings	Zhang et al.	Powder coatings	Ag nanoparticle carriers	Antibacterial
[15]	Effect of Electrodeposited Gold Coatings on Micro-Gaps, Surface Profile and Bacterial Leakage of Cast UCLA Abutments Attached to External Hexagon Dental Implants	Walton	Dental implants	Electrodeposited Au	Microgap sealing, bacterial leakage
[16]	Nanostructured Coatings for Spinal Fixation Screws: A Dual-Function Approach Against Biofilm Formation and Implant Failure	Panaitescu et al.	Stainless steel screws	Antibiofilm nanocoating	Antibiofilm, failure prevention
[17]	Silver Nanoparticles–Chitosan Nanocomposites as Protective Coatings for Dental Remineralization Treatment: An In Vitro Study	Aguirre et al.	Dental tissue	Ag–chitosan nanocomposites	Antibacterial + remineralization
[18]	Impact of Diamond-like Carbon Films on Reverse Torque: Superior Performance in Implant Abutments with Internal Conical Connections	Coelho et al.	Implant abutments	DLC coating	Friction, wear, stability
iv. Biological Response, Biomechanics, and Tissue–Implant Interface					
[19]	Biomechanical Behavior of Different Miniplate Designs for Skeletal Anchorage in the Anterior Open Bite Treatment	Ielpo et al.	Ti miniplates	Finite element analysis	Stress distribution
[20]	Potential of Cranberry to Stimulate Osteogenesis: An In Vitro Study	Bauer et al.	Cell culture	Phytochemical extract	Osteogenesis enhancement
[21]	Evaluation of Bone–Implant Interface: Effects of Angiotensin II Receptor Blockade in Hypertensive Rats	Mulinari-Santos et al.	Ti implants (in vivo)	Drug–implant interface	Osseointegration
[22]	Ultrasound-Assisted Acellular Spinal Cord Scaffold for Spinal Cord Injury Treatment	Deng et al.	Acellular scaffold	EDC-crosslinking + VEGF	SCI regeneration
[23]	Natural Selection on Hydroxyapatite Fiber Orientations for Resisting Damage of Enamel	Shen et al.	Enamel	Microstructural analysis	Wear/damage tolerance

2. Thematic Overview and Scientific Trends

The research articles included in this Special Issue span a spectrum of surface engineering approaches aimed at improving the biological, mechanical, and functional performance of biomaterials. For clarity and coherence, the content of this Special Issue has been grouped into four major scientific themes: (i) mechanical and physical surface treatments; (ii) bioactive and ceramic-based coatings; (iii) antibacterial and multifunctional hybrid surfaces; and (iv) biological response, biomechanics, and implant–tissue interface studies. Each theme represents a distinct yet interconnected dimension of modern surface engineering, reflecting ongoing efforts to address challenges such as osseointegration, infection control, degradation, wear, and long-term implant stability. In the following subsections, we highlight representative studies within each thematic group and critically discuss the mechanistic insights and emerging trends that arise from their combined findings.

2.1. Mechanical and Physical Surface Treatments

Mechanical and physical surface treatments remain central to the development of high-performance biomaterials, particularly for metallic implants where surface integrity, microstructural stability, and near-surface functionalization govern clinical outcomes [24]. Several contributions in this Special Issue demonstrate how severe plastic deformation, grain refinement, residual stress modulation, and controlled topographical changes can be strategically combined to enhance corrosion resistance, wear behavior, and biological performance.

Shot peening (SP) features prominently among these approaches, with multiple studies reporting its capacity to induce near-surface nanocrystallization, refine grains, and improve mechanical and electrochemical stability in titanium-based systems [25,26]. Conventional SP and fine-particle SP were comparatively investigated for biomedical Ti6Al4V, revealing substantial differences in roughness-driven wettability and antibacterial behavior, attributed to particle size-dependent impact energies and topology alterations [5]. These findings align with prior work on powder metallurgical Ti6Al4V [2] and Cp-Ti [4], where SP generated hardened subsurface layers and improved wear resistance through the formation of highly deformed α-Ti grains. Together, these studies reinforce the versatility of SP as a microstructural engineering tool for tailoring both mechanical durability and cell–surface interactions.

Physical surface modification beyond SP was also explored. Ion implantation of Zr/Nb into ZrO_2 films produced strengthened, wear-resistant surfaces, supported by first-principle calculations revealing improved bonding states and defect energetics [3]. Similarly, mechanical pre-tensioning via {332} twinning followed by aging treatment in Ti-15Mo altered grain boundary character and passivation behavior, concurrently enhancing corrosion resistance and wear stability through deformation-assisted microstructural reorganization [6]. These mechanistic insights highlight how controlled plastic deformation pathways, whether impact-driven or deformation-induced, can be exploited to optimize surface reactivity, oxide stability, and long-term implant robustness.

Overall, the mechanical and physical treatments presented across the Special Issue demonstrate a shared theme: altering surface and subsurface microstructure provides a powerful route to tuning implant performance, with SP-induced strain gradients, ion implantation-modified chemistries, and twinning-assisted microstructural evolution offering complementary strategies for advancing next-generation biomedical alloys.

2.2. Bioactive and Ceramic-Based Coatings

Bioactive and ceramic-based coatings represent one of the most intensively explored strategies for enhancing osseointegration, corrosion resistance, and biological performance of metallic biomaterials. Across the contributions in this Special Issue, hydroxyapatite (HA), calcium–phosphate (CaP), bioactive glass, and ceramic–oxide systems are engineered through sol–gel, micro-arc oxidation (MAO), electrophoretic deposition (EPD), liquid-phase deposition, and atomic layer deposition, each offering a distinct pathway for surface functionalization.

Low-temperature synthesis routes were successfully employed to deposit substituted HA coatings incorporating antibacterial ions such as Cu and Zn, yielding surfaces that combine apatite-like bioactivity with improved microbial suppression [7]. Similarly, ZnO-modified CaP layers fabricated via MAO displayed tailored zeta potential and enhanced antibacterial capability, underscoring the synergy between ceramic chemistry, nanoscale surface charge, and microbial interactions [9]. These findings highlight the importance of chemical doping and ionic substitution for modulating both biological and antibacterial responses in HA-based systems.

Complementary strategies using composite approaches were also widely represented. Bioactive glass coatings with ZnO nanoparticles deposited on stainless steel demonstrated enhanced bioactivity and structural integrity, driven by nanoparticle-mediated control of dissolution kinetics and apatite formation [8]. In parallel, Ga-doped HA coatings on β-Ti–45Nb alloys provided dual benefits by supporting osteogenic potential while introducing Ga-mediated antibacterial effects, illustrating how multi-ion engineering can drive multifunctional performance [10].

Magnesium-based biomaterials received particular attention, with several studies addressing their intrinsic challenges related to degradation. Superhydrophobic HA coatings produced through MAO followed by liquid-phase deposition significantly reduced corrosion rates and improved in vitro bioactivity on AZ31 Mg alloys [11]. A complementary approach using MAO ceramic layers combined with ALD TiO_2 demonstrated that ultrathin, conformal oxide films can effectively seal MAO porosity, yielding superior corrosion resistance and improved cytocompatibility under physiological conditions [12]. A systematic study on MAO CaP coatings further showed that increasing TiO_2 nanoparticle concentration alters surface morphology and oxide layer composition, enhancing passivation and supporting favorable cell responses [13].

Collectively, these contributions reveal a consistent mechanistic theme: bioactive ceramic coatings serve as tunable platforms where chemistry, crystallography, and hierarchical microstructure work in concert to regulate degradation behavior, protein adsorption, cellular attachment, and antibacterial activity. The integration of dopants, composite layers, and hybrid processing routes (e.g., MAO + ALD, MAO + LPD) demonstrates the shift toward multimodal coating architectures that bridge mechanical durability with biological functionality.

2.3. *Antibacterial and Multifunctional Hybrid Surfaces*

Preventing bacterial adhesion and biofilm formation remains one of the most persistent challenges in implantology. As demonstrated across several studies in this Special Issue, hybrid antibacterial surfaces, engineered through metallic ion incorporation, polymer–nanoparticle composites, or carbon-based lubricating films, offer promising routes to simultaneously enhance antimicrobial activity, interfacial stability, and tribological performance.

Silver-based systems are particularly prominent due to their broad-spectrum antibacterial activity. Powder coatings incorporating silver nanoparticles within different carrier matrices showed that antimicrobial effectiveness can be highly dependent on nanoparticle dispersion and carrier–silver compatibility, highlighting the need for controlled microstructural integration within polymeric systems [14]. Silver–chitosan nanocomposites developed for dental remineralization applications further illustrated multifunctionality, combining strong antibacterial action with the ability to promote surface remineralization, a dual benefit attributed to chitosan's inherent bioactivity and silver's bactericidal properties [17].

Gold coatings offer an additional design strategy. Electrodeposited gold layers on dental abutments effectively reduced microgaps and bacterial leakage, demonstrating that even nanoscale modifications to implant–abutment interfaces can influence microbial colonization and mechanical sealing behavior [15]. Similarly, ZnO-containing CaP coatings produced through micro-arc oxidation offered antibacterial function coupled with favorable surface charge characteristics, demonstrating a synergistic response where ionic release, microstructure, and surface energy collectively determine bactericidal efficacy [9]

Carbon-based thin films also emerged as a powerful antibacterial and tribological platform. Diamond-like carbon (DLC) coatings applied to implant abutments improved reverse torque performance while reducing wear and friction at internal conical connections, factors that directly affect implant longevity and mechanical stability [18]. The low surface energy and dense carbon network of DLC not only inhibit bacterial adhesion but also act as a solid lubricant, representing a distinct mechanistic pathway compared to metal-ion or ceramic-based antibacterial strategies.

Further multifunctional behavior was demonstrated in nanostructured antibacterial surfaces for spinal fixation screws, where engineered nano-architectures effectively suppressed biofilm formation and improved mechanical robustness under physiological loading conditions [16]. These results emphasize the importance of integrating textural and chemical cues to disrupt bacterial colonization on load-bearing orthopedic devices.

Taken together, the antibacterial and multifunctional hybrid surfaces presented here highlight a clear trend toward multi-mechanism antimicrobial strategies, where ion release, polymer–nanoparticle synergy, surface energy reduction, and nanoscale architectural control converge to combat biofilm formation while supporting mechanical and biological stability.

2.4. Biological Response, Biomechanics, and Tissue–Implant Interface

Biological interactions at the implant surface ultimately determine the success of biomedical devices, influencing early osseointegration, inflammatory response, long-term stability, and tissue regeneration. Several contributions in this Special Issue address these aspects through experimental, in vivo, and computational approaches, providing mechanistic insight into how surface modifications and biological environment jointly regulate tissue response.

Biomechanics-oriented analyses reveal how implant geometry and stress distribution affect clinical performance. Finite element simulations applied to different miniplate designs demonstrated that geometric optimization can substantially reduce stress concentrations at skeletal anchorage sites, with direct implications for implant stability and patient-specific orthodontic planning [19]. Such models highlight the fundamental link between surface treatment, structural design, and mechanical loading pathways in vivo.

Biochemical and pharmacological modulation of the tissue–implant interface was also explored. An in vivo study examining losartan, an angiotensin II receptor blocker commonly prescribed for hypertension, showed that systemic medication can influence bone–implant integration, enhancing osseointegration in hypertensive rat models [21]. This work underscores the need to consider patient-specific systemic factors when evaluating implant performance and demonstrates that implant success cannot be isolated from systemic biological context.

Biologically driven surface responses were further examined through natural extracts and tissue-level structural analyses. Cranberry extract exhibited osteogenesis-stimulating potential in vitro, attributed to phenolic and anthocyanin compounds that promote osteoblastic activity and modulate oxidative responses [20]. Similarly, the hierarchical orientation of hydroxyapatite fibers in enamel was shown to play a critical role in resisting wear and crack propagation, providing valuable biomimetic insights for engineering high-durability ceramic and CaP coatings [23]. These findings collectively emphasize the importance of microstructural anisotropy and natural hierarchical design principles in informing the next generation of bioinspired surface treatments.

Tissue engineering approaches further expanded the biological scope. An acellular spinal cord scaffold functionalized with VEGF and processed using ultrasound-assisted crosslinking demonstrated enhanced regenerative potential, offering a multifunctional platform for neuroregenerative applications [22]. The ability to modulate vascularization and support axonal growth highlights the broader relevance of surface engineering beyond classical orthopedic and dental implants.

Overall, these studies reinforce that biological response cannot be decoupled from surface chemistry, mechanical environment, or systemic physiology. By integrating biomechanical modeling, pharmacological effects, natural biomolecules, and tissue engineering strategies, this Special Issue provides a multidimensional perspective on how surface engineering influences, and is influenced by, the complex biological milieu surrounding implanted materials.

2.5. Review Articles: Current Landscape and Future Challenges

The Special Issue includes some impactful review papers that collectively frame the broader scientific context of surface engineering for biomedical applications. These reviews synthesize emerging trends, identify persistent challenges, and outline future research directions that resonate strongly with the findings reported in the research articles.

A comprehensive review on surface modification techniques for 3D-printed titanium implants emphasized the transformative role of additive manufacturing in biomedical device design and the necessity of surface treatments to overcome issues such as microstructural heterogeneity, surface roughness artifacts, and reduced osseointegration associated with as-printed components [27]. The authors highlighted a wide range of post-processing strategies, from mechanical treatments to biochemical coatings, thereby aligning closely with many of the experimental studies reported in this Special Issue. Their synthesis reinforces the view that AM-derived implant surfaces require tailored multi-step modification routes to achieve clinically reliable performance.

A second systematic review focused on multifunctional coatings across orthopedic, dental, and soft-tissue implant applications, consolidating the current state of antibacterial, osteogenic, and corrosion-resistant surface technologies [28]. This work underscored the rising importance of multifunctionality, a theme echoed in hybrid coatings incorporating ZnO, Ag–chitosan, DLC films, and complex ceramic dopants across several contributions in this Special Issue. The review argues for integrated solutions capable of addressing the simultaneous demands of mechanical durability, biological compatibility, and infection control, thereby providing a conceptual framework for the multifunctional surface systems identified in Section 2.3.

In addition, an editorial-style review summarizing advances in implant alloys and coatings provided a broader materials-oriented perspective, discussing emerging alloy systems, bioactive ceramics, and next-generation functional coatings [29]. This framework complements the mechanistic insights in Sections 2.1 and 2.2, reinforcing the need for surface engineering strategies that integrate microstructural control, chemical functionality, and biomimetic design principles.

Together, these reviews exemplify the multidimensional landscape of surface engineering research. Their collective perspective highlights several future challenges, including the need for predictive modeling of biological outcomes, improved synergistic multifunctional coatings, and integrated processing–performance correlations, that are reflected upon and further advanced in the research articles featured in this Special Issue.

3. Emerging Trends and Future Perspectives

Across the two volumes of this Special Issue, several overarching trends have emerged, collectively defining the current direction of surface engineering research for biomedical applications. The first major trend is the shift toward multifunctional and hierarchical surface systems, where a single coating or treatment integrates antibacterial, osteogenic, corrosion-resistant, and tribological functions. Hybrid approaches, such as MAO combined with ALD sealing layers, multi-ion substituted HA coatings, and polymer–nanoparticle composites, reflect a growing emphasis on synergistic mechanisms that simultaneously address the diverse clinical demands placed on implants.

A second trend is the increasing focus on microstructural and chemical precision, made possible through techniques such as ion implantation, ALD, fine-particle shot peening, and controlled ceramic deposition routes. These methods enable nanometer-to-micrometer scale control over grain structure, oxide chemistry, and surface energy, offering predictable pathways for tuning cell behavior, antibacterial activity, and mechanical reliability. As demonstrated in both titanium- and magnesium-based systems, optimized subsurface deformation and oxide stabilization are becoming key parameters for extending implant longevity. Recent progress in plasma-based ceramic oxidation (PEO) processes, particularly electrochemical oxidation (ECO) variants designed to suppress micro-arc instability and reduce coating porosity, further illustrates this trend. Such approaches have demonstrated improved adhesion, reduced surface defects, and enhanced corrosion resistance in magnesium-based systems, positioning ECO/PEO technologies as promising candidates for the next generation of structurally controlled oxide coatings [30,31].

Biologically oriented approaches continue to gain momentum. The incorporation of bioactive molecules, natural extracts, and pharmacological agents points to a growing recognition that implant performance is strongly influenced by the surrounding biological and systemic environment. Studies examining osteogenic stimulation, drug–implant interactions, and bioinspired hierarchical architectures illustrate the field's increasing movement toward bio–materials integration, where biological signaling and surface design are treated as interconnected components rather than separate domains.

Finally, several contributions highlight the need for predictive and personalized implant design, driven by biomechanics-informed evaluations and computational simulation. As implants become more complex and applications more patient-specific, integrating predictive modeling with advanced surface treatments will become essential for designing next-generation devices that balance mechanical robustness with favorable biological outcomes.

Together, these trends indicate a rapidly evolving landscape in which multi-functionality, microstructural precision, biological integration, and predictive modeling converge to shape future research. The insights collected through this Special Issue underscore the need for continued interdisciplinary efforts aimed at bridging materials science, biology, and biomechanics to engineer surfaces capable of meeting the increasingly complex requirements of modern biomedical implants.

4. Conclusions

The two-volume Special Issue *"Advances in Surface Engineering and Biocompatible Coatings for Biomedical Applications"* brings together a diverse yet thematically coherent collection of studies that collectively advance our understanding of how engineered surfaces can transform the performance of biomedical materials. Across 26 research and review articles, the contributions demonstrate that the interplay between surface chemistry, microstructure, mechanical integrity, and biological response is central to the success of modern implants.

The research findings highlight several overarching points. Mechanical and physical treatments continue to provide powerful pathways for tailoring near-surface microstructures, modulating residual stress, and enhancing mechanical durability. Bioactive and ceramic-based coatings offer finely tunable chemical and structural platforms for improving corrosion resistance, osteoconductivity, and antibacterial function, particularly through hybrid, doped, and multi-layer architectures. Antibacterial and multifunctional surfaces illustrate a broader trend toward integrated strategies that combine chemical, topographical, and nanostructural cues to control microbial behavior while enhancing tribological performance. Complementing these materials-focused advances, biological and biomechanical investigations demonstrate that systemic factors, cellular signaling, hierarchical structural organization, and mechanical loading pathways must be considered as integral components of implant design.

Together, the collective insights gained through this Special Issue underscore a unifying message: effective surface engineering requires a holistic, interdisciplinary approach that bridges materials science, mechanics, chemistry, and biology. As biomedical applications become increasingly demanding—spanning load-bearing orthopedic devices, bioactive dental interfaces, biodegradable magnesium systems, and regenerative scaffolds—the need for multifunctional, biologically informed, and predictively designed surface solutions will only continue to grow.

We hope that the research and perspectives presented across these two volumes will serve as a valuable resource for the community and stimulate further innovation in the engineering of advanced surfaces for next-generation biomedical implants and devices.

Author Contributions: Conceptualization, E.A. and M.G.; writing—original draft preparation, E.A.; writing—review and editing, E.A., M.G., Y.Y.A. and B.Y.; visualization, E.A.; supervision, E.A. and M.G.; project administration, E.A.; funding acquisition, E.A. and M.G. All authors have read and agreed to the published version of the manuscript.

Funding: The Guest Editors thank Kocaeli University Scientific Research Projects Coordination Unit (BAP) for project support under Project ID: 3749, Project Code: FKA-2024-3749.

Acknowledgments: The Guest Editors would like to express their sincere appreciation to all authors who contributed their high-quality research and review articles to this two-volume Special Issue, as well as to the reviewers for their timely and constructive evaluations. We also extend our gratitude to the Coatings editorial team for their continuous support throughout the organization and publication process. We would additionally like to acknowledge Mustafa Armağan and Eray Abakay for their assistance during the coordination of the second volume of the Special Issue. Their support in managing submissions and communication workflows is gratefully recognized.

Conflicts of Interest: The authors declare no conflicts of interest. The funders had no role in the design of the study; in the collection, analyses, or interpretation of data; in the writing of the manuscript; or in the decision to publish the results.

Abbreviations

The following abbreviations are used in this manuscript:

SP	Shot Peening
Ti6Al4V	Titanium Alloy Grade 5 (Ti-6Al-4V)
PM	Powder Metallurgy/Powder Metallurgical
Cp-Ti	Commercially Pure Titanium
ZrO_2	Zirconium Dioxide
HA	Hydroxyapatite
CaP	Calcium Phosphate
ZnO	Zinc Oxide
MAO	Micro-Arc Oxidation
EPD	Electrophoretic Deposition
ALD	Atomic Layer Deposition

BG	Bioactive Glass
DLC	Diamond-Like Carbon
VEGF	Vascular Endothelial Growth Factor
SCI	Spinal Cord Injury
FEA	Finite Element Analysis
TiO_2	Titanium Dioxide
β-Ti	Beta-Phase Titanium Alloy
HAp	Hydroxyapatite (crystallographic notation, often interchangeable with HA)
Mg	Magnesium
ARB	Angiotensin II Receptor Blocker
EDC	1-Ethyl-3-(3-dimethylaminopropyl)carbodiimide (crosslinking agent)

References

1. Badji, C.; Allal, A.; Dupin, J.-C.; Léonardi, F. Impact of Sterilization on the Adhesion Properties of a Polyamide 11 Coating on Textured Metal Substrates. *Coatings* **2024**, *14*, 424. [CrossRef]
2. Avcu, E.; Abakay, E.; Yıldıran Avcu, Y.; Çalım, E.; Gökalp, İ.; Iakovakis, E.; Koç, F.G.; Yamanoglu, R.; Akıncı, A.; Guney, M. Corrosion Behavior of Shot-Peened Ti6Al4V Alloy Produced via Pressure-Assisted Sintering. *Coatings* **2023**, *13*, 2036. [CrossRef]
3. Gao, Y.; Wang, L.; Li, D. The Surface Modification of ZrO_2 Film by Zr/Nb Ion Implantation and First-Principles Calculation. *Coatings* **2023**, *13*, 1696. [CrossRef]
4. Avcu, Y.Y.; Iakovakis, E.; Guney, M.; Çalım, E.; Özkılınç, A.; Abakay, E.; Sönmez, F.; Koç, F.G.; Yamanoğlu, R.; Cengiz, A.; et al. Surface and Tribological Properties of Powder Metallurgical Cp-Ti Titanium Alloy Modified by Shot Peening. *Coatings* **2023**, *13*, 89. [CrossRef]
5. Avcu, E.; Guney, M.; Yıldıran Avcu, Y.; Sulak, M.; Uzuner, H.; İlçe Bahadır, M.; Abakay, E.; Armağan, M.; Yamanoğlu, R.; Elibol, C.; et al. Comparative Effects of Fine and Conventional Shot Peening on Surface Morphology, Topography, Wettability, and Antibacterial Activity of Biomedical Ti6Al4V Alloy. *Coatings* **2025**, *15*, 1071. [CrossRef]
6. Zhang, S.; Yang, F.; Sun, W.; Cui, N.; Xu, T. Enhancing Passivation Behaviors and Wear Resistance of Biomedical Ti-15Mo Alloy via {332} Twinning Pre-Tension and Aging. *Coatings* **2024**, *14*, 1332. [CrossRef]
7. Lukaviciute, L.; Karciauskaite, J.; Grigoraviciute, I.; Vasiliauskiene, D.; Sokol, D.; Kareiva, A. Calcium Hydroxyapatite Coatings: Low-Temperature Synthesis and Investigation of Antibacterial Properties. *Coatings* **2023**, *13*, 1991. [CrossRef]
8. Heidari Laybidi, F.; Bahrami, A.; Abbasi, M.S.; Rajabinezhad, M.; Heidari Beni, B.; Karampoor, M.R.; Mousavi Anijdan, S.H. Electrophoretic Deposition of ZnO-Containing Bioactive Glass Coatings on AISI 316L Stainless Steel for Biomedical Applications. *Coatings* **2023**, *13*, 1946. [CrossRef]
9. Chebodaeva, V.; Sedelnikova, M.; Khimich, M.; Bakina, O.; Tolmachev, A.; Miller, A.; Golohvast, K.; Zakharenko, A.; Egorkin, V.; Vyaliy, I.; et al. Antibacterial Calcium Phosphate Coatings with Zinc Oxide Nanoparticles. *Coatings* **2023**, *13*, 1921. [CrossRef]
10. Vishnu, J.; Voss, A.; Hoffmann, V.; Alberta, L.A.; Akman, A.; Shankar, B.; Gebert, A.; Calin, M. Designing Gallium-Containing Hydroxyapatite Coatings on Low Modulus Beta Ti-45Nb Alloy. *Coatings* **2023**, *13*, 1817. [CrossRef]
11. Hu, Y.; Liang, X.; Yuan, Y.; Jian, F.; Tang, H. Preparation of Superhydrophobic Hydroxyapatite Coating on AZ31 Mg Alloy by Combining Micro-Arc Oxidation and Liquid-Phase Deposition. *Coatings* **2025**, *15*, 675. [CrossRef]
12. Han, X.; Wang, Y.; Ma, J.; Ma, X. Corrosion Resistance and In Vitro Biological Properties of TiO_2 on MAO-Coated AZ31 Magnesium Alloy via ALD. *Coatings* **2024**, *14*, 1198. [CrossRef]
13. Zhassulan, A.; Rakhadilov, B.; Baizhan, D.; Kengesbekov, A.; Kakimzhanov, D.; Musataeva, N. Influence of TiO_2 Nanoparticle Concentration on Micro-Arc Oxidized Calcium–Phosphate Coatings: Corrosion Resistance and Biological Response. *Coatings* **2025**, *15*, 1142. [CrossRef]
14. Zhang, H.; Cui, J.; Yang, J.; Yan, H.; Zhu, X.; Shao, Y.; Zhang, H.; Zhu, J. Effect of Carrier Materials for Active Silver in Antibacterial Powder Coatings. *Coatings* **2024**, *14*, 297. [CrossRef]
15. Walton, T.R. Effect of Electrodeposited Gold Coatings on Micro-Gaps, Surface Profile and Bacterial Leakage of Cast UCLA Abutments Attached to External Hexagon Dental Implants. *Coatings* **2023**, *13*, 1976. [CrossRef]
16. Panaitescu, T.G.; Niculescu, A.-G.; Grumezescu, V.; Costăchescu, B.; Bircă, A.C.; Balaure, P.C.; Oprea, O.C.; Voinea, I.C.; Stan, M.S.; Holban, A.M.; et al. Nanostructured Coatings for Spinal Fixation Screws: A Dual-Function Approach Against Biofilm Formation and Implant Failure. *Coatings* **2025**, *15*, 584. [CrossRef]
17. de Rodríguez, K.A.A.; de González, W.Y.E.; Castañeda Monroy, V.; Murphy, S.; Martínez-Castañón, G.-A.; Bach, H.; Niño-Martínez, N. Silver Nanoparticles–Chitosan Nanocomposites as Protective Coatings for Dental Remineralization Treatment: An In Vitro Study. *Coatings* **2025**, *15*, 40. [CrossRef]

8. Coelho, A.V.P.; Figueiredo, V.M.G.d.; Ferreira, L.L.; Silva, A.d.M.; Oliani, M.G.; Queiroz, J.R.C.d.; Sobrinho, A.S.d.S.; Nogueira Junior, L.; Prado, R.F.d. Impact of Diamond-like Carbon Films on Reverse Torque: Superior Performance in Implant Abutments with Internal Conical Connections. *Coatings* **2024**, *14*, 1168. [CrossRef]
9. Ielpo, A.P.M.; de Matos, J.D.M.; Noritomi, P.Y.; da Rocha Scalzer Lopes, G.; Queiroz, D.A.; Borges, A.L.S.; Nascimento, R.D. Biomechanical Behavior of Different Miniplate Designs for Skeletal Anchorage in the Anterior Open Bite Treatment. *Coatings* **2022**, *12*, 1898. [CrossRef]
10. Bauer, Y.G.; Magini, E.B.; Farias, I.V.; Della Pasqua Neto, J.; Fongaro, G.; Reginatto, F.H.; Silva, I.T.; Cruz, A.C.C. Potential of Cranberry to Stimulate Osteogenesis: An In Vitro Study. *Coatings* **2024**, *14*, 1352. [CrossRef]
11. Mulinari-Santos, G.; Santos, J.S.d.; de Souza Batista, F.R.; Pitol-Palin, L.; Silva, A.C.E.d.; Botacin, P.R.; Antoniali, C.; Okamoto, R. Evaluation of Bone–Implant Interface: Effects of Angiotensin II Receptor Blockade in Hypertensive Rats. *Coatings* **2025**, *15*, 73. [CrossRef]
12. Deng, X.; Liu, Y.; Xu, Z.; Yin, H. Ultrasound-Assisted Acellular Spinal Cord Scaffold for Spinal Cord Injury Treatment. *Coatings* **2024**, *14*, 1137. [CrossRef]
13. Shen, J.; Xin, H.; Li, X.; Kong, Y.; Zhu, S.; Zhou, Y.; Fan, Y.; Xia, J. Natural Selection on Hydroxyapatite Fiber Orientations for Resisting Damage of Enamel. *Coatings* **2024**, *14*, 1122. [CrossRef]
14. Avcu, E.; Avcu, Y.Y.; Armağan, M.; Abakay, E.; Yousif, B.F.; Guney, M. Editorial: Tribological behavior of biomaterials. *Front. Mater.* **2025**, *12*, 1549972. [CrossRef]
15. Abakay, E.; Armağan, M.; Yıldıran Avcu, Y.; Guney, M.; Yousif, B.F.; Avcu, E. Advances in improving tribological performance of titanium alloys and titanium matrix composites for biomedical applications: A critical review. *Front. Mater.* **2024**, *11*, 1452288. [CrossRef]
16. Yildiran Avcu, Y.; Yetik, O.; Guney, M.; Iakovakis, E.; Sinmazcelik, T.; Avcu, E. Surface, Subsurface and Tribological Properties of Ti6Al4V Alloy Shot Peened under Different Parameters. *Materials* **2020**, *13*, 4363. [CrossRef]
17. Long, S.; Zhu, J.; Jing, Y.; He, S.; Cheng, L.; Shi, Z. A Comprehensive Review of Surface Modification Techniques for Enhancing the Biocompatibility of 3D-Printed Titanium Implants. *Coatings* **2023**, *13*, 1917. [CrossRef]
18. Vishwakarma, V.; Kaliaraj, G.; Amirtharaj Mosas, K. Multifunctional Coatings on Implant Materials—A Systematic Review of the Current Scenario. *Coatings* **2022**, *13*, 69. [CrossRef]
19. Vishwakarma, V.; Kaliaraj, G.; Kirubaharan, A. Advanced Alloys and Coatings for Bioimplants. *Coatings* **2022**, *12*, 1525. [CrossRef]
20. Yavuzyegit, B.; Karali, K.; Avcu, E.; De Mori, A.; Quizon, D.; Hacıosmanoğlu, M.; Hekimoğlu, A.P.; Smith, N.; Usov, S.; Shashkov, P.; et al. Corrosion and mechanical performance of novel electrochemical oxidation coatings on AZ31 magnesium alloys for biomedical applications. *Surf. Coat. Technol.* **2025**, *507*, 132151. [CrossRef]
21. Yavuzyegit, B.; Karali, K.; Davis, S.; Morrison, B.; Karabal, S.; Balandiz, K.; Smith, N.; Usov, S.; Shashkov, P.; Bonithon, R.; et al. High-resolution DIC analysis of in situ strain and crack propagation in coated AZ31 magnesium alloys under mechanical loading. *J. Mater. Sci.* **2025**, *60*, 14708–14730. [CrossRef]

coatings

MDPI

Article

Natural Selection on Hydroxyapatite Fiber Orientations for Resisting Damage of Enamel

Junfu Shen [1,†], Haiyan Xin [2,†], Xiaopan Li [1], Yiyun Kong [1], Siqi Zhu [1], Yuankai Zhou [1], Yujie Fan [1] and Jing Xia [1,*]

[1] College of Mechanical Engineering, Jiangsu University of Science and Technology, Zhenjiang 212000, China; 221210201219@stu.just.edu.cn (X.L.)

[2] Department of Oral and Maxillofacial Surgery, Jinan Stamotological Hospital, Jinan 250001, China; xhy1987@stu.sdp.edu.cn

* Correspondence: 201800000005@just.edu.cn

† These authors contributed equally to this work.

Abstract: Teeth have excellent mechanical properties, with high wear resistance and excellent fracture resistance. This is due to their well-organized multilevel hierarchical structure. While a number of studies in the last decades have revealed the relationship between tooth structure and mechanical properties, there is still no general agreement on how different orientations of hydroxyapatite (HAp) fibers affect the mechanical properties of enamel. With a scanning electron microscope and nanoindenter, the orientations of HAp fibers and their properties were investigated. HAp fibers have two different orientations: parallel and perpendicular to the surface. Fibers oriented parallel to the surface exhibited higher hardness, elastic modulus and wear resistance. Under applied force, fibers oriented perpendicular to the surface suffered deeper shearing in the protein along the long axis, resulting in lower mechanical properties. Teeth resist damaging fractures by combining hard and soft structures. This study may lead to new insights into how nature selects for tooth structure and provide a theoretical basis for the bioinspired design.

Keywords: enamel structure; dental microwear; fiber orientation; biomaterial; biomechanics of human tissues

1. Introduction

The human tooth is a natural mineral substance that plays a very important role in human life. Teeth are among the hardest tissues in the human body, and are directly exposed to the mouth environment. Even after millions of chewing cycles, teeth disperse high chewing contact pressure with relatively limited damage rather than failure. The excellent mechanical properties of teeth have a close relationship with their complex chemical composition and orderly structure. The study of the relationship between tooth structure and properties is helpful for understanding the internal mechanism of natural materials to resist external damage, and provides a new idea for the construction and performance optimization of bionic materials.

From the overall structure, a tooth crown is mainly composed of the outer enamel and the inner dentin, with the dentin–enamel junction (DEJ) between them. It has been noted that enamel, dentin, and the DEJ play important roles in the resistance to fractures. Enamel is harder and more resistant to wear, while dentin is less hard and tougher, making teeth both strong and tough [1]. Borero-Lopez et al. [2] suggested that tooth enamel is a functionally graded material with anticracking properties. When cracks move from the outer enamel to the inner enamel, they are dissipated by the HAp column, deviating from their original path. Additionally, when faced with harder enamel at the boundary and softer underlying dentin, the cracks are further resisted. Also, research has shown the presence of numerous parallel microcrack structures in teeth at the boundary between enamel and dentin. These microcracks can self-repair through continuous absorption of organic matter. Enamel can alleviate stress concentration and act as a stress shield when impacted [3]. The role of the DEJ in tooth cracking cannot be ignored. The width and gradient structure of the DEJ vary at occlusal and cervical sites, with a wider and smoother transition at occlusal sites, promoting more efficient stress transfer between enamel and dentin without affecting the overall stiffness of the teeth [4]. In addition, the viscoelasticity of the DEJ functions as a loading attenuation mechanism between enamel and dentin [5]. Under a tensile load perpendicular to the DEJ, failure does not occur in the DEJ; this helps to stop the crack from extending into the dentin, thus preventing a potentially dangerous rupture [6,7].

Enamel is composed of keyhole-like enamel rods and inter-rods enamel [8]. The interfacial area between enamel rods and inter-rods enamel, termed sheaths, has a high organic content and is about 100–200 nm wide [9]. Enamel rods are arranged in different ways across the entire enamel. In the outer layer enamel, enamel rods almost arrange in parallel, revealing a radial pattern. In the inner enamel, enamel rods appear as bands of approximately cross-sectioned arrangement [10]. The varying mechanical properties from the outer layer of enamel to the inner layer have attracted a lot of attention. Cuy et al. [11] found a significant decrease in elastic modulus and hardness from the outer layer to the inner layer. However, the inner layer enamel demonstrated higher toughness and better ability to prevent crack propagation. In another study, it was found that the outer enamel displayed better wear resistance, whereas the inner enamel exhibited better creep and stress redistribution ability. The inner enamel also demonstrated better fracture toughness, dispersing stress from the outer enamel, and contributing to the overall structural integrity of the enamel in terms of wear and fracture resistance [12]. In the microscale, enamel rod and inter-rod consist of Hydroxyapatite (HAp) fibers with different orientations and compactness [13,14]. These fibers are roughly hexagonal and are bonded together by a protein layer [15]. The fibers reveal anisotropic mechanical properties in the occlusal surface and the axial-sectional plane, and it was found that the occlusal surfaces had better microscopic mechanical properties [16,17]. The change in the orientation of the HAp fibers enhanced the fracture toughness of enamel. Additionally, the transverse fibers effectively inhibit crack propagation [18,19]. The more the orientation and loading direction of HAp fibers converge, the better the bearing capacity and wear resistance of tooth enamel [17]. The exceptional properties of tooth enamel are not only due to its perfect structure, but also because of the effective binding of organic proteins and inorganic minerals. Proteins are mainly found in the sheath between the enamel rod and inter-rod enamel. They are considered the main reason for the visco-plastic and visco-elastic properties of enamel [20]. He et al. [15] found that the protein undergoes shear deformation when the enamel is subjected to an external load. During chewing, the enamel consumes energy through microdamage of proteins, thereby increasing the fracture toughness of enamel. An et al. [21] found that under the action of the indentation damage mode, the protein binding HAp particles undergo shear deformation and the macromolecular chains in the protein extend. With the increase in load, the degree of shear deformation of protein intensifies, and the macromolecular chains in the protein will break, leading to irreversible deformation of tooth enamel.

This study aims to investigate the effect of HAp fibers orientations on the mechanical properties of enamel. First, the orientation of HAp fibers in enamel rod and inter-rod enamel is observed. Then, the influence of HAp fiber orientations on hardness, elastic modulus, friction, and wear properties is analyzed. Finally, the factors contributing to these performance differences are analyzed.

2. Materials and Methods

2.1. Specimen Preparation

The specimens used in the study were prepared from healthy human third molar teeth obtained with the consent of the subjects. These teeth were stored in deionized water at 4 °C before use. The tooth roots were cut with a saw at low speed under constant water irrigation, while the other eight teeth were embedded with the occlusal surface facing down in epoxy resin, and the other eight teeth were cut into two halves along the axial section and embedded into the epoxy resin with the longitudinal section facing down. Subsequently, all specimens were ground sequentially using 800, 1200, and 1500 grit sandpaper, followed by polishing with 0.5 μm diamond paste. The polished samples were then cleaned with deionized water before testing. The samples were etched for 1 min with a 0.001 M citric acid solution to reveal the orientations of HAp fibers.

2.2. Nanoindentation Tests

To understand how the orientation of HAp fibers affects the mechanical properties of enamel, the mean elasticity modulus and hardness values of the different aligned HAp fibers were obtained using a nanoindenter (T750, Hysitron Inc., Minneapolis, MN, USA) with a Berkovich diamond tip. Four occlusal surface samples and four longitudinal section samples from different people were taken, respectively, for the nanoindentation tests. As shown in Figure 1A, the experiments were conducted on the outer enamel of the occlusal surface and longitudinal section samples, respectively. On the occlusal surface, approximately 7 × 7 indentations were placed with a spacing of about 5 μm under a normal load of 2 mN for each specimen. On the longitudinal section, about 10 × 5 indentations were placed with a spacing of about 2 μm under a normal load of 1 mN for each specimen. For each indentation, the loading function was adjusted to maintain 5 s of loading time, 2 s of delay at peak load, and 5 s of unloading time. The hardness and elastic modulus of the material are obtained by calculating the load–displacement curve drawn by the system in real time according to the Oliver–Pharr method [22]. For standard diamond tips, the determined elastic modulus is 1140 GPa and the hardness is 0.07. The Poisson's ratio of enamel used in the analysis was assumed to be 0.28 based on the literature [23]. After the nanoindentation tests, the topographies of the indentations on the enamel surface were scanned using an atomic force microscope (AFM, SPI 3800N, Seiko, Tokyo, Japan) with a standard Si_3N_4 tip, having a nominal radius of 10 nm. The indentations on the sample that were completely located in the enamel rod or the inter-rod enamel were selected to be included in the calculation of the mean hardness and elasticity modulus, and the indentations located in the sheath of the enamel were abandoned to avoid errors.

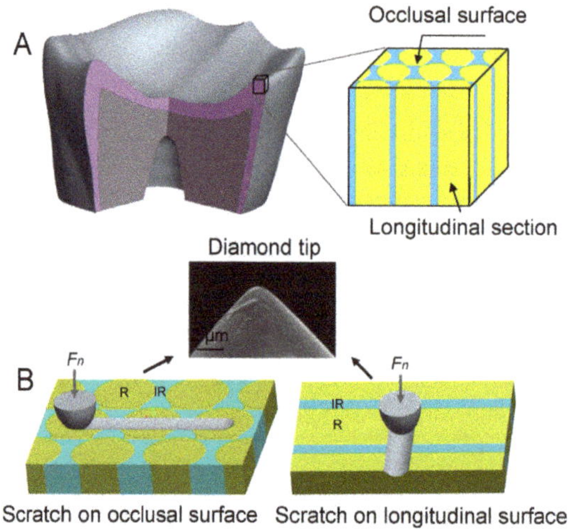

Figure 1. Schematic diagram of the test area and nanoscratch test. (**A**) Schematic diagram of the occlusal surface and longitudinal section of enamel. (**B**) Schematic diagram of the preparation direction of the scratch, with the upper illustration showing the SEM topography of the tip (IR: inter-rod enamel; R: enamel rod).

2.3. Nanoscratch Tests

To simulate the shearing movement of opposing teeth parallel to the occlusal surface and longitudinal section, scratch tests were conducted. The wear behavior and coefficient of friction were studied using a nanoindenter (T750, Hysitron Inc., Minneapolis, MN, USA) by a diamond tip with a radius of curvature of 1 μm. As depicted in Figure 1B, scratches were created on four specimens of the occlusal surface and four specimens of the longitudinal section. The scratches were made vertically to the inter-rod enamel in the occlusal surface, and perpendicular to the long axes of enamel rod in the longitudinal section. The scratches were made using normal loads of 0.5 mN, with a sliding velocity of 1 μm/s and sliding displacements of 15 μm. The friction coefficient of scratches was obtained in real time during the experiment. Following the nanoscratch tests, the topographies of the scratch-induced damage on the enamel were scanned using an Si_3N_4 tip in the AFM to indicate the depth of the scratch.

2.4. Statistical Analysis

The statistical software SPSS25 (International Business Machines Corporation, Armonk, NY, USA) was utilized to conduct a test aimed at identifying any significant differences between the samples. If $p < 0.05$, it signifies a difference between the samples, denoted by an asterisk (*). If $p < 0.01$, it indicates a significant difference between the samples, denoted by double asterisks (**).

3. Results and Discussion

3.1. HAp Fiber Orientations in Occlusal Surface and Longitudinal Section

HAp fibers are long, roughly flat hexagonal structures, approximately 20–25 nm thick and 50–70 nm wide, with lengths ranging from about 200 to 500 nm [24]. Human enamel is composed of these HAp fibers, with their long axes aligned at different orientations within enamel rod and inter-rod enamel. Using a high-resolution scanning electron microscope (SEM, JSM-7001F, JEOL Ltd., Tokyo, Japan), the orientations of HAp fibers in enamel rod and inter-rod enamel on the occlusal surface and the longitudinal section of the outer layer enamel were clearly observed. As shown in Figure 2A, the inter-rod enamel surrounds the enamel rod to form a typical keyhole shape. Two different orientations of HAp fibers were found; the long axes of HAp fibers in enamel rod (R) were approximately perpendicular to the occlusal surface, while in the inter-rod enamel (IR), the hydroxyapatite fibers aligned with their long axes parallel to the occlusal surface. Sheaths were observed with a width of about 100–200 nm between enamel rod and inter-rod enamel. In the longitudinal section displayed in Figure 2B, these two different aligned fibers were also observed. The long axes of HAp fibers in enamel rod were parallel to the longitudinal section, while in the inter-rod enamel, they were approximately perpendicular to the surface. Figure 1A,B both confirm the orientations of HAp fibers in the outer enamel—parallel or perpendicular to the surface. Several studies have noted that HAp fibers in the enamel rod are perpendicular to the occlusal surface, while the fiber orientation in the inter-rod enamel is parallel to the occlusal surface [25,26], which aligns with the findings of this experiment. The enamel rods in the outer and inner layers are arranged in different directions. In the outer layer enamel, enamel rods are almost arranged in parallel, revealing a radial pattern. The inner enamel is made up of enamel rods with different orientations arranged in a Hunter–Schreger band structure. This structure consists of bundles of hydroxyapatite crystals arranged at different angles to the surface [10]. In this paper, we only studied two different fiber orientations of the outer enamel. Further research is needed to understand the impact of the angle of deflection of HAp fibers on their properties.

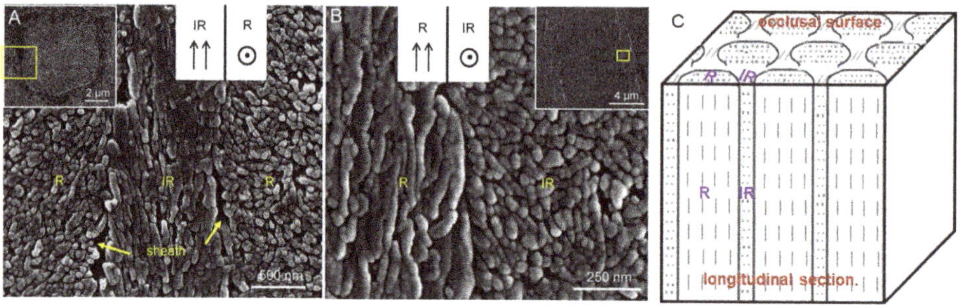

Figure 2. The orientations of HAp fibers in the inter-rod enamel (IR) and the enamel rod (R) in the outer enamel. (**A**) Occlusal surface. (**B**) Longitudinal section. (**C**) Stereoscopic diagram of the orientations of HAp fibers.

3.2. Effect of Fiber Orientation on Mechanical Properties of Enamel

As shown in Figure 3A, the topography of the enamel indentations in the occlusal surface was observed by AFM. The indentations completely located in the enamel rod (shown in the black block) and inter-rod enamel (shown in the red circle) were selected to calculate the mean hardness and elastic modulus. The enamel sheath between enamel rod and inter-rod enamel has much lower hardness and elastic modulus [27,28]. Therefore, the indentations located in the enamel sheath area were excluded to enhance the accuracy of the calculation. Based on the criteria mentioned above, 20 points in the enamel rod and inter-rod enamel were chosen from four samples to calculate the average hardness, elastic modulus, and standard deviation. The amplifying images in Figure 3A clearly show the differences in the indentations, with those in enamel rods displaying larger impression areas compared to those in inter-rod enamel. The load–displacement curves in Figure 3B demonstrate that the indentations in enamel rod exhibited greater depth displacement. The mean hardness and elastic modulus of the inter-rod enamel were found to be higher than those of the enamel rods, with indentations in the inter-rod enamel yielding a mean hardness of 4.99 ± 0.29 GPa and an elastic modulus of 102.01 ± 2.20 GPa. In comparison, the enamel rod exhibited a mean hardness of 4.06 ± 0.55 GPa and an elastic modulus of 98.94 ± 6.50 GPa. According to the statistical analysis, the hardness and elastic modulus of the indentations in the enamel rod and inter-rod enamel are different ($p < 0.05$). In combination with HAP fibers orientations as shown in Figure 2, HAp fibers with long axes parallel to the occlusal surface displayed superior mechanical behavior.

In Figure 4, the mechanical behaviors of enamel rod and inter-rod enamel in a longitudinal section are depicted. The mean hardness elastic modulus and standard deviation were calculated by 20 indentations completely located in the enamel rod (shown in the black block) and inter-rod enamel (shown in the red circle) on each of the four samples, respectively. As shown in Figure 4A, indentations located in enamel rods show lower impression areas and displacements compared to those located in the inter-rod enamel. The mean hardness and elastic modulus of the inter-rod enamel are lower than those of the enamel rods in the outer enamel. Specifically, the hardness and elastic modulus increase from H = 3.70 ± 0.45 GPa and E = 85.67 ± 0.19 GPa, where the indentations are located in the inter-rod enamel to H = 4.60 ± 0.27 GPa and E = 90.86 ± 5.63 GPa for enamel rods. According to the statistical analysis, the hardness and elastic modulus of the indentations in the enamel rod and inter-rod enamel are different ($p < 0.05$). Taking the HAp fiber orientations of enamel rod and inter-rod enamel in the longitudinal section into account, the results further demonstrate that HAp fibers with long axes parallel to the surface exhibit better mechanical behavior. The relationship between the orientation of HAp fibers and their mechanical properties is still a subject of debate. Staine et al. [29] discovered that when the force applied to the HAp fibers is perpendicular to the direction of its long axis, the measured hardness and elastic modulus are higher compared to when the force is applied parallel to the long axis. However, Sha and Katz et al. [30] obtained opposite results, suggesting that parallel loading leads to higher mechanical properties. These findings were based on comparing the hardness and elastic modulus of the occlusal surface and the longitudinal section. They suggested that the long axis direction of the HAp fibers on the occlusal surface is perpendicular to the surface, so the loading direction of the indentation tip is parallel to the long axis of the HAp fibers. On the longitudinal section, the long axis direction of the HAp fibers is parallel to the surface, and the loading direction of the indentation tip is perpendicular to the long axis of the HAp fibers. It is important to note that there are HAp fibers with different orientations in the enamel rod and inter-rod enamel in the occlusal surface and longitudinal section, leading to inconsistencies in the study process. Therefore, further exploration of the relationship between particle orientation and mechanical properties based on the specific arrangement orientation of hydroxyapatite particles is necessary.

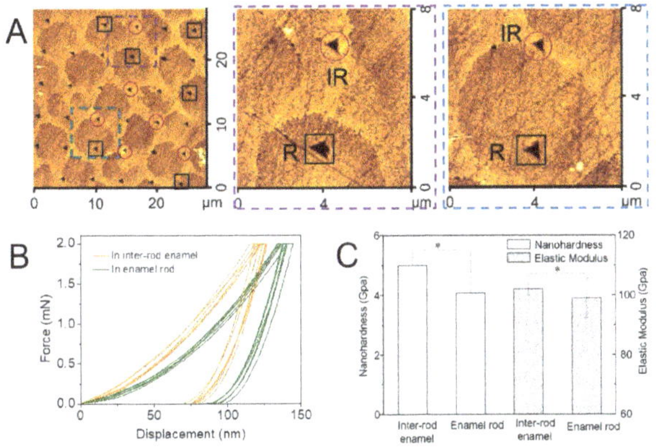

Figure 3. Observation of mechanical behaviors in enamel rod and inter-rod enamel on the occlusal surface under indentation tests. (**A**) Images of indentations in enamel surface. (**B**) Force–displacement curves of the indentations located in enamel rods and inter-rods enamel in (**A**). (**C**) Comparison of hardness and elastic modulus of the enamel rods and inter-rods enamel (* $p < 0.05$).

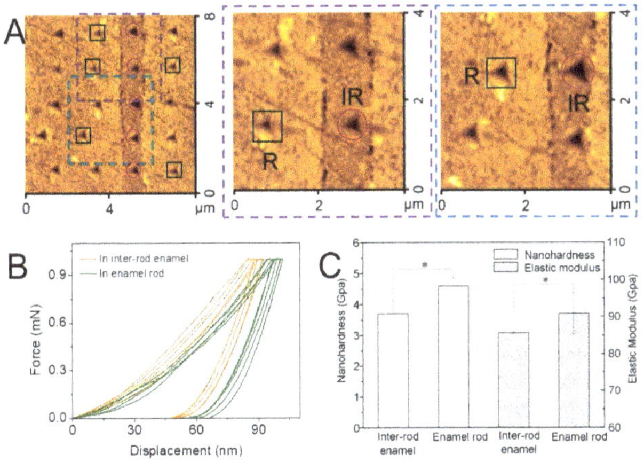

Figure 4. Observation of mechanical behaviors in enamel rod and inter-rod enamel on the longitudinal section under indentation tests. (**A**) Images of indentations in enamel surface. (**B**) Force–displacement curves of the indentations located in enamel rod and inter-rod enamel in (**A**). (**C**) Comparison of hardness and elastic modulus of enamel rod and inter-rod enamel (* $p < 0.05$).

3.3. Effect of Fiber Orientation on Microfriction Behavior of Enamel

Figure 5 illustrates the images of the scratches and the coefficient–distance curves. The friction curve corresponds to the scratch damage morphology and the cross-section profile curve of tooth enamel. It can be observed that on the occlusal surface, the coefficient of friction on the enamel rod is about 0.144, while the coefficient of friction on the inter-rod enamel is about 0.087. The coefficient of friction in the inter-rod enamel region is lower than that in the enamel rod region. On the longitudinal section, the coefficient of friction on the inter-rod enamel is about 0.158, while that on the enamel rod is about 0.108. The coefficient of friction in the inter-rod enamel region is higher than that in the enamel rod region. The enamel sheath shows the highest friction coefficient, which is consistent with its poor mechanical properties. Considering the HAp fiber orientations of the enamel rod and inter-rod enamel on these two surfaces, it can be concluded that HAp fibers with long axes parallel to the surface exhibited better microfriction behavior.

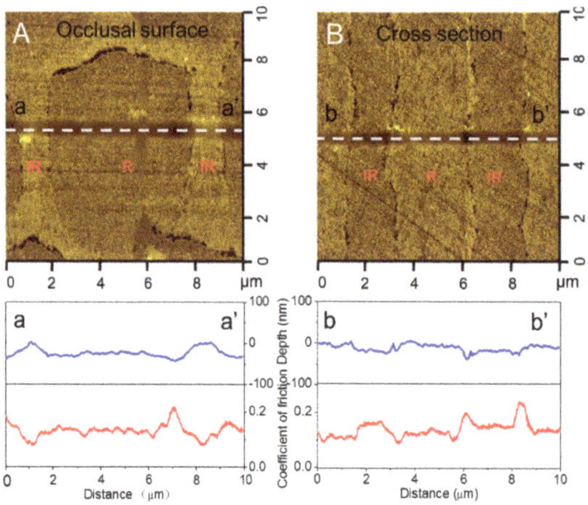

Figure 5. Friction–displacement curve of tooth enamel corresponding to scratch damage. (**A**) Occlusal surface. (**B**) Longitudinal section.

3.4. Effect of Fiber Orientation on the Wear Behavior of Enamel

In Figure 6, the wear behavior in the enamel rod and inter-rod enamel on the occlusal surface is compared. The scratches were made across the inter-rod enamel and enamel rod with a constant normal load of 0.5 mN. It was observed that wear in the enamel rod was more severe than in the inter-rod enamel. Based on the cross-sectional profile of the scratch in inter-rod enamel and enamel rod, shown in Figure 6D, the maximum wear depth in inter-rod enamel is about 15.15 nm, and in the enamel rod, it is 23.8 nm. The maximum wear depth decreased by more than 36.3% when moving from the inter-rod enamel to the enamel rod under the same load. Additionally, the inter-rod enamel showed a relatively smaller wear width compared with the enamel rod. When combined with the fiber orientations, it was evident that HAp fibers with long axes parallel to the surface exhibited better wear behavior.

The image in Figure 7 depicts the wear behavior in enamel rod and inter-rod enamel in a longitudinal section. Similar to the adverse fiber arrangements observed in the occlusal surface and longitudinal section, both surfaces exhibited adverse wear behavior of enamel rod and inter-rod enamel. The scratching was performed across the direction perpendicular to the long axes of inter-rod enamel and enamel rod with a constant normal load of 0.5 mN. Wear in the inter-rod enamel was found to be more severe compared to the enamel rod, with the maximum wear depth in the inter-rod enamel measuring about 21.4 nm, and 11.8 nm in the enamel rod. The maximum wear depth decreased by more than 44.8% when moving from the enamel rod to the inter-rod enamel under the same load. Considering the fiber orientations of enamel rod and inter-rod enamel in the longitudinal section, the results further demonstrate that HAp fibers with long axes parallel to the surface exhibited better wear behavior. In terms of structure, horizontally arranged HAp fibers have higher density, while there are more gaps between vertically arranged HAp fibers perpendicular to the occlusion surface. Studies have shown that the densification of fibers gives teeth good mechanical properties [31]. Horizontally arranged HAp fibers have better wear resistance than longitudinally arranged HAp fibers. Moreover, the change in the orientation of the fibers enhances the fracture toughness of enamel. Transverse fibers can effectively inhibit crack propagation [18,19]. It was suggested that the more similar the orientation of HAp fibers is to the direction of loading, the better the bearing capacity and wear resistance of tooth enamel [17]. HAp fibers align in various orientations to disperse energy and prevent stress concentration, thereby enhancing the overall toughness of teeth.

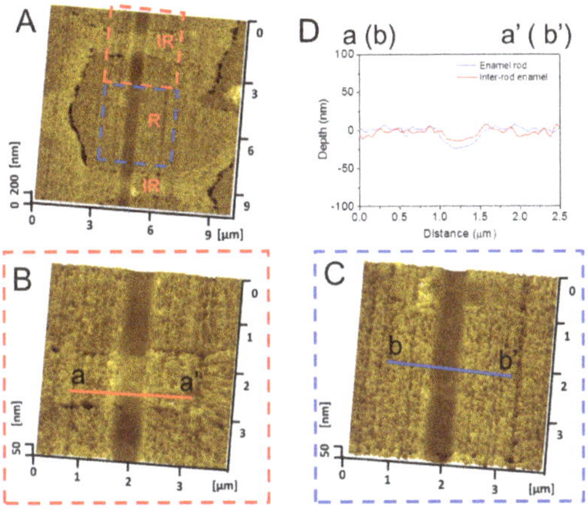

Figure 6. Observation of wear behavior in enamel rod and inter-rod enamel on the occlusal surface by scratch tests under normal load of 0.5 mN. (**A**) Image of scratch-induced damage in occlusal surface. (**B**,**C**) Details of the scratch in (**A**). (**D**) Cross-sectional profile of the scratches in (**B**,**C**).

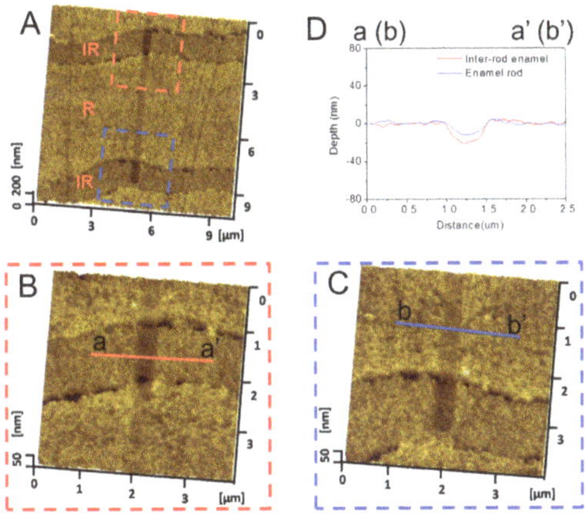

Figure 7. Observation of wear behaviors in enamel rod and inter-rod enamel on the longitudinal section by scratching tests under normal load of 0.5 mN. (**A**) Image of scratch-induced damage in longitudinal section. (**B**,**C**) Details of the scratch in (**A**). (**D**) Cross-sectional profile of the scratches in (**B**,**C**).

3.5. Influence Mechanism of Hydroxyapatite Alignment on Tooth Enamel Properties

The mechanical properties and microtribological behavior of HAp fibers vary with different orientations. Human enamel possesses a well-organized microstructure with a hierarchical assembly [10]; an analysis of how enamel reacts under load should consider its unique structure and inner bonding manner. HAp fibers are long, roughly flat hexagonal structures, approximately 20–25 nm thick and 50–70 nm wide, with lengths ranging from 200 to 500 nm [23]. These fibers are self-assembled from smaller nanoparticles, each about 20 nm in diameter, connected by rigid interfacial molecular bonds, including hydrogen bonds mediated by surface-hydrated water molecules [28]. The HAp fibers are bonded together by a thin layer of hydrated protein [15]. Proteins play a crucial role in the mechanical properties of teeth. Human tooth enamel, containing a small amount of protein, is about three times harder than dense HAp [8]. When the protein is removed, the fracture toughness of tooth enamel decreases by about 40% [32]. The protein "glue" layer between the HAp fibers experiences shear strain during enamel contact loading, which is 16 times higher than the contact strain. This facilitates the enamel to dissipate the contact deformation energy [15]. Additionally, the memory behavior of protein components makes tooth enamel exhibit a stress–strain relationship and creep behavior similar to metal [33–35]. In this paper, the theory of protein stress shear deformation is used to explain the difference between the two orientations of HAp fibers.

According to the structure and bonding manner of HAp fibers, the force analysis of the different aligned HAp fibers under indentation tests was analyzed, as shown in Figure 8. The protein matrix bonded to the HAp fibers has a yield stress of 0.4 GPa and an elastic modulus of 4.3 GPa, which are much lower than the mineral HAp fibers [36]. Therefore, when an indentation load is applied, it would initially result in a large shear strain in the protein layer. When the HAp fibers with long axes perpendicular to the surface experience force in a direction coaxial with the long axes of the fibers, the protein layer would be sheared along the long axes of the fiber (200–500 nm). The HAp fibers would be pressed into the underside due to sufficient space within the enamel matrix. On the other hand, when the HAp fibers with long axes parallel to the surface experience force in a direction perpendicular to the long axes of the fibers, the protein layer at the end of the fiber would be sheared (50–70 nm) first. However, further shear in the protein layer would be impeded by the fiber below. With an increase in the indentation load, the lower fiber would be bent, and during unloading, the bent fiber would be restored to some extent. The fiber elements are separated by continuous protein layers leading to a strong anisotropy between axial and transverse moduli, the latter being substantially lower due to the deformation of proteins [37,38]. As a result, HAp fibers with long axes parallel to the surface survived with lower indentation depth and had a better ability to resist elastic–plastic deformation, resulting in higher hardness and elastic modulus.

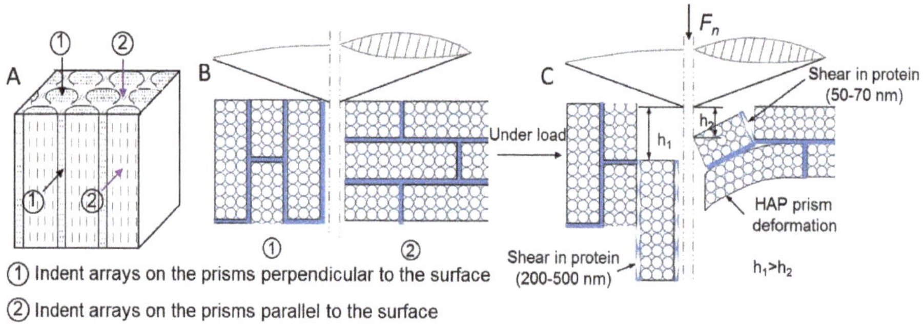

Figure 8. The force analysis of the different aligned HAp fibers under indentation tests. (**A**) Stereoscopic diagram of indentation positions. (**B**) Indentation direction diagram. (**C**) HAp fibers deformation diagram.

The different fiber orientations led to anisotropic microtribological behavior. Figure 9 explains the phenomenon through force analysis of the differently aligned HAp fibers under applied scratching load. The HAp fiber was made up of smaller nanoparticles with a diameter of approximately 20 nm and bonded by rigid interfacial molecular bonds, including H-bonds [28]. During scratch tests, the HAp fiber would be refined and fragmented into smaller ones. When the HAp fibers' long axes were perpendicular to the surface, under applied load, the protein layer would first be sheared along the depth direction (200–500 nm), and then the fiber would be bent forward along the scratch direction and refined into smaller ones (20 nm). When the HAp fibers were aligned with their long axes parallel to the surface, the protein that glued the fibers below would first be sheared along the long axes, and the fiber would be bent forward along the scratch direction and refined into smaller ones. As a result, the shear in the protein and the refinement of the fiber occurred along the long axes. HAp fibers with long axes parallel to the surface would endure a lower wear depth.

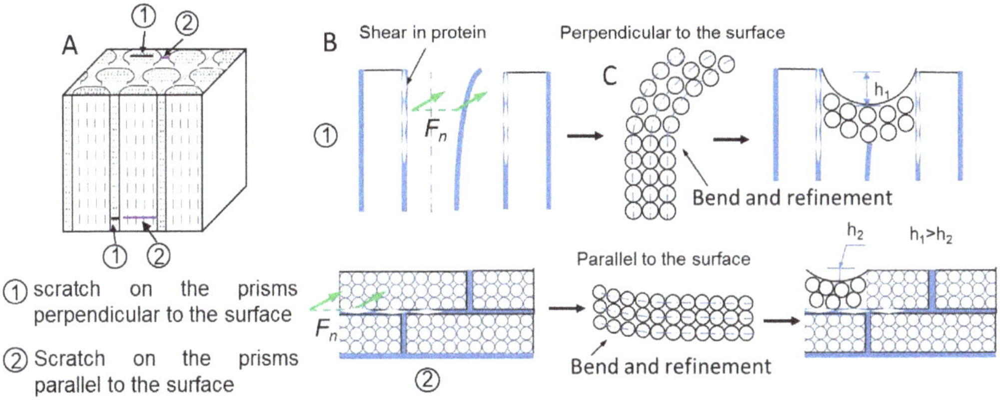

Figure 9. The force analysis of the different aligned HAp fibers under scratch tests. (**A**) Stereoscopic diagram of scratch positions. (**B**) Scratch direction diagram. (**C**) HAp fibers deformation diagram.

In this study, only the differences in the two different aligned HAp fibers in the outer layer of enamel of different healthy third molars were observed, and mechanical analysis was conducted on the observed fiber orientations. The orientations of HAp fibers at different layers of enamel showed great differences. In addition, different diets, oral environment, and different trace elements also had potential differences in tooth structural properties. These differences have not been studied in detail and are worth further exploration in the future.

The human enamel has different orientations of HAp in the outer layer, resulting in anisotropic properties in enamel rod and inter-rod enamel. The soft layer of the enamel helps to dissipate energy by limiting abrasion or deformation under large contact stresses. The combination of soft and hard structure limits the growth of cracks and prevents catastrophic failure. Studying the relationship between HAp fiber orientation and properties of enamel has broad implications, ranging from materials science to evolutionary biology to clinical dentistry. Structural anisotropy could serve as a fundamental design principle for self-assembly materials. For example, embedding hard materials vertically into soft materials or arranging nanoparticles with different orientations can optimize the mechanical properties of materials. Zhao et al. [39] synthesized HAp nanowires similar to natural tooth enamel using the hydrothermal method. The uniform growth of amorphous zirconia ceramics was achieved on the surface of the HAp nanowires by regulating material growth and nucleation. This resulted in the composite of amorphous/crystal nanowire and polymer and allowed for controllable assembly at a macro scale. The artificial tooth enamel possesses good mechanical properties. Controlling the direction of growth of the hydroxyapatite nanowires based on the orientation of the hydroxyapatite crystals in teeth, and assembling the nano-HAp nanowires into composite hydroxyapatite nanowires at different angles, may further enhance the mechanical properties of artificial tooth enamel. Furthermore, researchers have drawn inspiration from the fracture resistance mechanism of natural mother of pearl to develop mother-of-pearl imitation structural ceramics, which significantly improves the fracture toughness of ceramics [40]. The complex multistage structure of teeth also demonstrates strong fracture resistance, and the relevant theories can provide a theoretical reference for the toughening mechanism of composite ceramics.

4. Conclusions

With the development of nanoindentation and microscope techniques, the experimental accuracy was greatly improved. The main conclusions of this study can be summarized as follows:

1. There are two types of orientation of HAp fibers in the outer layer of tooth enamel: parallel and perpendicular to the occlusal surface.
2. The arrangements of HAp fibers significantly influenced the mechanical and microtribological behavior of human tooth enamel. HAp fibers with long axes parallel to the surface had higher hardness and elastic modulus and showed better microtribological behavior.
3. Teeth resist damaging fractures by combining hard and soft structures.

This research provides new insights into how nature selects for tooth structure. Animals have evolved different HAp fiber orientations to resist tooth damage. Additionally, research on the wear behavior of enamel can provide a theoretical basis for clinical dental prevention. In future studies, exploring the dynamic response of enamel under different loading conditions or different microstructures and the molecular mechanisms behind the observed anisotropic properties is worth further investigation. In addition, biomimetic design or optimization of materials based on the structural properties of teeth will be a hot focus of research.

Author Contributions: Conceptualization, J.S., H.X. and J.X.; data curation, J.S., H.X., X.L., S.Z. and Y.K.; funding acquisition, H.X. and J.X.; methodology, J.S. and J.X.; supervision, J.S., Y.Z., Y.F. and J.X.; validation, J.S., H.X. and J.X.; writing—original draft, J.S. and H.X.; writing—review and editing, J.X., J.S. and H.X. J.S. and H.X. contributed equally to this work. All authors have read and agreed to the published version of the manuscript.

Funding: This research was funded by National Science Foundation of China (51805226), Postgraduate Research & Practice Innovation Program of Jiangsu Province (KYCX24-4081), Hospital level clinical research project of Zhenjiang First People's Hospital in 2022, Y2022012, Jiangsu Provincial Health Commission's 2023 Guiding Project, 307.

Institutional Review Board Statement: Not applicable.

Informed Consent Statement: Not applicable.

Data Availability Statement: The original contributions presented in the study are included in the article; further inquiries can be directed to the corresponding author.

Acknowledgments: The authors thank all reviewers for their remarkable guidance on this article.

Conflicts of Interest: The authors declare that they have no known competing financial interests or personal relationships that could have appeared to influence the work reported in this paper.

References

1. Carreon, A.H.; Funkenbusch, P.D. Nanoscale properties and deformation of human enamel and dentin. *J. Mech. Behav. Biomed. Mater.* **2019**, *97*, 74–84. [CrossRef] [PubMed]
2. Borrero-Lopez, O.; Constantino, P.J.; Bush, M.B.; Lawn, B.R. On the vital role of enamel prism interfaces and graded properties in human tooth survival. *Biol. Lett.* **2020**, *16*, 20200498. [CrossRef] [PubMed]

3. Chai, H.; Lee, J.J.-W.; Constantino, P.J.; Lucas, P.W.; Lawn, B.R. Remarkable resilience of teeth. *Proc. Natl. Acad. Sci. USA* **2009**, *106*, 7289–7293. [CrossRef] [PubMed]
4. Wang, Z.; Wang, K.; Xu, W.; Gong, X.; Zhang, F. Mapping the mechanical gradient of human dentin-enamel-junction at different intratooth locations. *Dent. Mater.* **2018**, *34*, 376–388. [CrossRef]
5. Hasegawa, M.; Tanaka, R.; Zhong, J.; Kobayashi, M.; Manabe, A.; Shibata, Y. Deciphering load attenuation mechanisms of the dentin-enamel junction, Insights from a viscoelastic constitutive model. *Acta Biomater.* **2023**, *171*, 193–201. [CrossRef]
6. Yamada, Y.; Inoue, T.; Saito, M.; Nishimura, F.; Miyazaki, T. Anisotropic ultimate strength and microscopic fracture patterns during tensile testing in the dentin–enamel junction region. *Dent. Mater. J.* **2020**, *39*, 272–277. [CrossRef]
7. Zhu, X.; Wang, S.; Ye, J.; Guo, H.; Wang, R. Investigation of Fracture-resistance of Human Teeth at the Dentin-enamel Junction Using the J-integral Calculation of Finite Element Analysis. *J. Mech. Med. Biol.* **2022**, *22*, 2250069. [CrossRef]
8. White, S.; Luo, W.; Paine, M.; Fong, H.; Sarikaya, M.; Snead, M. Biological organization of hydroxyapatite crystallites into a fibrous continuum toughens and controls anisotropy in human enamel. *J. Dent. Res.* **2001**, *80*, 321–326. [CrossRef]
9. Xie, Z.-H.; Mahoney, E.; Kilpatrick, N.; Swain, M.; Hoffman, M. On the structure–property relationship of sound and hypomineralized enamel. *Acta Biomater.* **2007**, *3*, 865–872. [CrossRef]
10. Cui, F.Z.; Ge, J. New observations of the hierarchical structure of human enamel, from nanoscale to microscale. *J. Tissue Eng. Regen. Med.* **2007**, *1*, 185–191. [CrossRef]
11. Cuy, J.L.; Mann, A.B.; Livi, K.J.; Teaford, M.F.; Weihs, T.P. Nanoindentation mapping of the mechanical properties of human molar tooth enamel. *Arch. Oral Biol.* **2002**, *47*, 281–291. [CrossRef]
12. He, L.H.; Swain, M.V. Enamel—A functionally graded natural coating. *J. Dent.* **2009**, *37*, 596–603. [CrossRef]
13. Beniash, E.; Metzler, R.A.; Lam, R.S.; Gilbert, P. Transient amorphous calcium phosphate in forming enamel. *J. Struct. Biol.* **2009**, *166*, 133–143. [CrossRef] [PubMed]
14. Habelitz, S.; Marshall, S.J.; Marshall Jr, G.W.; Balooch, M. Mechanical properties of human dental enamel on the nanometre scale. *Arch. Oral Biol.* **2001**, *46*, 173–183. [CrossRef] [PubMed]
15. He, L.H.; Swain, M.V. Contact induced deformation of enamel. *Appl. Phys. Lett.* **2007**, *90*, 171916. [CrossRef]
16. Jia, Y.F.; Xuan, F.Z. Anisotropic fatigue behavior of human enamel characterized by multi-cycling nanoindentation. *J. Mech. Behav. Biomed. Mater.* **2012**, *16*, 163–168. [CrossRef] [PubMed]
17. Jeng, Y.R.; Lin, T.T.; Hsu, H.M.; Chang, H.J.; Shieh, D.B. Human enamel rod presents anisotropic nanotribological properties. *J. Mech. Behav. Biomed. Mater.* **2011**, *4*, 515–522. [CrossRef]
18. Ang, S.F.; Schulz, A.; Fernandes, R.P.; Schneider, G.A. Sub-10-micrometer toughening and crack tip toughness of dental enamel. *J. Mech. Behav. Biomed. Mater.* **2011**, *4*, 423–432. [CrossRef]
19. Xie, Z.; Swain, M.; Munroe, P.; Hoffman, M. On the critical parameters that regulate the deformation behaviour of tooth enamel. *Biomaterials* **2008**, *29*, 2697–2703. [CrossRef]
20. Rathsam, C.; Farahani, R.M.; Hains, P.G.; Valova, V.A.; Charadram, N.; Zoellner, H.; Swain, M.; Hunter, N. Characterization of inter-crystallite peptides in human enamel rods reveals contribution by the Y allele of amelogenin. *J. Struct. Biol.* **2018**, *204*, 26–37. [CrossRef]
21. An, B.; Wang, R.; Zhang, D. Role of crystal arrangement on the mechanical performance of enamel. *Acta Biomater.* **2012**, *8*, 3784–3793. [CrossRef] [PubMed]
22. Oliver, W.C.; Pharr, G.M. An improved technique for determining hardness and elastic modulus using load and displacement sensing indentation experiments. *J. Mater. Res.* **1992**, *7*, 1564–1583. [CrossRef]
23. Waters, N.E. Some mechanical and physical properties of teeth. *Symp. Soc. Exp. Biol.* **1980**, *34*, 99–135. [PubMed]
24. Moradian-Oldak, J. Protein-mediated enamel mineralization. *Front. Biosci. A J. Virtual Libr.* **2012**, *17*, 1996. [CrossRef]
25. Yilmaz, E.D.; Koldehoff, J.; Schneider, G.A. On the systematic documentation of the structural characteristics of bovine enamel, A critic to the protein sheath concept. *Dent. Mater.* **2018**, *34*, 1518–1530. [CrossRef] [PubMed]
26. Wang, C.; Li, Y.; Wang, X.; Zhang, L.; Tiantang Fu, B. The enamel microstructures of bovine mandibular incisors. *Anat.Rec. Adv. Integr. Anat. Evol. Biol.* **2012**, *295*, 1698–1706. [CrossRef]
27. Ge, J.C.F.Z.; Cui, F.Z.; Wang, X.M.; Feng, H.L. Property variations in the prism and the organic sheath within enamel by nanoindentation. *Biomaterials* **2005**, *26*, 3333–3339.
28. Mann, S. The biomimetics of enamel: A paradigm for organized biomaterials synthesis. In *Ciba Foundation Symposium 205-Dental Enamel: Dental Enamel: Ciba Foundation Symposium 205*; John Wiley & Sons, Ltd.: Chichester, UK, 2007; pp. 261–274.
29. Staines, M.; Robinson, W.H.; Hood, J.A.A. Spherical indentation of tooth enamel. *J. Mater. Sci.* **1981**, *16*, 2551–2556. [CrossRef]
30. Sha, M.C.; Li, Z.; Bradt, R.C. Single-crystal elastic constants of fluorapatite, $Ca_5F(PO_4)_3$. *J. Appl. Phys.* **1994**, *75*, 7784–7787. [CrossRef]
31. Lacruz, R.S.; Habelitz, S.; Wright, J.T.; Paine, M.L. Dental enamel formation and implications for oral health and disease. *Physiol. Rev.* **2017**, *97*, 939–993. [CrossRef]
32. Yahyazadehfar, M.; Arola, D. The role of organic proteins on the crack growth resistance of human enamel. *Acta Biomater.* **2015**, *19*, 33–45. [CrossRef]
33. He, L.H.; Fujisawa, N.; Swain, M.V. Elastic modulus and stress–strain response of human enamel by nano-indentation. *Biomaterials* **2006**, *27*, 4388–4398. [CrossRef]
34. He, L.H.; Swain, M.V. Enamel—A "metallic-like" deformable biocomposite. *J. Dent.* **2007**, *35*, 431–437. [CrossRef]

35. He, L.H.; Swain, M.V. Nanoindentation creep behavior of human enamel. *J. Biomed. Mater. Res.* **2009**, *91*, 352–359. [CrossRef] [PubMed]
36. Makrodimitris, K.; Masica, D.L.; Kim, E.T.; Gray, J.J. Structure prediction of protein–solid surface interactions reveals a molecular recognition motif of statherin for hydroxyapatite. *J. Am. Chem. Soc.* **2007**, *129*, 13713–13722. [CrossRef]
37. Katz, J.L. Hard tissue as a composite material—I. Bounds on the elastic behavior. *J. Biomech.* **1971**, *4*, 455–473. [CrossRef] [PubMed]
38. Spears, I.R. A three-dimensional finite element model of prismatic enamel: A re-appraisal of the data on the Young's modulus of enamel. *J. Dent. Res.* **1997**, *76*, 1690–1697. [CrossRef] [PubMed]
39. Zhao, H.; Liu, S.; Wei, Y.; Yue, Y.; Gao, M.; Li, Y.; Zeng, X.; Deng, X.; Kotov, N.A.; Guo, L.; et al. Multiscale engineered artificial tooth enamel. *Science* **2022**, *375*, 551–556. [CrossRef]
40. Meng, Y.F.; Zhu, Y.B.; Zhou, L.C.; Meng, X.S.; Yang, Y.L.; Zhao, R.; Xia, J.; Yang, B.; Lu, Y.J.; Wu, H.A.; et al. Artificial Nacre with High Toughness Amplification Factor: Residual Stress-Engineering Sparks Enhanced Extrinsic Toughening Mechanisms. *Adv. Mater.* **2022**, *34*, 2108267. [CrossRef]

coatings

Article

Ultrasound-Assisted Acellular Spinal Cord Scaffold for Spinal Cord Injury Treatment

Xi Deng [1,2], Yun Liu [3], Zhongsheng Xu [3,*] and Hong Yin [1,*]

1 Department of Orthopedics, Xinqiao Hospital, Army Medical University, Chongqing 400037, China; dx1114315@tmmu.edu.cn
2 Department of Ultrasound, Xinqiao Hospital, Army Medical University, Chongqing 400037, China
3 Second Affiliated Hospital, Chongqing Medical University, Chongqing 400010, China; yunliu@cqmu.edu.cn
* Correspondence: zhongshengxu@cqmu.edu.cn (Z.X.); yh123456@tmmu.edu.cn (H.Y.)

Abstract: Spinal cord injury (SCI) treatment remains challenging globally, with limited breakthroughs. Tissue engineering offers promise, particularly using acellular spinal cord scaffolds. This study developed a 1-ethyl-(3-dimethylaminopropyl) carbodiimide hydrochloride (EDC)-crosslinked vascular endothelial growth factor (VEGF)-modified acellular spinal cord scaffold for sustained VEGF release. The results show sustained VEGF release over 20 days without altering the scaffold's properties. Enhanced stability and mechanical properties were observed without increased cytotoxicity. In a rat SCI model, the system improved motor function, reduced glial scarring, and restored spinal cord morphology and histology, indicating potential for SCI therapy.

Keywords: acellular spinal cord scaffold; sustained-release system; spinal cord injury

1. Introduction

Spinal cord injury (SCI) has high rates of disability and mortality, posing a heavy burden on society and families. Currently, no effective treatment measures have been found for SCI [1]. The challenges in SCI repair and treatment research include primary and secondary neuronal cell death, lack of trophic factors in the local microenvironment of the injury, and the formation of glial scars at the injury site [2]. In recent years, various strategies have been explored for SCI repair. These include pharmacological treatments, such as the combination of ceftriaxone and methylprednisolone [3]; cellular transplantation therapies, where stem cells and other cell types are used to promote tissue repair [4]; and gene therapy approaches, like the use of microRNA markers for diagnosis and treatment [5]. Existing tissue-engineering scaffolds include natural degradable materials, like chitosan; synthetic polymer materials, such as poly(lactic-co-glycolic acid); and composite materials. However, these scaffolds often face limitations, such as insufficient three-dimensional biomimetic structures or suboptimal biocompatibility, which restrict their application in SCI regeneration and repair. A recent trend in tissue engineering is the use of natural decellularized tissues to construct advanced scaffolds. This approach has shown promise across various applications, including engineered hearts [6], blood vessels [7], tendons [8], corneas [9], and bones [10]. With the development of tissue engineering in recent years, there is hope that scaffold materials can be used as carriers to load seed cells and neurotrophic factors and implanted into injured sites to achieve the purpose for repairing damaged spinal cord tissue [11]. Among the three essential elements of tissue engineering, the construction of scaffold materials is a critical factor. Our research group has previously constructed allogeneic acellular spinal cord scaffold materials. These scaffolds possess a three-dimensional mesh structure that is highly biomimetic and possesses good biomechanical properties, low immunogenicity, and good biocompatibility [12]. However, these scaffolds face the challenge of inadequate vascularization, which is critical for tissue survival and regeneration, especially in large cell masses. To overcome this, we have developed a novel approach by introducing the neurotrophic factor VEGF (vascular endothelial growth factor) to the acellular spinal cord scaffold. This innovation aims to enhance vascularization by constructing a VEGF-modified acellular spinal cord tissue-engineering composite.

However, with further research, it has been found that the use of acellular spinal cord scaffolds alone for the regenerative repair of SCI faces the problem of the insufficient vascularization of the regenerated tissue in the transplant area. In spinal cord regeneration research, vascularization is crucial. Studies have shown that scaffolds and seed cells in tissue-engineered composites significantly impact tissue regeneration because of inadequate blood supply. For effective regeneration, seed cells must remain within 150–200 μm of blood vessels; cells beyond 200 μm from capillaries cannot survive because of a lack of nutrients and oxygen. Cell masses larger than 1 mm^3 are prone to necrosis if no vascular ingrowth occurs [13,14]. The EDC-crosslinked VEGF-modified acellular spinal cord scaffold developed in this study represents a significant advancement over existing approaches by addressing these vascularization challenges and ensuring a stable and sustained release of VEGF within the transplant microenvironment. This novel scaffold is expected to provide a more effective nutritional supply and a metabolite transport system at the transplantation site, thereby improving the prospects of SCI repair [15]. Simple physical methods, such as soaking or injection, cannot maintain the stable and sustained release of the VEGF in the transplant microenvironment. Studies have shown that 1-ethyl-3-(3-dimethylaminopropyl)carbodiimide hydrochloride (EDC) can covalently bind amino and carboxyl groups on protein molecules to form stable amide bonds and can slowly release bioactive proteins bound to three-dimensional porous collagen scaffolds into the microenvironment surrounding the scaffold [16].

Therefore, this study aims to use EDC as a crosslinker to stably bind VEGF to the collagen molecules of the acellular spinal cord scaffold through covalent bonds, enabling the slow release of the VEGF into the microenvironment surrounding the scaffold, thereby constructing a VEGF-modified acellular spinal cord scaffold sustained-release system. In vivo and in vitro experiments will be conducted to demonstrate the therapeutic effect of this sustained-release system on SCI repair, providing a theoretical basis for the design of novel tissue-engineered spinal cord scaffolds.

2. Materials and Methods

2.1. Materials

Sodium deoxycholate and Triton X-100 were purchased from Beijing Aoboxing Technology Co., Ltd. (Beijing, China). Rat neurotrophic factor-3 was purchased from BioVision (Milpitas, CA, USA), and 1-ethyl-(3-dimethylaminopropyl) carbodiimide hydrochloride (EDC), N-hydroxy succinimide (NHS), and VEGF were purchased from Thermo Scientific (Waltham, MA, USA). Anhydrous ethanol, a streptavidin–biotin–peroxidase complex (SABC) kit, and a CCK-8 (cell-counting kit 8) assay kit were purchased from Beyotime (Shanghai, China). Trypsin was purchased from Amresco (Solon, OH, USA). An HE (hematoxylin and eosin) staining kit was purchased from Beyotime. The anti-CD4 (anti-cluster of differentiation 4) antibody, anti-CD8 (anti-cluster of differentiation 8) antibody, and fluorescent secondary antibody were purchased from EarthOx Technology Co., Ltd. (Burlingame, CA, USA). Pentobarbital sodium was sourced from the Educational and Scientific Support Office of the Army Medical University. SD rats were obtained from the Animal Experiment Center of the Army Medical University. All the above chemical reagents were used as received without further purification.

2.2. Methods

2.2.1. Construction of VEGF-Modified Decellularized Spinal Cord Scaffold Slow-Release System

Adult Sprague–Dawley (SD) rats, weighing 230–260 g, regardless of sex, were provided by the Experimental Animal Center of the Third Affiliated Hospital of the Army Medical

University. A 0.3% sodium pentobarbital solution was prepared, and the SD rats were intraperitoneally anesthetized with a dosage of 40–50 mg/kg. After successful anesthesia, the SD rats were placed in a prone position, with their limbs fixed on a rat dissection board. The back was shaved, and, after routine disinfection and draping, the skin was longitudinally incised along the midline using dissection scissors. The paraspinal muscles were dissected away from the spinous processes to expose the spine. The spinal column was transected at the atlanto-occipital joint and the lumbosacral region. The laminae were progressively removed to expose the spinal cord, which was then extracted and cut into uniform 2 cm segments.

Freeze–Thaw + Ultrasonication + Chemical Extraction Method: The spinal cord specimens were stored in a deep freezer at −80 °C for 6 h and then thawed gradually in a −20 °C freezer, a 4 °C fridge, and a 37 °C water bath. The specimens were then soaked in distilled water for 6 h at room temperature. They were placed in a 2% Triton X-100/double-distilled water solution and treated with an ultrasonic disruptor (Sonics, Newtown, CT, USA, VCX-750) for 5 min, followed by continuous oscillation in a constant-temperature shaker for 3 h. After soaking and oscillating in PBS for 3 h, the specimens were placed in a 2% sodium deoxycholate/double-distilled water solution and treated with an ultrasonic disruptor for another 5 min, followed by continuous oscillation in the shaker for another 3 h. This process was repeated, and the spinal cord tissue was then dried in a freeze-dryer. The decellularized spinal cord scaffold was thus prepared [17].

The acellular spinal cord scaffold treated with freeze-drying was soaked in a PBS solution containing 1-ethyl-(3-dimethylaminopropyl) carbodiimide hydrochloride (EDC) and N-hydroxy succinimide (NHS) at room temperature for 20 min, with a solution ratio of (EDC:NHS:PBS = 24 mg:60 mg:1 mL). The scaffold was then removed and rinsed in PBS buffer for 5 min. After rinsing, the scaffold was soaked in a VEGF-containing PBS solution for 1 h for coupling treatment, with a VEGF solution concentration of 1 μg/mL. A VEGF-releasing acellular spinal cord scaffold system was prepared, and its physicochemical properties were compared with those of the pure acellular spinal cord scaffold.

2.2.2. Testing the Sustained-Release Characteristics of VEGF-Modified Decellularized Spinal Cord Scaffolds

The VEGF content in the VEGF-modified decellularized spinal cord scaffold sustained-release system was determined under different experimental conditions using ELISA (Beyotime, Nantong, China). The impacts of different preparation protocols on the VEGF-loading capacity of the modified decellularized spinal cord scaffolds were evaluated. The data were fit based on the loading capacity and time; the loading capacity curve was drawn, and an optimized modification scheme that can effectively increase the loading capacity of bioactive proteins was established.

The balance point between a high VEGF-loading capacity and good sustained-release characteristics of the VEGF-modified decellularized spinal cord scaffolds was explored, and the preparation protocol of the VEGF-modified decellularized spinal cord scaffold sustained-release system was further optimized as follows: The in vitro drug release properties of the microspheres were determined using the paddle method. Six equally modified decellularized spinal cords were placed in the dissolution cups of a D-800 intelligent drug dissolution tester (Tianjin University Precision Instrument Factory, Tianjin, China), and 500 mL of double-distilled water was added to each cup. The mixtures were stirred at a constant speed of 100 rpm at 37 ± 0.5 °C. Then, 2 mL of the sample solution was collected from each dissolution cup at each time point from 12 h to day 21 and promptly replenished with an equal volume of double-distilled water. The amount of the VEGF released from the sample solution was measured using the Elisa method; the cumulative release rate of the scaffold was calculated; the data were fit based on the cumulative release rate and time, and the sustained-release curve was drawn. The decellularized spinal cords that were not treated with EDC were directly immersed in VEGF and used as a control.

2.2.3. Biocompatibility Study of VEGF-Modified Decellularized Spinal Cord Scaffolds In Vivo and In Vitro

In vitro: Three groups of spinal cord scaffolds (Group A: normal spinal cord tissue, Group B: decellularized spinal cord tissue, and Group C: EDC-crosslinked decellularized spinal cord tissue) were prepared by slicing into sections approximately 200 μm thick, as shown in Figures 1–3. These sections were placed in EP tubes for sealed storage and then sterilized using Cobalt-60 irradiation (conducted at the Irradiation Center of the Army Medical University, with a dose of 20 kGy). A VEGF cell suspension with a concentration of 5×10^5 cells/mL was prepared. The sterilized scaffolds from Groups A, B, and C were placed into sterile cell culture plates, and 100 μL of the cell suspension was seeded onto each scaffold. The scaffolds with the cell complexes were then incubated for 2 h, after which endothelial cell culture medium was added to each well. The medium was replaced every 2 days, and the culture was maintained for a total of 7 days. To assess the relative growth rates of the endothelial cells on the scaffolds, a CCK-8 assay was performed. The control group consisted of endothelial cells cultured without scaffolds. Each culture plate was removed, and 110 μL of a CCK-8 and serum-free medium mixture (1:10) was added. The plates were incubated at 37 °C for 2 h, and the optical density (OD) values at 450 nm were measured using a microplate reader (Thermo Scientific).

In vivo: Healthy SD rats (Third Military Medical University) were randomly divided into four groups, with six rats in each group. Group A was the normal spinal cord tissue group, Group B was the decellularized-scaffold-only group, Group C was the EDC-crosslinked decellularized-scaffold group, and Group D was the negative control group. The SD rats were anesthetized by the intraperitoneal injection of 2% pentobarbital sodium. After successful anesthesia, the backs of the rats were shaved and prepared for surgery and then fixed on the surgical operation board for routine disinfection and draping. A longitudinal incision was made along the posterior midline spinous process, and the subcutaneous and deep fascia were bluntly separated. The scaffolds of Groups A, B, and C were implanted under the deep fascia, and the wounds were closed layer by layer. In Group D, only surgical exposure was performed, and the wound was directly sutured. All the rats received an intramuscular injection of penicillin postoperatively. Specimens were collected from each group at four time points: one week, two weeks, seven weeks, and fourteen weeks. The paraffin sections of the scaffolds from Groups A, B, and C at each time point were stained with HE for histological observation under a microscope (Leica, Wetzlar, Germany). The immunogenicity of the scaffolds in Groups A, B, C, and D was evaluated by the number of CD4+ and CD8+ T-lymphocytes. Specimens from each group at 4 weeks after implantation were taken for CD4 and CD8 immunohistochemical staining (using the SABC kit as the secondary antibody and 3,3′-diaminobenzidine (DAB) for color development). The CD4+ and CD8+ T-lymphocytes (cluster-of-differentiation-4-positive and cluster-of-differentiation-8-positive T-lymphocytes) in each group were observed under a microscope. Four random 400× fields were selected to measure the number of positive cells, and the mean value was calculated.

2.2.4. VEGF-Modified Decellularized Spinal Cord Scaffold Sustained-Release System Transplantation for Spinal Cord Injury

A rat T10 semi-transection model was prepared, and the EDC-crosslinked VEGF-modified decellularized spinal cord scaffold sustained-release system was implanted (anatomical, isometric transplantation) to observe its repair effect on rat spinal cord injury and explore its possible mechanisms. SD rats were randomly divided into four groups: Group A (laminectomy group), Group B (simple spinal cord semi-transection group), Group C (simple decellularized spinal cord scaffold transplantation group), and Group D (VEGF-modified decellularized spinal cord scaffold transplantation group). The Basso, Beattie, and Bresnahan (BBB) scoring of the motor function; motor-evoked potential (MEP) measurements; gross morphological observations of postoperative specimens; HE-staining observations of postoperative specimens; and biotinylated dextran amine (BDA) anterograde tracing were performed at corresponding time points.

2.2.5. Porosity, Hydration Rate, and Enzymatic Degradation Rate

Porosity refers to the ratio of the total volume of small pores within a porous scaffold to the total volume of the scaffold. It is expressed by the formula φ = (pore volume within the scaffold (V1)/total volume of the scaffold (V2)) × 100%. To measure the porosity, the normal spinal cord tissue (Group A), decellularized spinal cord scaffold (Group B), and EDC-crosslinked decellularized spinal cord scaffold (Group C) were subjected to freeze-drying. Each scaffold was placed in a graduated test tube, and anhydrous ethanol was added (initial volume (V1)). After ensuring that the scaffolds were fully immersed, the new volume was recorded as V2. The volume of each scaffold was, thus, V2 − V1. The scaffolds were then removed, and the volume of the ethanol after removing the scaffolds was noted as V3. The volume of the pores within each scaffold was V1 − V3. The total volume of the scaffold was calculated as (V2 − V1) + (V1 − V3) = V2 − V3. The porosity of each scaffold was then determined using the following formula: Porosity = ((V1 − V3)/V2 − V3) × 100%.

The hydration rate refers to the abilities of the freeze-dried normal spinal cord tissue (Group A), simple acellular spinal cord scaffold (Group B), and EDC-crosslinked acellular spinal cord scaffold (Group C) to bind water molecules. After freeze-drying, Groups A, B, and C were weighed, i.e., dry weight (T1), and immersed in a PBS buffer solution at room temperature for 2 h. Then, the brackets of each group were placed on the absorbent filter paper and weighed after absorbing more water, that is, wet weight (T2). The hydration rate is as follows: α = (T2 − T1)/T1 × 100%.

The enzymatic hydrolysis rate refers to the freeze-dried normal spinal cord tissue (Group A), simple acellular spinal cord scaffold (Group B), and EDC-crosslinked acellular spinal cord scaffold (Group C) weighed (T1), immersed in a trypsin–PBS solution (concentration: 1 mg/mL) at 37 °C (room temperature), and placed in a constant-temperature generator. They were taken out at 5 phase points: 12 h, 24 h, 2 days, 4 days, 8 days, and 12 days and freeze-dried and weighed (T2). The formula of the enzymatic hydrolysis rate is as follows: β = ((T1 − T2)/T1) × 100%.

2.2.6. Biomechanical Property Testing

In the Biomechanics Laboratory at Chongqing University, a biomechanical testing device (Instron model 1011) was used to evaluate the normal spinal cord tissue (Group A), decellularized spinal cord scaffold (Group B), and EDC-crosslinked decellularized spinal cord scaffold (Group C). The freeze-dried scaffolds from Groups A, B, and C were fully hydrated by immersing them in a PBS buffer solution for 12 h. After hydration, the scaffolds were removed and fixed at both ends to the testing apparatus at room temperature. The initial tensile force was set at zero Newtons, and a caliper was used to measure the length and diameter of each scaffold. A maximum tensile force of 50 Newtons was applied at a constant rate of 1 mm/min. After the scaffold ruptured, the computer system automatically generated the maximum tensile stress, maximum tensile strain, and elastic modulus.

2.2.7. In Vivo Subcutaneous Implantation Experiment

The SD rats were anesthetized with an intraperitoneal injection of 2% sodium pentobarbital. Once successful, the rats were shaved on their backs, fixed on a surgical board, and prepared with routine disinfection and draping. A longitudinal incision was made along the midline of the back, and blunt dissection was performed to separate the subcutaneous tissue from the deep fascia. Scaffolds from Groups A, B, and C were implanted in a subfascial manner, and the incision was closed in layers. For Group D, only a surgical exposure was performed, and the wound was directly sutured. All the rats received intramuscular injections of penicillin postoperatively. Samples from each group were collected at four time points: one week, two weeks, seven weeks, and fourteen weeks. After retrieval, all the specimens were fixed in formalin, embedded in paraffin, and sectioned. And the specimens were collected for hematoxylin and eosin (HE) staining and to evaluate the therapy's efficacy.

2.2.8. Immunohistochemical Observations

To evaluate the immunogenicity of scaffolds from Groups A, B, C, and D, the numbers of CD4+ and CD8+ T-lymphocytes were assessed. Sections from the subcutaneous implantation experiment at 4 weeks were used for this analysis. CD4 and CD8 immunohistochemical staining was performed using a secondary antibody from the SABC kit, with DAB as the chromogen. Microscopic (Leica) examination was conducted to observe the CD4+ and CD8+ T-lymphocyte distributions in each group. Four random fields at 400× magnification were selected, and the numbers of positive cells were counted and averaged. The paraffin sections were deparaffinized and hydrated and then treated with a 3% hydrogen peroxide solution for 10 min to quench the endogenous peroxidase activity. The sections were then rinsed with distilled water. Next, the sections were placed in a citrate buffer and heated to boiling in an oven and then cooled and washed in PBS buffer. The sections were blocked with 5% fetal bovine serum for 20 min, and excess liquid was removed. Primary antibodies (rabbit anti-CD4 and anti-CD8) were added to the slides and incubated overnight. The slides were washed with PBS for 2 min and then incubated with anti-rabbit IgG for 20 min, followed by another 2 min PBS wash. The slides were then incubated with SABC for 20 min and washed with PBS for 5 min. The DAB chromogen was added to the slides, and, after achieving satisfactory staining under the microscope, staining was terminated by rinsing the slides with distilled water. The slides were counterstained with hematoxylin, rinsed with distilled water, dehydrated through an alcohol gradient, cleared with xylene, and mounted with neutral resin.

2.2.9. Analysis of Statistical Results

The data were processed using the SPSS 13.0 statistical software package. The specimens were randomly selected from each group of stent data for measurements (N = 6). The results were expressed as means ± standard deviations ($x \pm s$). The comparisons between the groups were performed using one-way analysis of variance and independent sample *t*-tests for statistical analysis. The test level was set at $\alpha = 0.05$; ($p < 0.05$ indicates a significant difference, and $p < 0.01$ indicates a highly significant difference, both with statistical significance).

3. Results

3.1. Construction of VEGF-Modified Decellularized Spinal Cord Scaffold Sustained-Release System and Measurement of Its Physical and Chemical Properties

3.1.1. General Morphology

The general morphology of the normal spinal cord tissue is cylindrical, with a creamy white color and sparsely distributed capillary networks on the surface (Figure 1a). The simple scaffold group appears milky white and cylindrical, with a length slightly shorter than the normal spinal cord tissue (Figure 1b). The EDC-crosslinked decellularized spinal cord scaffold group appears yellowish brown and cylindrical, with a length slightly shorter than the normal spinal cord tissue (Figure 1c).

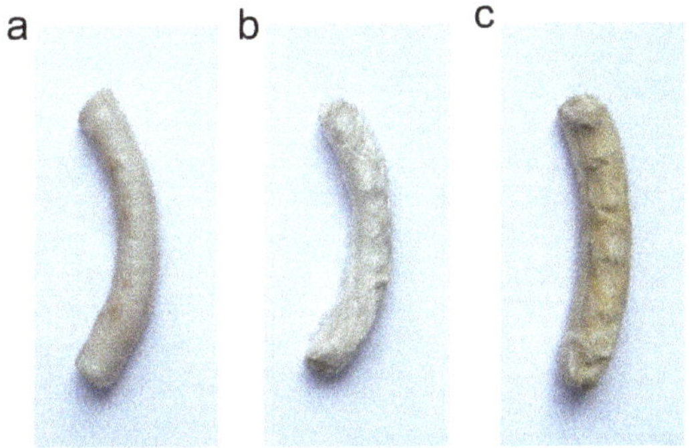

Figure 1. (**a**) Normal spinal cord tissue, magnification 2×; (**b**) purely acellular spinal cord tissue; (**c**) EDC-crosslinked acellular spinal cord scaffold.

3.1.2. Histology

HE staining: The HE-staining sections of the normal spinal cord tissue in Group A showed a large number of nerve cells, myelin sheaths, and an extracellular matrix closely integrated, as shown in Figure 2a,b (Figure 2a is a cross-section; Figure 2b is a longitudinal section). The HE-staining sections of the acellular spinal cord scaffolds in Group B and EDC-crosslinked acellular spinal cord scaffolds in Group C showed that both scaffolds exhibited a network-like structure under optical microscopy, with the complete removal of cells and a relatively intact retention of the extracellular matrix components (Figure 2c–f).

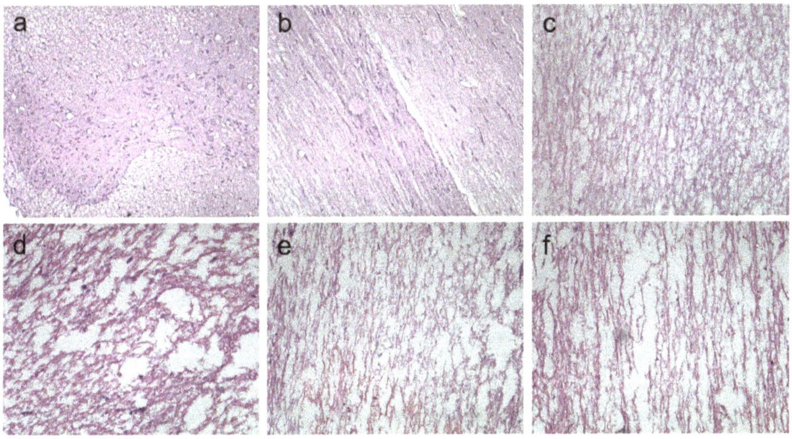

Figure 2. (**a**) Normal spinal cord tissue from Group A (transverse section); (**b**) normal spinal cord tissue from Group A (longitudinal section); (**c**) simple acellular spinal cord scaffold in Group B (cross-section); (**d**) simple acellular spinal cord scaffold of Group B (longitudinal section); (**e**) cross-section of EDC-crosslinked acellular spinal cord scaffold in Group C; (**f**) crosslinked acellular spinal cord scaffold of Group C (longitudinal section); magnification: 400×.

3.1.3. Scanning Electron Microscopy and Measurement of Pore Size

Under scanning electron microscopy observation (Crossbeam 340, ZEISS, Jena, Germany), we can see in Group A (normal spinal cord scaffold) that the longitudinal section (Figure 3a) shows significant wrinkling of the tissue, with no visible three-dimensional reticulate fibrous structure. The cross-section of the normal spinal cord scaffold (Figure 3b) has a compact structure, with no obvious porous reticulate structure visible. In contrast, comparing the longitudinal sections of Group B (purely decellularized spinal cord scaffold) and Group C (EDC-crosslinked decellularized spinal cord scaffold) (Figure 3c,d), there is no significant difference observed under scanning electron microscopy. Both groups show an internal structure, with irregularly intersecting pores that are interconnected to form a complex three-dimensional network. The cross-sections of Group B and Group C (Figure 3e,f) show similar scanning electron microscopy results, with the extracellular collagen matrix of the decellularized scaffolds preserved. Through measurements using the built-in analysis software of the scanning electron microscope, the average pore size in Group A was 0 μm, while the average pore sizes in Group B and Group C were 47.3 ± 18.3 μm and 42.8 ± 13.7 μm, respectively (Figure 3g). There was no statistically significant difference between the pore sizes of Group B (purely decellularized spinal cord scaffold group) and Group C (EDC-crosslinked decellularized spinal cord scaffold group) ($p > 0.05$).

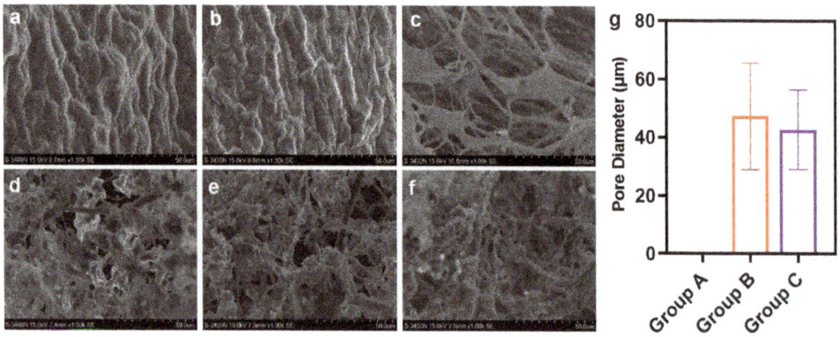

Figure 3. (**a**) Longitudinal section of normal spinal cord tissue in Group A; (**b**) cross-section of normal spinal cord tissue in Group A; (**c**) longitudinal section of the simple acellular spinal cord scaffold in Group B; (**d**) longitudinal section of the EDC-crosslinked decellularized spinal cord scaffold in Group C; (**e**) cross-section of the acellular spinal cord scaffold in Group B; (**f**) cross-section of EDC-crosslinked decellularized spinal cord scaffold from Group C; (**g**) comparison of pore sizes in Groups A, B, and C; magnification: 400×.

3.1.4. Porosity, Hydration Rate, and Enzymatic Hydrolysis Rate

Porosity refers to the ratio of the total volume of tiny pores within the porous scaffold to the total volume of the porous scaffold. The calculation formula is φ = (pore volume within the scaffold (V1)/total volume of the scaffold (V2)) × 100%. Upon measurement, the porosity of Group A (normal spinal cord tissue group) ranged from a maximum of 14.8% to a minimum of 9.7%, with an average of (12.1 ± 2.2)%. Group B (acellular spinal cord scaffold group) had a porosity ranging from a maximum of 94.8% to a minimum of 82.5%, with an average of (89.4 ± 4.3)%. Group C (EDC-crosslinked acellular spinal cord scaffold group) had a porosity ranging from a maximum of 92.1% to a minimum of 80.4%, with an average of (87.4 ± 5.1)%. There was a statistically significant difference between Group A and Group B ($p < 0.01$), as well as between Group A and Group C ($p < 0.01$). However, there was no statistically significant difference between Group B and Group C ($p > 0.05$). It can be seen that the porosities of the acellular spinal cord scaffold group and the EDC-crosslinked acellular spinal cord scaffold group are significantly higher than that of the normal spinal cord group, and EDC crosslinking has no significant effect on the porosity of the acellular spinal cord scaffold (Figure 4a). The hydration rate refers to the ability of scaffolds in each group, including normal spinal cord tissue after the freeze-drying treatment (Group A), decellularized spinal cord scaffolds alone (Group B), and EDC-crosslinked decellularized spinal cord scaffolds (Group C), to bind water molecules. The calculation formula is as follows: $\alpha = (T2 - T1)/T1 \times 100\%$. The measured results showed that the hydration rate of Group A (normal spinal cord tissue group) ranged from a minimum of 42.1% to a maximum of 62.5%, with an average of (52.3 ± 7.8)%. The hydration rate of Group B (decellularized spinal cord scaffold group alone) ranged from a minimum of 190.4% to a maximum of 212.3%, with an average of (199.4 ± 8.4)%. The hydration rate of Group C (EDC-crosslinked decellularized spinal cord scaffold group) ranged from a minimum of 170.1% to a maximum of 192.8%, with an average of (182.2 ± 8.3)%. It can be seen that the hydration rates of the Group B and Group C scaffolds were significantly higher than that of Group A, and this difference was statistically significant ($p < 0.05$). The hydration rate of the Group C scaffolds was slightly lower than that of Group B, and there was no statistical significance between the two groups ($p > 0.05$). It can be seen that the hydration rates of the decellularized spinal cord scaffolds and EDC-crosslinked decellularized spinal cord scaffolds were higher than that of the normal spinal cord scaffolds, and EDC crosslinking can cause a slight decrease in the hydration rate of the decellularized spinal cord scaffolds (Figure 4b). The enzymatic degradation rate refers to the rate of the dissolution and digestion of scaffolds from Groups A, B, and C after freeze-drying them in a trypsin solution. The calculation formula is $\beta = ((t1 - t2)/t1) \times 100\%$. The average enzymatic degradation rates of Group A (normal spinal cord tissue group) at various time points of 12 h, 24 h, 2 days, 4 days, 8 days, and 12 days under room-temperature conditions at 37 °C were as follows: (5.1 ± 1.3)%, (9.7 ± 0.7)%, (12.9 ± 1.5)%, (20.3 ± 1.5)%, (24.5 ± 1.0)%, and (31.1 ± 2.1)%, respectively. For Group B (purely acellular spinal cord scaffold group), the average enzymatic degradation rates at the same time points were (43.9 ± 5.9)%, (76.0 ± 7.8)%, (97.1 ± 3.3)%, 100%, 100%, and 100%, respectively. For Group C (EDC-crosslinked acellular spinal cord scaffold group), the average enzymatic degradation rates at the same time points were (5.8 ± 0.9)%, (8.3 ± 0.6)%, (11.1 ± 0.9)%, (16.1 ± 0.8)%, (21.8 ± 1.7)%, and (27.6 ± 1.4)%, respectively. As shown in the figure below (Figure 4c), the enzymatic degradation rate of Group B was significantly lower than those of Groups A and C, with statistical significance ($p < 0.01$). The enzymatic degradation rate of Group C was slightly lower than that of Group A, and there was no statistical significance between the two groups ($p > 0.05$). This suggests that EDC crosslinking can significantly improve the enzymatic resistance of the purely acellular spinal cord scaffolds, making these scaffolds' enzymatic resistance close to that of the normal spinal cord tissue.

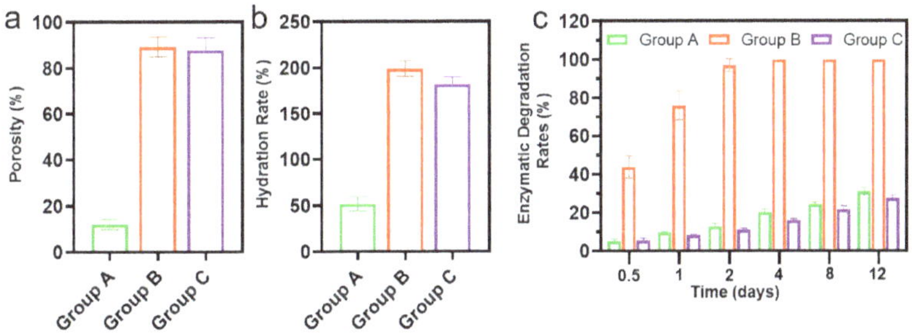

Figure 4. (**a**) Comparison of porosities among scaffolds in Groups A, B, and C; (**b**) comparison of hydration rates among scaffolds in Groups A, B, and C; (**c**) comparison of enzymatic degradation rates among scaffolds in Groups A, B, and C.

3.1.5. Biomechanical Characterization

The biomechanical results of Group A (normal rat spinal cord tissue), Group B (purely acellular spinal cord scaffold), and Group C (EDC-crosslinked acellular spinal cord scaffold) are shown in Figure 5. According to the maximum tensile stress diagrams of the three groups of scaffolds, the maximum tensile stress of both Group B and Group C scaffolds is lower than that of Group A, with statistically significant differences ($p < 0.01$). The maximum tensile stress of the Group C scaffolds is greater than that of Group B, with statistically significant differences ($p < 0.05$). This indicates that the maximum tensile stress of the normal spinal cord tissue is significantly greater than those of both the purely acellular spinal cord scaffold and the EDC-crosslinked acellular spinal cord scaffold. The maximum tensile stress of the EDC-crosslinked acellular spinal cord scaffold is significantly improved compared to that of the purely acellular spinal cord scaffold (Figure 5a). The maximum tensile strain diagrams of the three groups of scaffolds show that there is no significant difference in the maximum tensile strain among Groups A, B, and C, with no statistically significant differences ($p > 0.05$). This suggests that the maximum tensile strain of the normal spinal cord tissue does not change significantly after acellular treatment, and, after EDC-crosslinking, the maximum tensile strain of the Group C scaffolds does not change significantly compared to that of the normal spinal cord tissue (Figure 5b). The elastic modulus diagrams of the three groups of scaffolds reveal that the elastic moduli of the Group B and Group C scaffolds are lower than that of Group A, with statistically significant differences ($p < 0.01$). The elastic modulus of the Group C scaffolds is greater than that of Group B, with statistically significant differences ($p < 0.05$). This indicates that the elastic modulus of the normal spinal cord tissue is significantly greater than those of both the purely acellular spinal cord scaffold and the EDC-crosslinked acellular spinal cord scaffold. The elastic modulus of the EDC-crosslinked acellular spinal cord scaffold is significantly improved compared to that of the purely acellular spinal cord scaffold (Figure 5c).

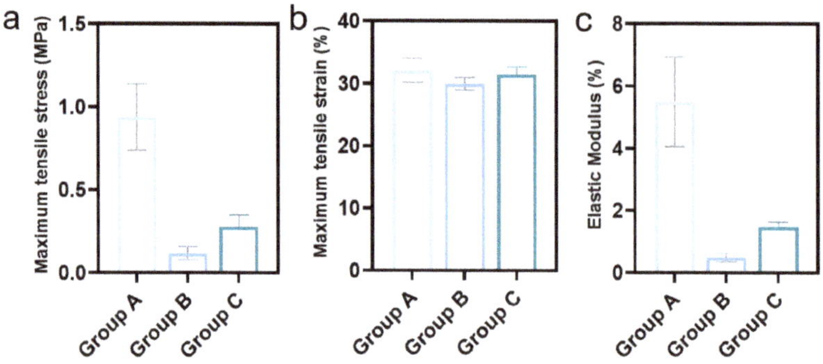

Figure 5. (**a**) Comparison of the maximum tensile stress among scaffolds in Groups A, B, and C; (**b**) comparison of the maximum tensile strain of the stents in Groups A, B, and C; (**c**) comparison of the elasticity modulus of the stents in Groups A, B, and C.

3.2. *Sustained-Release Characteristics of VEGF-Modified Decellularized Spinal Cord Scaffolds*

From the single sustained-release curve graphs of the two scaffold groups (Figure 6a), it can be seen that Group A (EDC-crosslinked decellularized spinal cord scaffold group) has the highest release amounts at 12 h and 24 h, with the concentration of the free VEGF in the sustained-release solution decreasing over time. Detected using ELISA kits, the sustained-release of the VEGF can last for more than 20 days. Group B (non-crosslinked decellularized spinal cord scaffold group) has significantly higher release amounts than Group A at 12 h and 24 h, with a sharp decrease in the release amount on the second day, which is significantly lower than that of Group A, forming a clear burst-release phenomenon. On the eighth day, almost no free VEGF can be detected in the sustained-release solution through ELISA kits.

From the cumulative sustained-release curve graphs of the two scaffold groups (Figure 6b), it can be seen that the curve of Group A (EDC-crosslinked decellularized spinal cord scaffold group) shows a steady upward trend in the cumulative release over time during the sustained-release process. The curve of Group B (non-crosslinked decellularized spinal cord scaffold group) reaches a plateau after 24 h, and negligible free VEGF is released later. It can be seen that almost all the VEGF in the Group B scaffolds is burst-released within the first day. Therefore, the EDC-crosslinked decellularized spinal cord scaffold can maintain a slow and sustained release of VEGF for a certain period of time, demonstrating certain sustained-release characteristics.

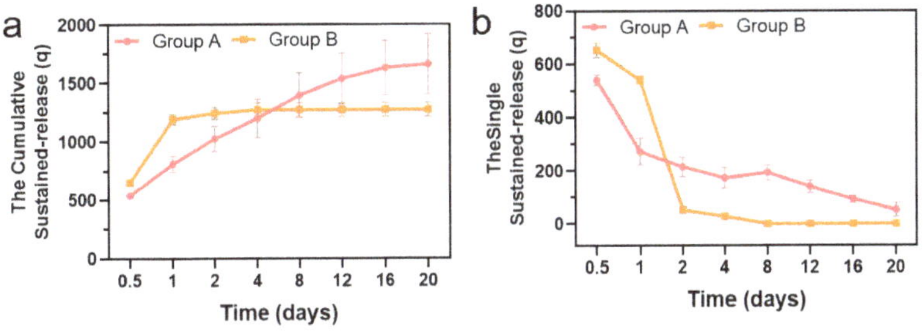

Figure 6. (**a**) Single-dose sustained-release curves; (**b**) cumulative sustained-release curves.

3.3. Biocompatibility and Biosafety of VEGF-Modified Decellularized Spinal Cord Scaffolds

The relative growth rates of the rat vascular endothelial cells cultured with scaffolds from Groups A, B, and C for one, three, five, and seven days were determined using the CCK-8 method. The results are presented in Figure 7a. It can be seen that after the coculture of the vascular endothelial cells with the normal spinal cord (Group A), decellularized spinal cord scaffolds alone (Group B), and EDC-crosslinked decellularized spinal cord scaffolds (Group C), the relative growth rates of Group A were slightly higher than those of Group B and Group C at each time point, but the differences were not statistically significant ($p > 0.05$). The relative growth rates of Group B were slightly higher than those of Group C at each time point, but the differences were also not statistically significant ($p > 0.05$). Therefore, both the pure decellularized spinal cord scaffolds and the EDC-crosslinked decellularized spinal cord scaffolds exhibited a cytotoxicity similar to that of normal spinal cord tissue cells.

The HE staining was observed in subcutaneous embedding experiments for the scaffolds in Groups A, B, and C. The specimens removed from the embedding experiments of Group A (normal spinal cord tissue group), Group B (acellular spinal cord scaffold group), and Group C (EDC-crosslinked acellular spinal cord scaffold group) at different time points were observed after paraffin sectioning and HE staining to observe the tissue responses of the different scaffolds under the skin. As shown in Figure 7b below, one week after the subcutaneous embedding of the scaffolds in Groups A, B, and C, many reactive inflammatory cells were observed under the microscope, similar to the body's foreign body response. And, two weeks after the subcutaneous embedding of the scaffolds in the three groups, the reactive inflammatory cells under the microscope were significantly reduced in Groups A, B, and C. The tissue structure of Groups A and B was loose, and proliferative fibroblasts and new blood vessels were visible. Similarly, proliferative fibroblasts and new blood vessels were observed around the scaffolds in Group C. Seven weeks after the subcutaneous embedding of the scaffolds in the three groups, the scaffolds in Groups A and B were partially organized and absorbed, and a large number of proliferative fibrous tissues separated and wrapped the scaffolds. This organized absorption of scaffolds in Group C was not obvious, and the tissue structure was still tight, but there was still fibrous tissue surrounding it. Fourteen weeks after the subcutaneous embedding of the scaffolds in the three groups, the scaffolds in Groups A and B were almost completely organized and absorbed and replaced by proliferative fibrous tissues. Some scaffolds in Group C also began to be organized and absorbed, separated, and wrapped by proliferative fibrous tissues and were completely organized and absorbed within half a year after embedding. It can be seen that the scaffolds in Groups A, B, and C showed a certain degree of foreign body response in vivo, but no significant rejection immune reaction occurred. The scaffolds in Groups A and B were absorbed rapidly in vivo, while the scaffolds in Group C, crosslinked with EDC, could slow down the absorption process of the scaffolds in vivo.

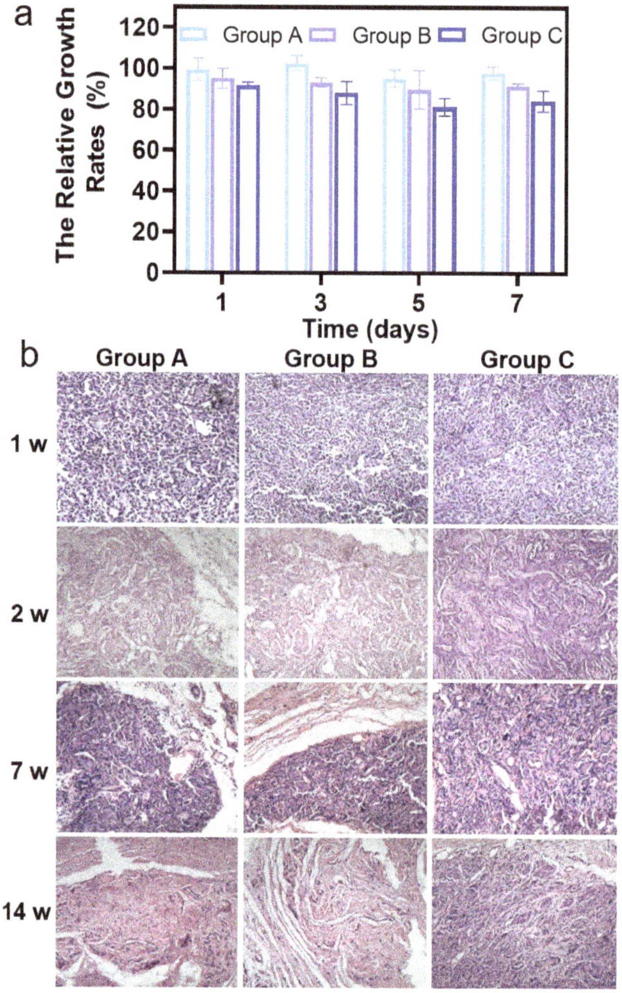

Figure 7. (**a**) Relative growth rates of vascular endothelial cells in the stent; (**b**) HE-staining images of subcutaneously embedded scaffolds of Groups A, B, and C in the stent after different times (1, 2, 7, and 14 weeks); magnification: 200×.

3.4. Immunohistochemical Observation of Subcutaneous Implantation of EDC-Crosslinked Decellularized Spinal Cord Scaffolds

Four groups of implantation experiments, including Group A (normal spinal cord tissue group), Group B (decellularized spinal cord scaffold group), Group C (EDC-crosslinked decellularized spinal cord scaffold group), and Group D (negative control group), were performed. Four weeks after surgery, the transplanted specimens from all four groups were retrieved and subjected to immunohistochemical detection of CD4 and CD8. The cytoplasm of the CD4- and CD8-positive T-cells appeared brownish. The immunohistochemical results of the normal spinal cord group (Group A) showed a large number of CD4- and CD8-positive T-cells (Figure 8a). In the decellularized spinal cord scaffold group (Group B), the numbers of CD4- and CD8-positive T-cells were significantly lower than in Group A (Figure 8a). The immunohistochemical results of the EDC-crosslinked decellularized spinal cord scaffold group were similar to those of the negative control group (Figure 8a), showing only a small number of CD4- and CD8-positive T-cells.

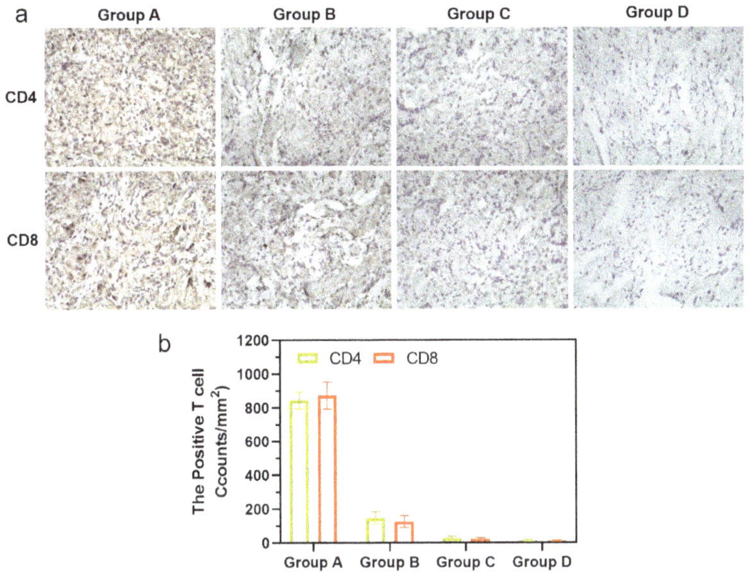

Figure 8. (**a**) Immunohistochemical results of T-cells in Groups A, B, C, and D after four weeks of subcutaneous embedding; magnification: 400×; (**b**) CD4- and CD8-positive T-cell counts in specimens from Groups A, B, C, and D.

The results of the CD4- and CD8-positive T-cell counts are shown in Figure 8b. It can be seen that the CD4- and CD8-positive T-cell counts in the acellular spinal cord scaffold group (Group B), the EDC-crosslinked acellular spinal cord scaffold group (Group C), and the negative control group (Group D) were significantly lower than those in the normal spinal cord tissue group (Group A), with statistically significant differences ($p < 0.01$). The CD4- and CD8-positive T-cell counts in the EDC-crosslinked acellular spinal cord scaffold group (Group C) were significantly lower than those in the acellular spinal cord scaffold group (Group B), with statistically significant differences ($p < 0.05$). This suggests that the immunogenicity of both the acellular spinal cord scaffold and the EDC-crosslinked acellular spinal cord scaffold is significantly lower than that of normal spinal cord tissue, and the immunogenicity of the EDC-crosslinked acellular spinal cord scaffold is lower than that of the acellular spinal cord tissue.

3.5. Experimental Study on VEGF-Modified EDC-Crosslinked Acellular Spinal Cord Scaffold Transplantation for the Treatment of Spinal Cord Injury

3.5.1. BBB Scores of Motor Function in Rats from Groups A, B, C, and D after Surgery

Observations were made from two weeks postoperatively to the fourth month postoperatively. All the rats in Group A (laminectomy group) survived, and there was no significant change in the BBB scores of the motor function in the right hindlimb before and after surgery, which were both normal. In Group B (simple spinal cord hemisection group, N = 24), one rat died on the day after surgery, and two rats died three weeks postoperatively. After death, the number of rats in this group was replenished using modeling. Except for two rats with slight hip flexion activity in the right lower limb, all the other rats in this group exhibited complete flaccid paralysis of the right lower limb. In Group C (simple acellular spinal cord scaffold transplantation group, N = 24), two rats died on the day after surgery, one rat died two weeks postoperatively, and three rats died six weeks postoperatively. After death, the number of rats in this group was replenished using modeling. In Group D (VEGF-modified acellular spinal cord scaffold transplantation group, N = 24), one rat died on the day after surgery, two rats died one week postoperatively, two rats died three weeks postoperatively, and three rats died eight weeks postoperatively. After death, the number of rats in this group was replenished using modeling. In Groups C and D, partial joint activity (mainly in the right hip joint) in the right lower limbs of some rats was observed two weeks postoperatively, and non-weight-bearing sliding of the right lower limb appeared over time. According to the BBB score results, the scores of Groups C and D were higher than those of Group B but lower than those of Group A, with statistical significance ($p < 0.05$). The results of Group C were lower than those of Group D, and there was no statistical significance at two weeks, one month, and two months postoperatively ($p > 0.05$). However, at four months postoperatively, the results of Group C were significantly lower than those of Group D, with statistical significance ($p < 0.05$). The results are shown in Figure 9a.

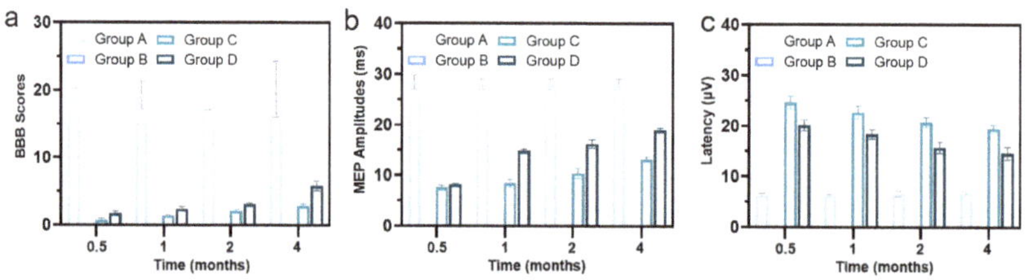

Figure 9. (**a**) BBB score results of rats in Groups A, B, C, and D; (**b**) MEP amplitude results of rats in Groups A, B, C, and D; (**c**) results of MEP latency in rats of Groups A, B, C, and D.

3.5.2. Measurement of Motor-Evoked Potentials (MEPs) in Rats of Groups A, B, C, and D after Surgery

Literature reports indicate that in rats with spinal cord injury, there is a decrease in the amplitude of motor-evoked potentials (MEPs) accompanied by the prolongation of latency [18]. Observations were made at two weeks, one month, two months, and four months postoperatively. All the rats in Group A (laminectomy group) survived, and the MEP amplitudes of the right lower limbs were (28.45 ± 1.47) ms, (27.92 ± 1.13) ms, (28.05 ± 1.05) ms, and (28.14 ± 0.98) ms, respectively, and the latency results were (6.24 ± 0.43) µV, (6.18 ± 0.27) µV, (6.58 ± 0.53) µV, and (6.69 ± 0.38) µV, respectively. In Group B (simple spinal cord hemisection group), no MEP amplitudes or latencies were elicited in the right lower limbs of the rats, with results of 0 ms and 0 µV. In Group C (simple acellular spinal cord scaffold transplantation group), the MEP amplitudes of the right lower limbs were (7.56 ± 0.48) ms, (8.47 ± 0.79) ms, (10.45 ± 0.97) ms, and (13.21 ± 0.54) ms, respectively, and the latency results were (24.59 ± 1.35) µV, (22.73 ± 1.27) µV, (20.47 ± 1.42) µV, and (19.25 ± 1.19) µV, respectively. In Group D (VEGF-modified acellular spinal cord scaffold transplantation group), the MEP amplitudes of the right lower limbs were (8.13 ± 0.21) ms, (14.78 ± 0.41) ms, (16.27 ± 0.88) ms, and (19.02 ± 0.48) ms, respectively, and the latency results were (20.19 ± 1.08) µV, (18.43 ± 0.92) µV, (15.69 ± 1.13) µV, and (14.58 ± 1.26) µV, respectively. The MEP amplitudes of the rats in Groups C and D were significantly lower than those in Group A and higher than those in Group B at each postoperative time point ($p < 0.01$) (Figure 9b). The MEP latencies of the rats in Groups C and D were significantly longer than those in Group A and shorter than those in Group B at each postoperative time point ($p < 0.01$) (Figure 9c). At each postoperative time point, the amplitudes in Group D were greater than those in Group C, and the latencies were slightly shorter in Group D compared to Group C, with statistical significance ($p < 0.05$).

3.5.3. Gross Morphological Observation of Rats in Groups A, B, C, and D after Surgery

Group A (laminectomy group): There were no abnormal gross morphological findings in the specimens at two months and four months after surgery (Figure 10). The spinal cord continuity was good, and no hyperemia or edema was observed on the surface of the dura mater;

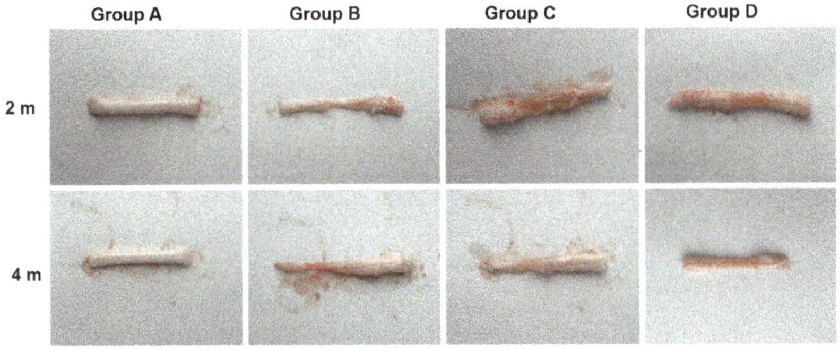

Figure 10. General morphologies of Groups A, B, C, and D two or four months (2 m or 4 m) after surgery.

Group B (simple spinal cord hemisection group): Gross morphological observation of the specimens at two months and four months after surgery showed that the defect in the right hemisection area persisted, and no spinal cord tissue growth was observed at the upper and lower ends of the transection side (Figure 10);

Group C (simple acellular spinal cord scaffold transplantation group): Gross morphological observation of the specimens at two months after surgery showed that the acellular spinal cord scaffold in the right hemitransplantation area had restored morphological connection with the spinal cord tissue at the upper and lower ends, and the scaffold was significantly adhered to the surrounding soft tissue, making separation difficult (Figure 10). Gross morphological observation of the specimens at four months after surgery showed that the right hemitransplantation area had restored morphological connection with the upper and lower ends, and the scaffold in the transplantation area appeared thinner, tougher, and darker in color compared to the spinal cord at the cephalic and caudal ends. The scaffold was not significantly adhered to the surrounding soft tissue, but separation was still difficult (Figure 10);

Group D (VEGF-modified acellular spinal cord scaffold transplantation group): Gross morphological observation of the specimens at two months after surgery showed that the acellular spinal cord scaffold in the right hemitransplantation area had restored morphological connection with the spinal cord tissue at the upper and lower ends, and the scaffold was not significantly adhered to the surrounding soft tissue (Figure 10). Gross morphological observation of the specimens at four months after surgery showed that the right hemitransplantation area had restored morphological connection with the upper and lower ends, and the scaffold in the transplantation area appeared thinner, tougher, and darker in color compared to the spinal cord at the cephalic and caudal ends. The scaffold was not significantly adhered to the surrounding soft tissue (Figure 10).

3.5.4. HE-Staining Observation of Rats in Groups A, B, C, and D after Surgery

Group A (laminectomy group): The HE-staining results of the specimens at two months and four months after surgery were the same as those of the normal spinal cord tissue, showing a large number of nerve cells, nuclei, myelin sheath tissue, and extracellular matrix closely combined (Figure 11). In Group B (simple spinal cord hemitransection group), HE staining of the specimens at two months and four months after surgery showed that there was still a defect in the right hemitransection area, and a large number of nerve cells, nuclei, myelin sheath tissue, and extracellular matrix were closely combined in the left non-transected area (Figure 11). In Group C (simple acellular spinal cord scaffold transplantation group), HE staining of the specimens at two months after surgery showed that the spinal cord stump had been combined with the scaffold, and inflammatory cell infiltration, liquefaction, and necrotic cavities were visible in the scaffold area, accompanied by fiber encapsulation (Figure 11). At four months after surgery, HE staining showed that the spinal cord stump had been combined with the scaffold, and a large number of glial scars and cavities were formed in the scaffold area (Figure 11). In Group D (VEGF-modified acellular spinal cord scaffold transplantation group), HE staining of the specimens at two months after surgery showed that the spinal cord stump had been combined with the scaffold, and a large number of vascular tissues and cellular components were visible in the transplant area, with some fiber encapsulation (Figure 11). At four months after surgery, the specimens showed that the spinal cord stump had been combined with the scaffold, and some fiber encapsulation and scar tissue were visible in the transplant area, with an area significantly smaller than that in Group C (Figure 11).

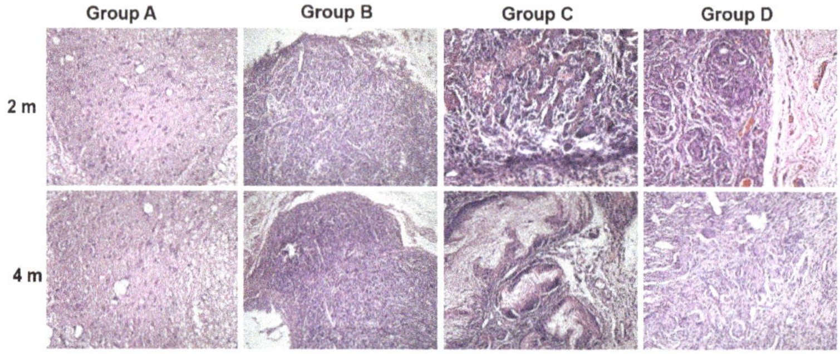

Figure 11. HE staining of Groups A, B, C, and D two or four months (2 m or 4 m) after surgery (magnification 100×).

3.5.5. Retrograde-Tracing Observation of Biotinylated Dextran Amine (BDA) in Rats of Groups A, C, and D after Surgery

In group A (laminectomy group), biotinylated dextran amine staining showed the general direction of the spinal cord nerves two months and four months after surgery, with red fluorescence indicating the direction of the nerve travel (Figure 12). In group C (simple acellular spinal cord scaffold transplantation group), scattered punctate red fluorescence was observed two months and four months after surgery (Figure 12). In group D (VEGF-modified acellular spinal cord scaffold transplantation group), a small amount of red fluorescence was visible two months and four months after surgery (Figure 12). Among the three groups, the fluorescence imaging of groups C and D was significantly weaker than that of group A, and the fluorescence imaging of group C was weaker than that of group D.

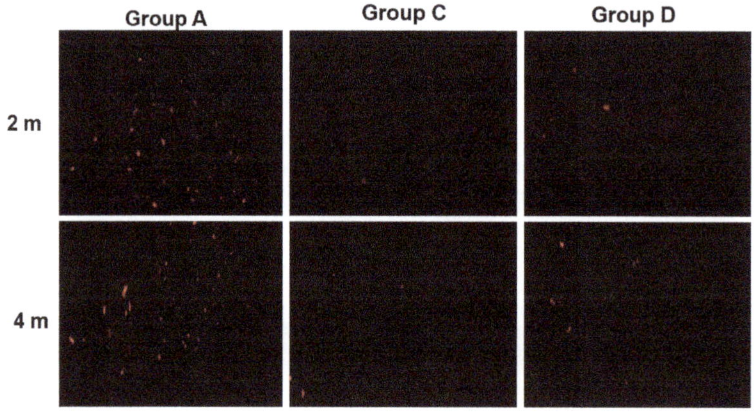

Figure 12. BDA staining in Groups A, C, and D two or four months (2 m or 4 m) after surgery (magnification 100×).

4. Discussion

In recent years, with the rapid development of modern industrial and transportation systems, the incidence of high-energy injuries, such as falls from heights and traffic accidents, has increased, leading to a gradual rise in the prevalence of spinal cord injuries [19]. The treatment of spinal cord injuries remains a global challenge, and because of the complexity and uniqueness of the pathophysiological changes that occur after spinal cord injury, there has been no breakthrough progress in research on the treatment of spinal cord injuries [20]. Basic research on the use of tissue-engineering techniques to repair spinal cord injuries is currently a hot topic [21]. Our research team has successfully prepared allogeneic acellular spinal cord scaffolds and optimized the preparation process. We have also conducted preliminary investigations into the physicochemical properties and biocompatibilities of these scaffolds [12,17]. However, as our research continues to deepen, we have found that the use of acellular spinal cord scaffolds for the regenerative repair of spinal cord injuries still faces the severe challenge of insufficient vascularization of the regenerated tissue [22]. In this part of the experiment, we inoculated vascular endothelial cells onto the EDC-crosslinked VEGF-modified acellular spinal cord scaffold slow-release system and transplanted this scaffold–cell complex into a rat model of spinal cord hemisection to observe the recovery of the spinal cord function in the rats. Our study explored the use of an EDC-crosslinked VEGF-modified acellular spinal cord scaffold, designed to enhance vascularization and promote spinal cord regeneration. The sustained release of the VEGF from the scaffold played a crucial role in improving the biological environment within the injury site, leading to better recovery outcomes. Specifically, VEGF's role in angiogenesis facilitated the formation of new blood vessels, ensuring an adequate supply of nutrients and oxygen to the regenerating tissue. This vascularization is critical because previous studies have shown that the survival and function of transplanted cells are severely limited without proper blood circulation, leading to necrosis and inadequate tissue regeneration.

The BBB motor function scoring is currently an objective and noninvasive method for studying the degree of recovery from spinal cord injury. This scoring system covers the behavioral changes in the recovery of the hindlimb motor function after surgery in rats and can reflect the details of spinal cord injury repair [23]. The results of this part of the experiment confirmed that four time points were selected: two weeks, one month, two months, and four months. We observed the recovery of the motor function in the right hindlimb at each time point in the following four groups: the simple laminectomy group (Group A), the simple spinal cord hemisection group (Group B), the simple acellular spinal cord scaffold group (Group C), and the VEGF-modified acellular spinal cord scaffold group (Group D). Starting from two weeks postoperatively, the motor function of the right hindlimb in Groups C and D recovered to a certain extent. When comparing Groups C and D, the scores in Group C were lower than those in Group D at two weeks, one month, and two months, but these differences were not statistically significant ($p > 0.05$). At four months postoperatively, the score in Group C was significantly lower than that in Group D, with statistical significance ($p < 0.05$). The scores in Groups B, C, and D were significantly lower than those in Group A, with statistical significance ($p < 0.05$). Group B, the simple spinal cord hemisection group, showed some recovery of the motor function in the right hindlimb after surgery in some rats. This recovery of the hindlimb motor function is attributed to the spinal cord compensatory function inherent in lower animals. Some scholars have reported that when performing spinal cord transection, if about 5% of the white matter fibers are preserved, some rats will still show recovery of the hindlimb motor function postoperatively [24]. Therefore, we disregarded the spontaneous recovery of the hindlimb motor function in Group B as an error. It can be seen that the VEGF-modified acellular spinal cord scaffold combined with vascular endothelial cells can promote the recovery of the spinal cord motor function to a certain extent and is significantly better than the simple acellular spinal cord scaffold group. This enhancement is likely related to the roles of VEGF in promoting angiogenesis, improving blood supply to the transplant area, and ensuring nutrient and oxygen supply to the regenerating tissue. Additionally, VEGF may further promote nerve regeneration by modulating the local inflammatory response and reducing glial scar formation.

Evoked-potential measurements are a commonly used auxiliary method to assess the recovery of spinal cord injuries, including somatosensory-evoked potential (SEP) and motor-evoked potential (MEP). SEP can be used to detect the pathway for the ascending spinal cord sensory fiber conduction, while MEP can be used to detect the pathway for the descending spinal cord motor fiber conduction [25]. Because of practical issues, such as operational difficulties and large errors in SEP detection in animals, in this part of the experiment, we chose to use MEP to detect the changes in the amplitude and latency at various time points in the simple laminectomy group (Group A), the simple spinal cord hemisection group (Group B), the simple acellular spinal cord scaffold group (Group C), and the VEGF-modified acellular spinal cord scaffold group (Group D). The amplitude changes of the MEP reflect the intensity of the potential, while the latency changes reflect the neural conduction function. After spinal cord injury, the amplitude of the MEP decreases, and the latency increases. In this part of the experiment, the amplitude and latency changes in rats in Groups A, B, C, and D were measured separately at two weeks, one month, two months, and four months postoperatively. The results showed that the amplitudes of Groups C and D were both lower than that of the normal group (Group A) and higher than that of the simple hemisection group, with statistically significant differences ($p < 0.01$). Among them, the amplitude of Group C was lower than that of Group D, with statistical significance ($p < 0.05$). It can be seen that the VEGF-modified acellular spinal cord scaffold combined with vascular endothelial cells can restore the conduction function of the spinal cord to a certain extent, and it is significantly better than the simple acellular spinal cord scaffold group. This suggests that the vascularization properties of the VEGF-modified decellularized spinal cord scaffold help to reduce tissue degeneration and glial scar formation in the transplantation area, improving the structural integrity of the spinal cord tissue.

Specimens from the simple laminectomy group (Group A), the simple acellular spinal cord scaffold group (Group C), and the VEGF-modified acellular spinal cord scaffold group (Group D) were removed for gross morphological observation at two months and four months postoperatively. In Group A, the specimens at two and four months postoperatively were similar to normal spinal cord tissue in rats, with good continuity of the spinal cord and no congestion or edema on the surface of the dura mater. In Group B, at two months postoperatively, the right half of the specimen still showed a defect; at four months postoperatively, the right half of the specimen still showed a defect, and no new spinal cord tissue growth was observed at the head and tail ends of the transection area. At the same time, the left half of the spinal cord and the tail end became thinner. In Group C, at two and four months postoperatively, the transplant area had restored the continuity between the head and tail ends of the transected spinal cord, and the transplant area was significantly thinner than the head and tail segments of the spinal cord. At two months postoperatively, there was severe adhesion between the right half of the transplanted spinal cord and the surrounding soft tissue, making separation difficult; at four months postoperatively, the adhesion between the right half of the transplanted spinal cord and the surrounding soft tissue had been reduced. In Group D, at two and four months postoperatively, the transplant area also restored the continuity between the head and tail ends of the transected spinal cord, and the transplant area was also thinner than the head and tail segments. However, the degree of thinning was less severe than in Group C. From the results of the HE staining, it can be seen that the transplant areas in Groups C and D have restored histological connections with the head and tail segments of the spinal cord. In Group C, a large amount of inflammatory infiltration and liquefaction necrosis cavities of varying sizes were observed in the transplant area at two months postoperatively; at four months postoperatively, a large number of glial scars and cavities were formed in the transplant area. In Group D, a large number of vascular tissues and cellular components were observed in the transplant area at two months postoperatively, along with partial fibrous encapsulation; at four months postoperatively, necrosis cavities, partial fibrous encapsulation, and scar tissue were observed, with significantly smaller areas and sizes compared to those in Group C. Therefore, transplantation of VEGF-modified acellular spinal cord scaffolds combined with vascular endothelial cells for the treatment of spinal cord injury can restore the gross morphological and histological continuity of the transected spinal cord. The areas and sizes of the liquefaction necrosis, fibrous encapsulation, and scar tissue in the transplant area are significantly smaller than those in the simple acellular spinal cord scaffold transplantation group. These results further support the critical role of VEGF-induced angiogenesis in SCI repair and suggest that regulating VEGF release can effectively reduce tissue degeneration and promote nerve regeneration.

Biotinylated dextran amine (BDA) is currently the most commonly used neural tracer, which has the advantages of long-term preservation and the ability to be combined with various immunohistochemical techniques or fluorescent tracers [26]. BDA can be injected extracellularly and taken up by neuronal cells and then propagated anterogradely or retrogradely along axons to distal sites. By performing corresponding immunofluorescence staining on the distal specimens, it is possible to observe whether there are nerve fibers in that area. This method is often used as an indicator to assess the recovery of the neural conduction in the injured area after spinal cord injury [26]. Therefore, in this part of the experiment, the recovery of the neural conduction in the postoperative specimens of rats in the four groups was observed through the technique of anterograde tracing using BDA in the simple laminectomy group (Group A), the simple acellular spinal cord scaffold group (Group C), and the VEGF-modified acellular spinal cord scaffold group (Group D). The results of the BDA fluorescent staining in this part showed that there was a large amount of red fluorescence along the corticospinal tract in the normal spinal cord specimens of Group A; sporadic punctate red fluorescence was visible in the transplant area specimens of Group C; a small amount of red fluorescence was also observed in the transplant-area specimens of Group D, with significantly more than in Group C. It can be seen that the transplantation of VEGF-modified acellular spinal cord scaffolds complexed with vascular endothelial cells can restore a certain degree of neural conduction in fiber bundles in the treatment of spinal cord injury, and the treatment effect is significantly better than that of the simple acellular spinal cord scaffold group. These results indicate that the VEGF-modified decellularized spinal cord scaffold promotes angiogenesis and nerve regeneration, showing significant advantages in enhancing SCI repair outcomes.

The broader implications of these findings suggest that the VEGF-modified acellular spinal cord scaffold not only enhances the structural repair of the spinal cord but also promotes functional recovery by creating a more conducive environment for neuronal regeneration. The ability of the scaffold to release VEGF in a controlled manner over time ensures that the regenerative processes are supported during the critical phases of healing. When considering the potential clinical applications, the VEGF-modified acellular spinal cord scaffold represents a promising approach for treating SCI. However, translating these findings into clinical practice requires further investigation into the long-term effects, scalability of scaffold production, and validation in larger animal models and human trials. Comparing this approach with other emerging therapies, such as gene therapy [5] or stem cell transplantation [4], our scaffold offers a unique advantage in its ability to simultaneously support vascularization and neuronal repair without the need for genetic modification or complex cell culture procedures.

5. Conclusions

In this study, simple acellular spinal cord scaffolds and VEGF-modified acellular spinal cord scaffolds were transplanted into a rat spinal cord hemitransection model to observe the recovery of the spinal cord function in rats. BBB scores were used to evaluate the recovery of the motor function in the right lower limbs of rats at different time points after surgery; motor-evoked potentials (MEPs) were also used to assess the motor function recovery. Specimens from the transplant areas were taken at different time points after surgery for gross morphological observation, histological observation (HE staining), and anterograde-tracing detection using BDA. The spinal cord repair situation in the VEGF-modified acellular spinal cord scaffold transplantation treatment of the rat spinal cord hemitransection model was comprehensively evaluated. The results showed that the VEGF-modified acellular spinal cord scaffold group had significantly better results in motor function BBB scores and MEP than the simple acellular spinal cord scaffold group and the negative control group. Moreover, the VEGF-modified acellular spinal cord scaffold group demonstrated superior restoration of both the gross morphological and histological connections of the spinal cord. This was evidenced by reduced liquefaction necrosis, smaller cavities, and less fibrous encapsulation, as seen in HE staining, as well as stronger BDA anterograde tracing compared to that of the simple acellular spinal cord scaffold group. These findings indicate that the VEGF-modified acellular spinal cord scaffold can better promote the repair of spinal cord injury, providing a new approach for tissue-engineered spinal cord research.

In addition to these key findings, this study highlights this scaffold's abilities to enhance vascularization, reduce glial scarring, and support neuronal regeneration, which are critical for functional recovery. Future research should focus on long-term studies to assess the sustained effects of the VEGF-modified acellular spinal cord scaffold, explore the scalability of scaffold production, and conduct trials in larger animal models or human subjects. These steps are essential to translate these promising findings into clinical applications.

Author Contributions: Conceptualization, X.D. and H.Y.; methodology, X.D. and Y.L.; resources, X.D. and H.Y.; writing—original draft preparation, X.D. and H.Y.; writing—review and editing, Z.X. and Y.L.; visualization, H.Y.; supervision, H.Y.; project administration, H.Y.; funding acquisition, H.Y. All authors have read and agreed to the published version of the manuscript.

Funding: This work was supported by the Natural Science Foundation of Chongqing (grant number: cstc2020jcyj-msxmX0800) and the Qingbo Program of Xinqiao Hospital Affiliated with the Army Military Medical University (grant number: 2022d030).

Institutional Review Board Statement: The animal studies were performed following the protocols approved by the Laboratory Animal Welfare and Ethics Committee of the Army Medical University (AMUWEC2020572).

Informed Consent Statement: Not applicable.

Data Availability Statement: The raw data supporting the conclusions of this article will be made available by the authors on request.

Conflicts of Interest: The authors declare no conflicts of interest.

References

1. Wang, J.Z.; Yang, M.; Meng, M.; Li, Z.H. Clinical characteristics and treatment of spinal cord injury in children and adolescents *Chin. J. Traumatol.* **2023**, *26*, 8–13. [CrossRef]
2. Shao, A.; Tu, S.; Lu, J.; Zhang, J. Crosstalk between stem cell and spinal cord injury: Pathophysiology and treatment strategies *Stem Cell Res. Ther.* **2019**, *10*, 238. [CrossRef] [PubMed]
3. Mirzaei, F.; Meshkini, A.; Habibi, B.; Salehpour, F.; Rafei, E.; Fathi, W.; Alavi, S.; Majdi, A.; Aghasan, S.; Seyed, A. Ceftriaxone Plus Methylprednisolone Combination Therapy Versus Methylprednisolone Monotherapy in Patients with Acute Spinal Cord Injury: A Randomized, Triple-Blind Clinical Trial. *Int. J. Spine Surg.* **2020**, *14*, 706–712. [CrossRef]
4. Zipser, C.M.; Cragg, J.J.; Guest, J.D.; Fehlings, M.G.; Jutzeler, C.R.; Anderson, A.J.; Curt, A. Cell-based and stem-cell-based treatments for spinal cord injury: Evidence from clinical trials. *Lancet Neurol.* **2022**, *21*, 659–670. [CrossRef]
5. Martirosyan, N.L.; Carotenuto, A.; Patel, A.A.; Kalani, M.; Yagmurlu, K.; Lemole, G.; Preul, M.; Theodore, N. The Role of microRNA Markers in the Diagnosis, Treatment, and Outcome Prediction of Spinal Cord Injury. *Front. Surg.* **2016**, *3*, 56 [CrossRef]
6. Delgado, A.L.J.; Carreira, A.C.O.; Hianka, J.C.; Palma, R.K.; Castro, S.; Carvalho, C.; León, M.; Barreto, R.; Miglino, M Development of a new decellularization protocol for the whole porcine heart. *J. Clin. Transl. Res.* **2021**, *7*, 563–574.
7. Li, P.; Jin, D.; Dou, J.; Wang, L.; Wang, Y.; Jin, X.; Han, X.; Kang, I.; Yuan, J.; Shen, J. Nitric oxide-releasing poly(ε-caprolactone)/S nitrosylated keratin biocomposite scaffolds for potential small-diameter vascular grafts. *Int. J. Biol. Macromol.* **2021**, *189*, 516–527 [CrossRef]

8. Baldwin, M.; Snelling, S.; Dakin, S.; Carr, A. Augmenting endogenous repair of soft tissues with nanofibre scaffolds. *J. R. Soc. Interface* **2018**, *15*, 20180019. [CrossRef] [PubMed]
9. Santillo, D.; Mathieson, I.; Corsi, F.; Göllner, R.; Guandalini, A. The use of acellular porcine corneal stroma xenograft (BioCorneaVet™) for the treatment of deep stromal and full thickness corneal defects: A retrospective study of 40 cases (2019–2021). *Vet. Ophthalmol.* **2021**, *24*, 469–483. [CrossRef]
10. Irfan, D.; Ahmad, I.; Patra, I.; Margiana, R.; Rasulova, M.T.; Sivaraman, R.; Kandeel, M.; Mohammad, H.J.; Al-Qaim, Z.H.; Jawad, M. Stem cell-derived exosomes in bone healing: Focusing on their role in angiogenesis. *Cytotherapy* **2023**, *25*, 353–361. [CrossRef] [PubMed]
11. Guo, S.Z.; Ren, X.J.; Wu, B. Preparation of the acellular scaffold of the spinal cord and the study of biocompatibility. *Spinal Cord.* **2010**, *48*, 576–581. [CrossRef] [PubMed]
12. Yaldiz, B.; Saglam-Metiner, P.; Yesil-Celiktas, O. Decellularised extracellular matrix-based biomaterials for repair and regeneration of central nervous system. *Expert Rev. Mol. Med.* **2021**, *23*, e25. [CrossRef]
13. Tran, K.; Partyka, P.; Jin, Y.; Bouyer, J.; Fischer, I.; Galie, P. Vascularization of self-assembled peptide scaffolds for spinal cord injury repair. *Acta Biomater.* **2020**, *104*, 76–84. [CrossRef]
14. Cassell, C.; Hofer, O.; Morrison, W.; Knight, K. Vascularization of tissue-engineered grafts: The regulation of angiogenesis in resconstructive surgery and in disease states. *Br. J. Plast. Surg.* **2002**, *55*, 603–610. [CrossRef] [PubMed]
15. Chiu, L.L.Y.; Radisic, M. Scaffolds with covalently immobilized VEGF and Angiopoietin-1 for vascularization of engineered tissues. *Biomaterials* **2010**, *31*, 226–241. [CrossRef]
16. Jiang, T.; Ren, X.; Tang, J.; Yin, H.; Wang, K.; Zhou, C. Preparation and characterization of genipin-crosslinked rat acellular spinal cord scaffolds. *Mater. Sci. Eng. C* **2013**, *33*, 3514–3521. [CrossRef] [PubMed]
17. Gong, W.; Zhang, T.; Che, M.; Wang, Y.; He, C.; Liu, L.; Lv, Z.; Xiao, C.; Wang, H.; Zhang, S. Recent advances in nanomaterials for the treatment of spinal cord injury. *Mater. Today Bio* **2023**, *18*, 100524. [CrossRef]
18. Xie, C.; Li, X.; Fang, L.; Wang, T. Effects of Athermal Shortwave Diathermy Treatment on Somatosensory Evoked Potentials and Motor Evoked Potentials in Rats with Spinal Cord Injury. *Spine* **2019**, *44*, 749–758. [CrossRef]
19. Fehlings, M.G.; Wilson, J.R.; Tetreault, L.A.; Aarabi, B.; Anderson, P.; Arnold, P.M.; Brodke, D.S.; Burns, A.S.; Chiba, K.; Joseph, R. A clinical practice guideline for the management of patients with acute spinal cord injury: Recommendations on the use of methylprednisolone sodium succinate. *Glob. Spine J.* **2017**, *7*, 203S–211S. [CrossRef]
20. Li, C.; Song, J.; Wang, Y.; Shi, Y.; Ji, J.; Lin, Q.; Liu, Y. Adhesion and proliferation of bone marrow stromal cells on acellular spinal cord scaffolds. *Int. J. Neurosci.* **2022**, *134*, 889–898. [CrossRef]
21. Qin, Y.; Zhang, K.; Sun, T.; Shen, P.; Wang, H.; Zhang, Z.; Ren, Q.G. Effect of miR-222 on the Angiogenesis of Bone Microvascular Endothelial Cells After Repair of Vascular Endothelial Growth Factor with Polylactide-Poly (ethylene glycol)-Polylactide (PELA) Microspheres. *J. Biomater. Tissue Eng.* **2022**, *12*, 2478–2482. [CrossRef]
22. Song, Z.; Ye, Y.; Zhang, Z.; Shen, J.; Hu, Z.; Wang, Z.; Zheng, J. Noninvasive, targeted gene therapy for acute spinal cord injury using LIFU-mediated BDNF-loaded cationic nanobubble destruction. *Biochem. Biophys. Res. Commun.* **2018**, *496*, 911–920. [CrossRef] [PubMed]
23. You, S.; Chen, B.; Liu, H.; Lang, B.; Xia, J.; Jiao, X.; Ju, G. Spontaneous recovery of locomotion induced by remaining fibers after spinal cord transection in adult rats. *Restor. Neurol. Neurosci.* **2003**, *21*, 39–45.
24. Simão, C.R.; de Holanda, L.J.; Urbini, L.F.; Lacerda, M.O.; Fernandes, K.; da Silva, P.M.; Morya, E.; Lindquist, A.R. Surface electromyography to identify top-down modulation in complete chronic spinal cord injury. *Eur. J. Phys. Med.* **2022**, *58*, 144. [CrossRef]
25. Ogurcov, S.; Shulman, I.; Garanina, E.; Sabirov, D.; Baichurina, I.; Kuznetcov, M.; Masgutova, G.; Kostennikov, A.; Rizvanov, A.; James, V. Blood serum cytokines in patients with subacute spinal cord injury: A pilot study to search for biomarkers of injury severity. *Brain Sci.* **2021**, *11*, 322. [CrossRef] [PubMed]
26. Albayar, A.A.; Roche, A.; Swiatkowski, P.; Antar, S.; Ouda, N.; Emara, E.; Smith, D.H.; Ozturk, A.K.; Awad, B.I. Biomarkers in spinal cord injury: Prognostic insights and future potentials. *Front. Neurol.* **2019**, *10*, 27. [CrossRef]

coatings

MDPI

Article

Impact of Diamond-like Carbon Films on Reverse Torque: Superior Performance in Implant Abutments with Internal Conical Connections

Arianne Vallim Pinto Coelho [1], Viviane Maria Gonçalves de Figueiredo [1], Leandro Lameirão Ferreira [2], Alecsandro de Moura Silva [3], Marcelo Gallo Oliani [4], José Renato Cavalcanti de Queiroz [5], Argemiro Soares da Silva Sobrinho [2], Lafayette Nogueira Junior [1] and Renata Falchete do Prado [6,*]

1 Department of Prosthodontic and Dental Material, Institute of Science and Technology, São Paulo State University (UNESP), São José dos Campos 12245-000, Brazil; arianne.odonto@gmail.com (A.V.P.C.); vivi_mfigueiredo@yahoo.com.br (V.M.G.d.F.); lafayette.nogueira-junior@unesp.br (L.N.J.)

2 Department of Physics, Technical Institute Aerospace (ITA), São José dos Campos 12228-900, Brazil; leandrofisic@hotmail.com (L.L.F.); argemiro@ita.br (A.S.d.S.S.)

3 Department of Dentistry, Taubaté University (UNITAU), Taubaté 12020-270, Brazil; alecsms@yahoo.com.br

4 Department of Prosthodontics, Faculdade de Pindamonhangaba, FUNVIC, Pindamonhangaba 12412-825, Brazil; marcelo.oliane@unesp.br

5 Department of Biotechnology, UnP—Laureate University, Natal 59056-450, Brazil; joserenatocq@hotmail.com

6 Department of Bioscience and Oral Diagnosis, Institute of Science and Technology, São Paulo State University (UNESP), São José dos Campos 12245-000, Brazil

* Correspondence: renata.prado@unesp.br; Tel./Fax: +55-12-3947-9010

Abstract: The loosening or fracture of the prosthetic abutment screw is the most frequently reported complication in implant dentistry. Thin diamond-like carbon (DLC) films offer a low friction coefficient and high wear resistance, functioning as a solid lubricant to prevent the weakening of the implant–abutment system. This study evaluated the effects of DLC nanofilms on the reverse torque of prosthetic abutments after simulated chewing. Abutments with 8° and 11° taper connections, with and without DLC or silver-doped DLC coatings, were tested. The films were deposited through the plasma enhanced chemical vapor deposition process. After two million cycles of mechanical loading, reverse torque was measured. Analyses with scanning electron microscopy were conducted on three samples of each group before and after mechanical cycling to verify the adaptation of the abutments. Tribology, Raman and energy-dispersive spectroscopy analyses were performed. All groups showed a reduction in insertion torque, except the DLC-coated 8° abutments, which demonstrated increased reverse torque. The 11° taper groups experienced the most torque loss. The nanofilm had no significant effect on maintaining insertion torque, except for the DLC8 group, which showed improved performance.

Keywords: dental implants; diamond-like carbon nanofilm; dental implant abutment connection; torque

1. Introduction

The internal conical connections have been developed to improve the biomechanical properties of implants [1] and to reduce the incidence of problems usually found in conventional connections. Stability may vary according to the conicity degree, geometries, and surface interface [2].

However, regardless of the connection type, screw-retained abutment systems present a gap in the abutment–implant interface. This space allows for possible micro infiltration and bacterial colonization [3] and may be responsible for abutment micromovement, leading to mechanical failures (loosening and fracture of the screw) [4] with consequent preload loss.

A screw system failure occurs in two stages. The first stage starts when external forces, such as mastication, act on the system. At this point, screw threads slide over the implant's internal threads and/or wear out, causing thread height to diminish, consequently leading to a loss of screw preload. The second stage occurs when the decrease in preload leads to torque loss in the threads and, consequently, to loss of direct contact between the screw and the implant's internal threads [5].

The compression and preload amounts are determined by several variables. These include the elasticity modulus of the materials used in the manufacturing of the fixing screw and the parts to be attached to it, the coefficient of friction of sliding surfaces [6], component adaptation, the lubrication system [7], applied torque and speed of application, the head and screw threads geometry, and their superficial roughness [8].

Studies have shown that most of the screwing strength, when applied to prosthetic screws, is lost due to the friction between micro-roughness and the metal surfaces, which limits screw rotation [9,10].

The preload maintenance will depend, in part, on the accommodation effect (settling effect) obtained between the surfaces [11], since they are not smooth and full contact between them does not occur. Additionally, such micro-roughness suffers wear and deformation, which leads to contact loss between the coupling threads and consequent preload loss [12].

Reverse torque, the force required to remove an implant abutment, is crucial for assessing the mechanical stability of implant–abutment connections. Studies have shown that both indexed and non-indexed abutments experience significant torque loss after insertion [13,14]. However, non-indexed abutments generally demonstrate better torque retention compared to indexed abutments [15]. The implant material also influences reverse torque values, with hydroxyapatite-coated implants exhibiting significantly higher torque removal values than titanium implants [16]. Additionally, mandibular implants tend to show greater torque resistance than maxillary implants. These findings suggest that factors such as abutment design, implant material, and placement location play important roles in maintaining implant stability. Understanding reverse torque values is essential for determining appropriate torque levels for abutment fastening and ensuring the long-term success of implant-supported restorations [13]. Hence, reducing friction between metallic surfaces could increase rotation and, consequently, the preload. The addition of a lubricant to this interface has been a proposed alternative [10]. Thin diamond-like carbon coating (DLC) offers a low friction coefficient and high resistance to wear [17], acting as a solid lubricant. The unique properties of DLC coatings depend on the ratio of sp3/sp2 phases, which can be tailored for specific applications [18]. These coatings are widely used in orthopedic, cardiovascular, and dental applications, and for various biomedical devices such as vascular stents, prosthetic heart valves, and surgical instruments [19]. Known for their hemocompatibility, they are applied to artificial heart surfaces in contact with blood and are effective in pressure sensors, chemical electrodes, and biochemical sensors. They promote rapid endothelization of implants, minimize platelet activation, and reduce thrombus formation. DLC coatings are considered highly promising for biomedical applications, with growing potential in automotive and adhesive uses as well [18,19].

In the literature, the coatings used in implant dentistry are intended for the prior preparation of the implant and consider their advantages in the osseointegration process. For example, common coating materials include carbon, bisphosphonates, bioactive glass, fluoride, hydroxyapatite (HA), and titanium nitride (TiN) [20]. HA remains the most biocompatible, while bioglass shows promise. TiN coatings may reduce bacterial colonization, enhance fibroblast proliferation, and improve mechanical properties [21].

The use of DLC film on screws offers another perspective on the use of coatings, serving as a solid lubricant in the fixation of prostheses and aiming for the perfect coupling of prosthetic components on implants.

The advantage of the use of this type of lubricant is the possibility of incorporating nanoparticles (V, Ti, F, H, Ag) into its microstructure, allowing for an improvement of this film with additional biological properties. The incorporation of silver particles may be particularly advantageous because of its known bactericidal effect [22]. To prevent supra- and sub-gingival biofilm formation and plaque accumulation, allowing antimicrobial function on implant abutments and/or prostheses would be advantageous for the prevention of peri-implantitis, which would subsequently increase the long-term success of the implant [23]. A suitable deposition method of DLC coating containing silver nanoparticles is the plasma enhanced chemical vapor deposition [24].

Implants with conical connections are generally manufactured with a difference in the angles of the conicities between the implant and the abutment. This fact directly influences the behavior of the implant system in general, the size of the microgap, the tension generated in the bone tissue, the loss of preload of the screw, and, in particular, the removal torque [25]. Tapered implants show advantages in low-density bone and immediate loading scenarios, increasing primary stability and minimizing stability loss at compression sites [26]. Research on dental implant taper angles suggests that increasing the abutment taper angle can significantly improve fracture resistance, with 8° and 10° angles showing higher fracture forces compared to 6° [27]. When comparing internal conical connections with different taper angles (5.4°, 12°, 45°, and 60°), no significant differences were found in bacterial leakage under dynamic loading conditions [28].

This study aimed to cover the prosthetic abutments of two different conical connections with DLC nanofilms to evaluate their effect on reverse torque after masticatory cycling.

The null hypotheses were:

- There is no influence of the presence of films on reverse torque after mechanical cycling. Reverse torque is a critical parameter in assessing the long-term stability of dental implants, as it reflects the ability of the abutment to resist loosening under functional loads. By testing this null hypothesis, the study aimed to determine whether the application of nanofilms introduces any beneficial or detrimental effects on reverse torque, which could influence clinical outcomes.
- DLC coating does not present different mechanical behavior compared to silver nanoparticle-doped DLC coating. This hypothesis was formulated to compare the mechanical performance of two different types of coatings: pure DLC and silver-doped DLC. Silver doping is known for its antimicrobial properties, but it is important to evaluate whether adding silver nanoparticles affects the mechanical integrity of the coating, particularly in terms of reverse torque after mechanical cycling. By examining this hypothesis, this study sought to clarify whether silver doping compromises the mechanical benefits of DLC or enhances them without negatively impacting reverse torque stability.
- The DLC films do not interfere with reverse torque regardless of abutment wall angulations (8° and 11°). The angulation of the abutment walls can influence the mechanical interaction between the abutment and the implant, potentially affecting the distribution of stresses and the overall stability of the connection. This hypothesis was crucial to determine whether the mechanical benefits or potential drawbacks of DLC films are consistent across different abutment designs, or if they are dependent on specific geometric factors.

2. Materials and Methods

2.1. Samples

Seventy-eight internal conical-type abutments were randomly distributed into six groups (n = 13) according to two variables: wall angulations (8° and 11°) and surface treatment (C, control group without coating; DLC, coated with DLC film; DLC-Ag, coated with silver nanoparticles doped DLC).

Thirty-nine solid abutments (internal conical 8°, Ø4.0 × 7.0 − thread M2.0) CMAC 7.0 − batch: 153012 (Titaniumfix—São José dos Campos, SP, Brazil) and thirty-nine direct abutments (11° internal conical, Ø4.0 × 5.5 – thread M1.6) 406.015 – batch: 098510 (Titaniumfix) were used.

Thirty-nine internal octagon implants (Ø4.0 × 13.0) with internal conical 8° abutments and thirty-nine double hexagon implants (Ø4.0 × 13.0) with internal conical 11° abuments, both manufactured by Titaniumfix, were used.

The implants were inserted in a polyurethane (elasticity modulus of 3.6 GPa—F16, AXSON, São Paulo, Brazil) cylindrical sample, positioned through a clamp and a metallic matrix to perform perforation and avoid movement. The samples were fixed at an inclination of 30°. A digital caliper (Digimess, São Paulo, Brazil) was used to ensure the height of the installation. All these steps were performed as recommended in the ISO 14801:2007 standard. Briefly, the implants were positioned such that the platform was precisely located at a height of 3.0 ± 0.5 mm above the surface of the fixation block. The installation process used specialized wrenches tailored for the specific implants (codes CMSWA and 612.305, Titaniumfix, São José dos Campos, Brazil), in conjunction with a calibrated torque wrench. A digital caliper (Digimess, Nikeypar, São Paulo, Brazil) was utilized to accurately verify the installation height. Subsequently, the abutments were affixed using a torque wrench explicitly designed for the installation and removal of the prosthetic abutments in use (code CMDTAC, Titaniumfix, São José dos Campos, Brazil), with the tightening torque applied via a digital torque wrench (TQ-680, Instrutherm, São Paulo, Brazil) featuring a precision of 0.01 Ncm. The tightening torque was administered following the manufacturer's specifications, applying 32 Ncm for the solid abutment and 20 Ncm for the direct abutment. Ten minutes after the initial tightening, the same torque values were reapplied. Upon achieving the desired tightening torque, the hemispherical loading devices were meticulously positioned on the abutments.

2.2. Film Deposition and Abutments Installation

The films were deposited over the abutments (threads and base) via pulsed DC-discharged plasma-enhanced vapor deposition (PECVD) at 20 kHz.

After the substrate cleansing process (argonium plasma, flow 20 sccm) for 40 min at 7.5×10^{-5} Torr), a silicon carbon-rich thin film was deposited over the samples as an interlayer [29], using hexamethyldisiloxane (HMDSO, $C_6H_{18}OSi_2$) as the precursor gas at 1.0×10^{-2} Torr for 30 min. After this period, a transition layer was deposited using a mixture of HMDSO and methane at a pressure of 1.6×10^{-1} Torr for 10 min.

For the deposition of the DLC nanofilm itself, the HMDSO flow was closed, and the methane flow was maintained at 13 sccm and the argon flow at 5 sccm for 1 h and 30 min, at a pressure of 1.1×10^{-1} Torr (groups DLC8 and DLC11).

Before the deposition of silver-doped DLC film, the silver–hexane solution (30–50 nm) was stirred in ultrasound for 15 min to disperse particles homogeneously in the solution. The same cleaning and transition layer deposition steps were performed. The silver-doped DLC film deposition started; the hexane with silver nanoparticles flow was maintained at 12 sccm and the argon flow at 5 sccm for 1 h and 30 min, at a pressure of 8×10^{-2} Torr (DLC-Ag8 and DLC-Ag11). The deposition power was maintained at around 200 w. The cathodic tension and the electric current were kept constant during the entire deposition process.

After all the previous steps, each abutment (C8, C11, DLC8, DLC11, DLC-Ag8, DLC-Ag11) was fixed over proper implants. The abutments were installed using a digital torque meter (TQ-680, Instrutherm, São Paulo, Brazil) according to the manufacturer's instructions (32 Ncm for 8° solid abutment and 20 Ncm for 11° direct abutment). Ten minutes later, additional screwing was performed with the same torque value.

Hemispheric loading devices were machined from stainless steel at A.S. Technology Comp. Esp. Ltda. (São José dos Campos, Brazil), according to the parameters described in ISO 14801:2007. The devices were custom-designed and machined for each abutment model to ensure a perfect fit and to avoid movement.

2.3. Mechanical Cycling and Reverse Torque Analysis

These hemispherical loading devices were positioned over the abutments, and the set was subjected to mechanical cycling (ER-11000 Plus, Erios, São Paulo, Brazil) at 2 Hz for 2×10^6 cycles, with samples immersed in distilled water (n = 13). The sample size was based on a previous study by Figueiredo et al. [30], which adopted a significance level of 5% and a beta power of 80% using the Bonding test data, and determined that a total of at least 8 samples were needed. The frequency and number of adopted cycles were followed according to the parameters described in the ISO 14801:2007 norm. The applied load was 120 Ncm, which is equivalent to the average physiological masticatory load. The tip used for load application had a flat end, which provided a stable contact point on the hemispherical loading device, as also specified by the ISO 14801:2007 norm. At the end of the cycling, the samples were removed from the cycler, and reverse torque values were registered using the same digital torque meter (TQ-680).

Two methods were adopted for data analysis. The first method considered the absolute difference (Ncm) by subtracting the reverse torque (Tr) from the insertion torque (Ti). The second method calculated the relative difference (%) by dividing the absolute difference by the Ti.

2.4. Morphological Analysis of Abutments Adaptation

The samples (C, DLC, and DLC-Ag non-cycled and C cycled, DLC cycled and DLC-Ag cycled) (n = 3) were analyzed via scanning electron microscopy (SEM Inspect S50, FEI Company, Orlando, FL, USA) in BSE mode with 1000× magnification. For this analysis, the implants with their respective fixed abutments were cut along their axis until the fixing screw was exposed. Then, they were sanded with an automatic sanding machine polisher using 400- and 1200-grit wet sandpaper and cleansed with alcohol in an ultrasound bath (5 min).

2.5. Characterization of Films

The film thicknesses were obtained by measuring the steps formed between the covered and uncovered surface of additional silicon samples positioned in the plasma deposition chamber, using an Alpha-Step 500 profilometer (Tencor, San Diego, CA, USA).

During the process of film deposition over the abutments, additional titanium plate samples (14 × 14 mm) were prepared and coated with the different nanofilms for analysis and characterization. The scratch test was performed using a tribometer (CETR Biomedical Tribometer, CETR-Bruker, Campbell, CA, USA). The adherence of the film to the substrate and the coefficient of friction were analyzed on two coated plates (n = 2) at 24.8 °C and 46% relative humidity. The test was carried out at the speed of 0.1 mm/s with a progressive load from 0.2 Ncm to 20 Ncm over a 100 mm area using a Rockwell-type diamond tip (Rockwell C, CETR-Bruker, Campbell, CA, USA). The initial failure point was determined as the point at which a crack occurred, or film delamination was observed. Three tests were carried out per sample, aiming to determine the critical load.

The coefficient of friction analysis was performed using a spherical steel tip (4 mmø) with a progressive load of 2 Ncm (5 mm/s, 5 mm during 1000 s).

Roughness analysis (Ra) was carried out using a digital optical profilometer (WYKI NT 1100, Vecco, New York, NY, USA) in VSI mode at 20× magnification. Five Ra measurements were made in each sample (n = 2) for each surface treatment (C, DLC, and DLC-Ag).

The chemical composition and structure of the films were analyzed via energy-dispersive spectroscopy (EDS) and Raman scattering spectroscopy (Renishaw 2000, Renishaw, New Mills, UK), which operates with a scattered argonium laser at 514 nm.

2.6. Statistical Analysis

A significance level of 5% was adopted. The GraphPad Prism 6 software program (GraphPad Software, San Diego, CA, USA) was used to perform statistical analysis. The data were tested for normality using the the D'Agostino–Pearson normality test. Reverse torque tests were analyzed using the Kruskal–Wallis test and Dunn's multiple comparison test to evaluate the effects of the nanofilms. The Mann–Whitney test was used to determine differences in the mechanical behavior of each wall angulation. The methodology was reviewed by an independent statistician.

3. Results

Some groups of data did not present a normal distribution according to the D'Agostino–Pearson normality test.

The results for absolute difference (Ncm) and relative difference (%) are presented in Table 1 and Figure 1. Positive values indicate torque loss and negative values indicate torque gain. The Kruskal–Wallis test (for groups 8° and 11°, respectively) demonstrated significant differences for both groups in terms of absolute difference ($p = 0.0396$; $p = 0.0017$) and relative difference ($p = 0.0355$; $p = 0.0014$).

Table 1. Medians from the statistical analysis of the absolute difference values (Ncm) and the relative values (%). *p*-values, obtained by searching for the effect of conicity values using the Mann–Whitney test are presented.

Group	Unit	Internal Conical 8°	Internal Conical 11°	*p*-Value
C		2.75^{A}	1.5^{A}	0.45
DLC	Ncm	-1.55^{B}	4.1^{B}	<0.0001 *
DLC-Ag		1.45^{AB}	3.2^{AB}	0.08
C		8.52^{a}	7.39^{a}	0.81
DLC	%	-4.81^{b}	20.45^{b}	<0.0001 *
DLC-Ag		4.52^{ab}	15.65^{ab}	0.0001 *

Values that do not share the same superscript letters are significantly different from each other ($p < 0.05$). Superscript uppercase letters are results from Dunn's test for absolute difference values; superscript minute letters are results from Dunn's test for the relative values. * Significant differences (−1.55 Ncm is lower than 4.1 Ncm; −4.81% is lower than 20.45%; and 4.52% is lower than 15.65%).

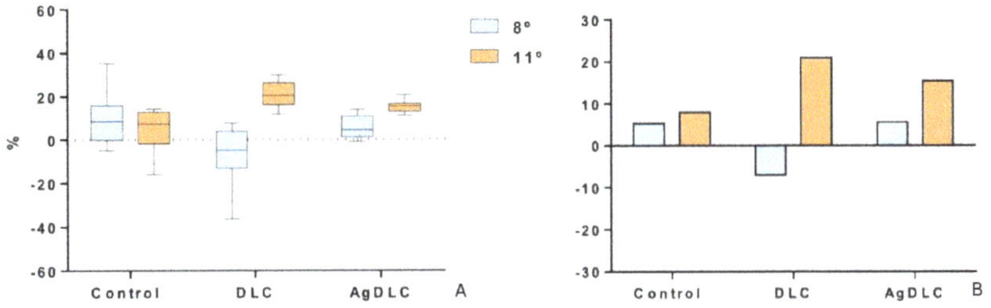

Figure 1. Boxplot of the relative difference (%) of the torque values (**A**). Column graph of the average relative difference (%) of the torque values (**B**).

Considering the coatings, according to Dunn's multiple comparisons test, differences occurred between C and DLC groups for both 8° and 11°. In the internal conical connection with an angle of 8°, the DLC group exhibited the lowest torque loss. In the internal conical connection with an angle of 11°, the DLC group exhibited the highest torque loss.

According to the Mann–Whitney test, torque loss values did not present differences between the control groups with 8° and 11° wall angulations. In the coated groups, the internal conical connection with an angle of 11°, coated with both nanofilms, demonstrated the highest torque loss values.

Morphological Analysis of Abutment Adaptation

Deformations observed on the threads of abutment screws and internal threads of implants, apart from the internal connections of implants, were verified. The samples presented a perfect adaptation of the pillar–implant interface for both cycled and non-cycled groups, and there were no qualitative differences in the thread images (Figure 2).

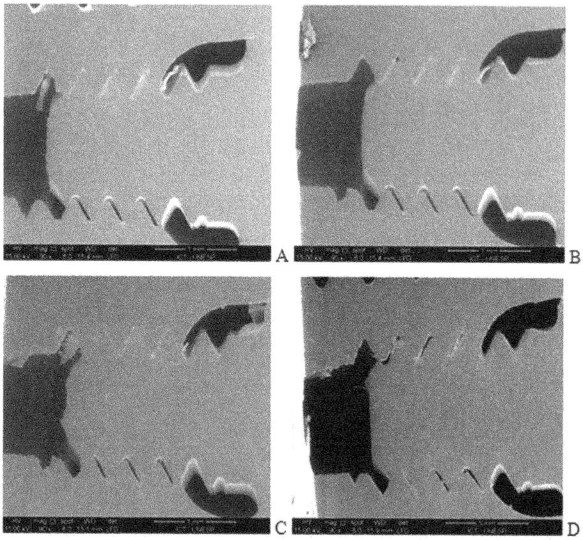

Figure 2. Thread images of C8 pre-cycling (**A**), post-cycling (**B**), DLC8 pre-cycling (**C**), and post-cycling (**D**) at 90× magnification in SE mode.

The thickness of the DLC nanofilm was 1.58 μm and the thickness of the DLC-Ag nanofilm was 2.91 μm. Since the deposition time was the same for both films, this difference in thickness is probably due to the fact that each molecule of methane, the precursor gas for the DLC film, has only one carbon atom, while hexane, the precursor gas for the AgDLC film, has six carbon atoms.

Table 2 presents data from the scratch test. The results demonstrated that, during the test, the film kept its adherence to the entire interior thread and was removed only in one sample of the DLC-Ag at an applied force of 9.7 N. Thread analysis revealed a shell-format film saturation, suggesting good adherence to the substrate. The mean values of the scratch test were lower for the DLC-Ag nanofilm compared to the DLC nanofilm.

Table 2. Scratch test values for DLC and DLC-Ag. The moment of the first crack was LC1 and the first delamination was LC2.

	DLC			DLC-Ag		
	LC_1 (N)	LC_2 (N)	Removal	LC_1 (N)	LC_2 (N)	Removal
	−5.8	−5.8	NO	−4.5	−5.5	NO
	−4.2	−6.0	NO	−4.6	−5.0	NO
	−4.7	−5.8	NO	−4.7	−6.1	9.7 N.
Mean	−4.9	−5.87		−4.6	−5.53	

Regarding the coefficient of friction (COF), performed using a spherical steel tip (4 mmø) with a progressive load of 2 Ncm, the control group presented the highest COF (0.46 ± 0.14), followed by DLC (0.15 ± 0.2) and DLC-Ag (0.12 ± 0.2). The low coefficient of friction and high hardness of the DLC layer offered some benefits in preventing the deformation of the implant top.

The average surface roughness measurements (Ra—mm), carried out using a digital optical profilometer, were 0.4 ± 0.02 (C), 0.4 ± 0.07 (DLC), and 0.3 ± 0.01 (DLC-Ag). Ra represents the average between the values of the peaks and valleys present in the tested area.

Figure 3 shows the Raman spectra of DLC and DLC-Ag. The Raman spectra revealed that the G band was larger in the DLC nanofilm, indicating a higher concentration of sp2 hybridization in the silver-doped film. The D bands did not present significant changes.

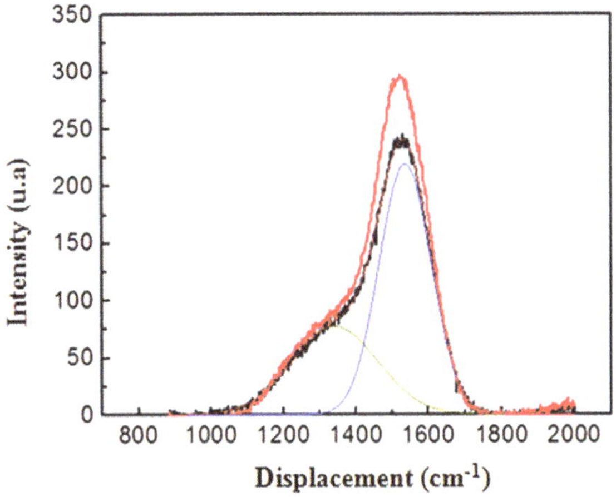

Figure 3. Raman spectra of silver doped Ag-DLC (in black) and undoped DLC (in red). The presence of silver on the film changes the position and the intensities of the bands G that compose each spectrum. Raman spectra revealed that the G band is larger in the DLC film. The spectra include the fitting curves. Blue curve indicates the Lorentzian curves for the G-band without Gaussian convolution. Yellow curve indicates the Lorentzian curves for the D-band without the same Gaussian convolution.

EDS quantitative analysis demonstrated an atomic percentage of 90.5% C, 2.4% Si, 7.1% Ti, and 0.5% Ag in the DLC-Ag group, and 89.0% C, 2.1% Si, and 8.5% Ti in the DLC group.

4. Discussion

This in vitro assessment evaluated the effect of taper angles on the internal conical connections of implant systems and two different DLC nanofilms on the reverse torque of abutment screws after load cycling. We obtained great results with both DLC and DLC-Ag nanofilms in the internal conical connection with an angle of 8°. Torque loss values were lower than the values of the control group. We only accepted the second null hypothesis, which stated that DLC coating does not present a different mechanical behavior compared to silver nanoparticle-doped DLC coating.

The results of the present study, particularly the differences in coating types (DLC vs. silver-doped DLC) and abutment angulations (8° vs. 11°), can have several potential impacts on clinical outcomes. The DLC8 group showed superior reverse torque compared to other groups, including those with silver-doped DLC. This suggests that the DLC coating might provide better mechanical stability, reducing the risk of abutment loosening. In clinical practice, this could translate to longer-lasting implant stability, less frequent need for adjustments or retightening, and a lower risk of complications related to screw loosening, ultimately reducing overall healthcare costs and improving patient satisfaction.

DLC coatings are known for their hardness and wear resistance [17], which can reduce material wear over time. In contrast, the addition of silver nanoparticles, which could offer antimicrobial properties [22] and reduce the risk of peri-implantitis, altered the mechanical properties of the coating. The silver-doped DLC demonstrates lower mechanical stability, which accelerates the wear and potentially shortens the lifespan of the abutment, increasing the need for earlier intervention or replacement. This study showed that 11° abutments experienced higher torque loss than 8° abutments. Clinically, this could mean that implants with a steeper taper (11°) might be more prone to loosening over time, especially if not coupled with an optimal coating. This could lead to a higher likelihood of abutment failure, necessitating more frequent follow-ups, retightening, or even implant replacement.

Bone density is a critical factor in the success of dental implants and the planning of treatment. Various preoperative methods exist for assessing jawbone density, many of which show strong correlations with clinical outcomes [31]. Computed tomography provides valuable insights into bone quality before implant placement, highlighting significant variations in density across different oral regions [32]. The anterior mandible typically exhibits the highest bone density, followed by the anterior maxilla, posterior mandible, and posterior maxilla [32]. Additionally, bone density varies significantly between individuals and even within different areas of the same jaw [33]. Factors such as bone quality, bone quantity, and overall patient health are critical to implant success [34]. While it is generally observed that women have a lower bone density than men, no significant correlation has been established between bone density and patient age [33]. These findings underscore the importance of carefully selecting the abutment angulation and conducting a patient-specific assessment to customize implant treatment according to each patient's unique anatomical and functional needs. For example, a patient with higher occlusal forces might benefit more from an 8° abutment with a diamond-like carbon (DLC) coating, which can minimize the risk of abutment loosening and enhance long-term stability.

We proved that the internal conical angle of 11° resulted in higher torque loss values, consequently enhancing the chance of implant failure. Although clinical studies have verified that a low torque (less than 30 Ncm) is sufficient to ensure dental implant osseointegration [35], and all studies reviewed by Lemos et al. [36] reported high survival rates irrespective of the insertion torque, we suspect that the torque loss in the 11° internal cone was related to the insertion torque of 20 Ncm used in our study, which is considered a low torque.

The thickness of DLC coatings plays a significant role in their performance, with thicker coatings providing better friction reduction under elastohydrodynamic lubrication conditions [37]. Additionally, DLC coatings can serve as an effective galvanic corrosion barrier between titanium abutments and nickel–chromium superstructures without compromising the fit and integrity of prosthetic assemblies [38]. While DLC coatings have shown potential to enhance the reliability of dental implant-supported restorations, their effectiveness may vary depending on the coating method and thickness [37,39].

A significant difference in the implant survival rates (14% failure rates for regular or low insertion torque as compared with 0% for high insertion torque) has been reported in implants with immediate loading [40]. Additionally, the torque loss observed in the 11° group here agrees with Ottoni et al. [41], who reported that low insertion torque (20 Ncm) is associated with a high risk of failure (9 of 10 implants failed) in implant-supported single crowns.

Based on previous studies [42,43], which showed that low-frequency cycling is more effective than continuous frequency increase cycling, a change in the ISO 14801 norm was adopted in the methods of the present study. As expected, the results demonstrated a decrease in the reverse torque values of all groups after mechanical cycling, which indicates a decrease in the torque of the screws. The DLC8 group had the lowest reverse torque loss.

The promising mechanical behavior of the DLC8 group in the present study is in accordance with previous studies [44]. Higher reverse torque values for abutment screws with DLC coating have been reported [43]. The main difference from the present study was the cycle type used. The authors performed cyclic loading using a thermomechanical testing device to represent approximately one year of clinical usage [44]. In contrast, this study used 2×10^6 cycles at 2 Hz and 120 N and water immersion, to simulate two and a half years of masticatory function [45]. In external-hex implants, the DLC coating did not influence torque loss when compared to the control group [46].

Research on the mechanical cycling of implant–abutment connections reveal significant effects on their stability and performance. Short-term cycling (36×10^4 cycles) leads to improved adaptation and sealing ability, with decreased microgap size in Morse taper connections [47]. Some studies employed long-term cyclic loading using five million cycles as the upper limit [47–49]. Fatigue limits varied across designs, ranging from 210–240 N for different connections to 300–800 N in another study [48]. Internal connections generally showed higher fatigue strength than external ones, while external hex designs exhibited higher static strength [49]. Zirconia abutments demonstrated higher fatigue strength than titanium in external hex configurations [49].

Some studies have evaluated the performance of coated dental prosthetic abutments under simulated masticatory loads and torque tests. DLC coatings on abutments have shown mixed results. One study found that DLC coating reduced screw loosening after mechanical cycling [46]. However, another study reported no improvement in torque maintenance for DLC-coated screws with or without diamond nanoparticles [50]. Bacchi et al. (2015) [51] observed that conventional titanium screws maintained higher loosening torque values than DLC-coated screws after cyclic loading. For zirconia abutments, a study found that ceramic coating did not significantly affect their mechanical behavior under static and dynamic loading conditions [52]. These studies demonstrate that while various coatings have been investigated for dental prosthetic abutments, their effectiveness in maintaining torque and mechanical stability varies, highlighting the need for further research in this area.

As a result of the abutment threads analysis, none of the groups presented changes in the adaptation, wear, or deformation after mechanical cycling, confirming results from previous literature [53]. Screw joint failure begins with external functional loading, which gradually reduces the preload in the joint. The DLC layer's lower friction coefficient and greater hardness compared to pure titanium may help prevent implant abrasion, erosion, and deformation [54]. The superior torque results in the DLC8-coated groups can also be explained by this. Dziedzic et al. [55] reported that the carbon coating on screw threads reduced interface friction and enhanced preload values, thereby increasing the clamping force between the abutment and the implant platform. The DLC's low friction coefficient enables greater preload with the same torque, thereby improving joint stability.

The factors that influence the settling effect and, consequently, screw loosening include preload, the coefficient of friction, and retightening [56]. The coefficient of friction in the DLC and DLC-Ag groups was lower than that obtained in the control group. In agreement, Raman spectra in the present study demonstrated higher sp2 hybridization concentrations in the DLC-Ag nanofilm. Tribological tests have proven that coatings with high sp2 content tend to have lower friction coefficients in dry conditions [57]. Despite differences in the coefficient of friction, our roughness measurements, expressed as Ra values, were similar in all groups. This coefficient depends not only on the roughness of the contacting surfaces, but also on loading, the hardness of the threads, and the presence, quantity, and quality of the lubricant on the surfaces. When decreased to appropriate values, the friction coefficient increases the preload and decreases the chance of screw loosening [56]. The DLC coating is an alternative method to prevent screw loosening [44].

Peri-implant diseases present a prevalence rate of 43% across Europe and 22% across South and North America [58]. Both peri-implant mucositis and peri-implantitis have an infectious etiology based on the accumulation of a biofilm composed of periodontal pathogens on the implant surface [59]. EDS analysis confirmed the successful deposition of a silver-doped film over our samples, thus our antibacterial coating on abutments could be advantageous in preventing peri-implantitis and implant failures.

This study presents certain methodological limitations that could be addressed in future research. Regarding the selection of implants, to standardize the clinical scenario, implants with the same external diameter (4.0 mm) were chosen. However, these implant models exhibit specific design characteristics that influence this standardization. The implant with an 8° taper (c-fix; Titaniumfix, Brazil) features an M2.0 internal thread (2 mm diameter) and an octagonal indexing system. In contrast, the implant with an 11° taper (b-fix; Titaniumfix, Brazil) has an M1.6 internal thread (1.6 mm diameter) and a dual hexagonal indexing system. The indexing systems impact the area occupied by the taper, which may affect the active area of each implant model. Consequently, the load distribution across each thread differs, given the variation in the threaded area and active taper area in each model. Future studies should consider standardizing the internal thread diameter and selecting implants with different diameters to determine if the active taper area associated with thread diameter influences reverse torque values after mechanical cycling. The tribological analyses conducted were qualitative only. The results for the coefficient of friction, average roughness, and scratch hardness were not subjected to statistical analysis. More detailed studies are necessary to investigate whether there is a correlation between tribological results and reverse torque values. Another analysis that was not statistically tested involved scanning electron microscopy images. The assessment of taper adaptation and thread wear was purely visual, and quantitative evaluations are required for more significant and conclusive results. Additionally, regarding the adaptation of the abutments, the application of DLC and AgDLC films was carried out across the entire active area of the abutments (taper and threads), but the effect of the films on abutment adaptation was only evaluated after fatigue testing. The differences in reverse torque values between coated and uncoated abutments may not have been as apparent due to potential misadaptation in the taper area of the abutments with DLC and AgDLC films, which could have influenced the reverse torque values. Future studies should examine the thickness of the films and their influence on taper adaptation before fatigue testing to achieve film thicknesses that do not affect adaptation.

5. Conclusions

Based on the findings of this in vitro study, it can be concluded that the use of diamond-like carbon (DLC) coatings, particularly with an 8° abutment angulation, significantly enhances the mechanical stability of abutment screws by minimizing torque loss after load cycling. The DLC coating proved more effective than silver-doped DLC, which exhibited reduced mechanical stability. These results suggest that DLC-coated abutments may offer greater implant longevity and fewer complications related to screw loosening in clinical settings. However, the study's limitations include the specific conditions under which the tests were conducted, which may not fully replicate the complexities of the oral environment. Further in vivo research is necessary to validate these findings and assess the long-term clinical performance of DLC coatings in various implant scenarios.

Author Contributions: A.V.P.C.: Methodology, Investigation, Formal analysis, Data curation, Writing—original draft, Visualization, Project administration; V.M.G.d.F.: Methodology, Data curation; Writing—original draft; L.L.F.: Methodology, Software, Investigation, Formal analysis, Data curation, Writing—original draft, Visualization, Project administration; A.d.M.S.: Conceptualization, Validation, Writing—review and editing, Project administration; M.G.O.: Conceptualization, Resources, Supervision; J.R.C.d.Q.: Investigation, Formal analysis, Writing—review and editing; A.S.d.S.S.: Conceptualization, Validation, Resources, Writing—review and editing, Supervision, Project administration, Funding acquisition; R.F.d.P.: Software, Investigation, Formal analysis, Writing—review and editing; L.N.J.: Conceptualization, Validation, Resources, Writing—review and editing, Supervision, Project administration, Funding acquisition. All authors have read and agreed to the published version of the manuscript.

Funding: Titaniumfix (São José dos Campos, Brazil) provided the implants of this study.

Institutional Review Board Statement: Not applicable.

Informed Consent Statement: Not applicable.

Data Availability Statement: Data will be made available upon request.

Conflicts of Interest: There are no conflicts of interest with this research.

References

1. Ceruso, F.M.; Barnaba, P.; Mazzoleni, S.; Ottria, L.; Gargari, M.; Zuccon, A.; Bruno, G.; Di Fiore, A. Implant-Abutment Connections on Single Crowns: A Systematic Review. *Oral Implantol.* **2017**, *10*, 349–353. [CrossRef]
2. Bagegni, A.; Spies, B.C.; Kern, M.; Hazard, D.; Kohal, R. The Influence of Prosthetic Crown Height and Implant-Abutment Connection Design Selection on the Long-Term Implant-Abutment Stability: A Laboratory Study. *J. Mech. Behav. Biomed. Mater.* **2021**, *113*, 104095. [CrossRef]
3. Passos, S.P.; Gressler May, L.; Faria, R.; Özcan, M.; Bottino, M.A. Implant-Abutment Gap versus Microbial Colonization: Clinical Significance Based on a Literature Review. *J. Biomed. Mater. Res. Part B Appl. Biomater.* **2013**, *101*, 1321–1328. [CrossRef]
4. Gois Filho, D.M.; de Gois-Santos, V.T.; Silva, R.S.; Marqueti, A.C.; Cortes, A.R.G.; Trento, C.L. Evaluation of Sealing between Abutment and Inner Connection of Cone Morse Dental Implant. *Clin. Lab. Res. Dent.* **2018**, *2018*, 1–6. [CrossRef]
5. Kim, B.J.; Yeo, I.S.; Lee, J.H.; Kim, S.K.; Heo, S.J.; Koak, J.Y. The effect of screw length on fracture load and abutment strain in dental implants with external abutment connections. *Int. J. Oral Maxillofac. Implant.* **2012**, *27*, 820–823.
6. Jo, J.Y.; Yang, D.S.; Huh, J.B.; Heo, J.C.; Yun, M.J.; Jeong, C.M. Influence of Abutment Materials on the Implant-Abutment Joint Stability in Internal Conical Connection Type Implant Systems. *J. Adv. Prosthodont.* **2014**, *6*, 491–497. [CrossRef] [PubMed]
7. Jörn, D.; Kohorst, P.; Besdo, S.; Rücker, M.; Stiesch, M.; Borchers, L. Influence of Lubricant on Screw Preload and Stresses in a Finite Element Model for a Dental Implant. *J. Prosthet. Dent.* **2014**, *112*, 340–348. [CrossRef]
8. dos Santos, M.V.; Elias, C.N.; Cavalcanti Lima, J.H. The Effects of Superficial Roughness and Design on the Primary Stability of Dental Implants. *Clin. Implant. Dent. Relat. Res.* **2011**, *13*, 215–223. [CrossRef]
9. Mangano, C.; Mangano, F.; Piattelli, A.; Iezzi, G.; Mangano, A.; La Colla, L. Prospective Clinical Evaluation of 1920 Morse Taper Connection Implants: Results after 4 Years of Functional Loading. *Clin. Oral Implant. Res.* **2009**, *20*, 254–261. [CrossRef] [PubMed]
10. Stüker, R.A.; Teixeira, E.R.; Beck, J.C.P.; Da Costa, N.P. Preload and Torque Removal Evaluation of Three Different Abutment Screws for Single Standing Implant Restorations. *J. Appl. Oral Sci.* **2008**, *16*, 55–58. [CrossRef]
11. Winkler, S.; Ring, K.; Ring, J.D.; Boberick, K.G. Implant Screw Mechanics and the Settling Effect: An Overview. *J. Oral Implant.* **2003**, *29*, 242–245. [CrossRef]
12. Al Jabbari, Y.S.; Fournelle, R.; Ziebert, G.; Toth, J.; Iacopino, A.M. Mechanical Behavior and Failure Analysis of Prosthetic Retaining Screws after Long-Term Use In Vivo. Part 1: Characterization of Adhesive Wear and Structure of Retaining Screws. *J. Prosthodont.* **2008**, *17*, 168–180. [CrossRef] [PubMed]
13. Cerutti-Kopplin, D.; Rodrigues Neto, D.J.; Lins Do Valle, A.; Pereira, J.R. Influence of Reverse Torque Values in Abutments with or without Internal Hexagon Indexes. *J. Prosthet. Dent.* **2014**, *112*, 824–827. [CrossRef] [PubMed]
14. Cerutti-Kopplin, D.; Rodrigues Neto, D.J.; Do Valle, A.L.; Pereira, J.R.; Pinho, L.G.N.D. Reverse Torque Values in Indexed Abutments. *J. Res. Dent.* **2013**, *1*, 192. [CrossRef]
15. Martins, C.M.; Ramos, E.V.; Kreve, S.; de Carvalho, G.A.P.; Franco, A.B.G.; de Macedo, L.G.S.; Silva, A.D.M.; Dias, S.C. Reverse Torque Evaluation in Indexed and Nonindexed Abutments of Morse Taper Implants in a Mechanical Fatigue Test. *Dent. Res. J.* **2019**, *16*, 110–116. [CrossRef]
16. Carr, A.B.; Larsen, P.E.; Papazoglou, E.; McGlumphy, E. Reverse torque failure of screw-shaped implants in baboons: Baseline data for abutment torque application. *Int. J. Oral Maxillofac. Implant.* **1995**, *10*, 167–174.
17. Watabe, T.; Amanov, A.; Tsuboi, R.; Sasaki, S. Friction and Fretting Wear Characteristics of Different Diamond-like Carbon Coatings against Alumina in Water-Lubricated Fretting Conditions. *J. Nanosci. Nanotechnol.* **2013**, *13*, 8167–8175. [CrossRef]
18. Malisz, K.; Świeczko-Żurek, B.; Sionkowska, A. Preparation and Characterization of Diamond-like Carbon Coatings for Biomedical Applications—A Review. *Materials* **2023**, *16*, 3420. [CrossRef]
19. Peng, Y.; Peng, J.; Wang, Z.; Xiao, Y.; Qiu, X. Diamond-like Carbon Coatings in the Biomedical Field: Properties, Applications and Future Development. *Coatings* **2022**, *12*, 1088. [CrossRef]
20. Xuereb, M.; Camilleri, J.; Attard, N. Systematic Review of Current Dental Implant Coating Materials and Novel Coating Techniques. *Int. J. Prosthodont.* **2015**, *28*, 51–59. [CrossRef]
21. del Castillo, R.; Chochlidakis, K.; Galindo-Moreno, P.; Ercoli, C. Titanium Nitride Coated Implant Abutments: From Technical Aspects And Soft Tissue Biocompatibility to Clinical Applications. A Literature Review. *J. Prosthodont.* **2022**, *31*, 571–578. [CrossRef] [PubMed]
22. Schwarz, F.P.; Hauser-Gerspach, I.; Waltimo, T.; Stritzker, B. Antibacterial Properties of Silver Containing Diamond like Carbon Coatings Produced by Ion Induced Polymer Densification. *Surf. Coat. Technol.* **2011**, *205*, 4850–4854. [CrossRef]
23. Odatsu, T.; Kuroshima, S.; Sato, M.; Takase, K.; Valanezhad, A.; Naito, M.; Sawase, T. Antibacterial Properties of Nano-Ag Coating on Healing Abutment: An in Vitro and Clinical Study. *Antibiotics* **2020**, *9*, 347. [CrossRef]
24. Marciano, F.R.; Bonetti, L.F.; Santos, L.V.; Da-Silva, N.S.; Corat, E.J.; Trava-Airoldi, V.J. Antibacterial Activity of DLC and Ag–DLC Films Produced by PECVD Technique. *Diam. Relat. Mater.* **2009**, *18*, 1010–1014. [CrossRef]
25. Aguirrebeitia, J.; Abasolo, M.; Vallejo, J.; Ansola, R. Dental Implants with Conical Implant-Abutment Interface: Influence of the Conical Angle Difference on the Mechanical Behavior of the Implant. *Int. J. Oral Maxillofac. Implant.* **2013**, *28*, e72–e82. [CrossRef]
26. Tang, C.L.; Zhao, S.K.; Huang, C. Features and Advances of Morse Taper Connection in Oral Implant. *Chin. J. Stomatol.* **2017**, *52*, 59–62. [CrossRef]
27. Wang, K.; Geng, J.; Jones, D.; Xu, W. Comparison of the Fracture Resistance of Dental Implants with Different Abutment Taper Angles. *Mater. Sci. Eng. C Mater. Biol. Appl.* **2016**, *63*, 164–171. [CrossRef]

28. Ozdiler, A.; Bakir-Topcuoglu, N.; Kulekci, G.; Isik-Ozkol, G. Effects of Taper Angle and Sealant Agents on Bacterial Leakage Along the Implant-Abutment Interface: An In Vitro Study Under Loaded Conditions. *Int. J. Oral Maxillofac. Implant.* **2018**, *33*, 1071–1077. [CrossRef]
29. Schauer, J.C.; Winter, J. Plasma Deposition of Elastic Wear Resistant Si-C Coatings on Nickel-Titanium for Biomedical Applications. *J. Appl. Phys.* **2008**, *103*, 113302. [CrossRef]
30. de Figueiredo, V.M.G.; de Silva, A.D.M.; Massi, M.; Sobrinho, A.S.D.S.; de Queiroz, J.R.C.; Machado, J.P.B.; Do Prado, R.F.; Junior, L.N. Effect of the Nanofilm-Coated Zirconia Ceramic on Resin Cement Bond Strength. *J. Dent. Res. Dent. Clin. Dent. Prospect.* **2022**, *16*, 170–178. [CrossRef]
31. Molly, L.; Molly, L. Bone Density and Primary Stability in Implant Therapy. *Clin. Oral Implant. Res.* **2006**, *17*, 124–135. [CrossRef]
32. Turkyilmaz, I.; Tözüm, T.F.; Tumer, C. Bone Density Assessments of Oral Implant Sites Using Computerized Tomography. *J. Oral Rehabil.* **2007**, *34*, 267–272. [CrossRef] [PubMed]
33. Di Stefano, D.A.; Arosio, P.; Pagnutti, S.; Vinci, R.; Gherlone, E.F. Distribution of Trabecular Bone Density in the Maxilla and Mandible. *Implant. Dent.* **2019**, *28*, 340–348. [CrossRef]
34. Venkatakrishnan, C.J.; Bhuminathan, S.; Chandran, C.A.R. Dental Implant Insertion Torque and Bone Density—Short Review. *Biomed. Pharmacol. J.* **2017**, *10*, 1305–1309. [CrossRef]
35. Maló, P.; Lopes, A.; de Araújo Nobre, M.; Ferro, A. Immediate Function Dental Implants Inserted with Less than 30 N·cm of Torque in Full-Arch Maxillary Rehabilitations Using the All-on-4 Concept: Retrospective Study. *Int. J. Oral Maxillofac. Surg.* **2018**, *47*, 1079–1085. [CrossRef]
36. Lemos, C.A.A.; Verri, F.R.; de Oliveira Neto, O.B.; Cruz, R.S.; Luna Gomes, J.M.; da Silva Casado, B.G.; Pellizzer, E.P. Clinical Effect of the High Insertion Torque on Dental Implants: A Systematic Review and Meta-Analysis. *J. Prosthet. Dent.* **2021**, *126*, 490–496. [CrossRef]
37. Björling, M.; Larsson, R.; Marklund, P. The Effect of DLC Coating Thickness on Elstohydrodynamic Friction. *Tribol. Lett.* **2014**, *55*, 353–362. [CrossRef]
38. Ozkomur, A.; Erbil, M.; Akova, T. Diamondlike Carbon Coating as a Galvanic Corrosion Barrier between Dental Implant Abutments and Nickel-Chromium Superstructures. *Int. J. Oral Maxillofac. Implant.* **2013**, *28*, 1037–1047. [CrossRef]
39. Bordin, D.; Coelho, P.G.; Bergamo, E.T.P.; Bonfante, E.A.; Witek, L.; Del Bel Cury, A.A. The Effect of DLC-Coating Deposition Method on the Reliability and Mechanical Properties of Abutment's Screws. *Dent. Mater.* **2018**, *34*, e128–e137. [CrossRef]
40. Cannizzaro, G.; Leone, M.; Torchio, C.; Viola, P.; Esposito, M. Immediate versus Early Loading of 7-Mm-Long Flapless-Placed Single Implants: A Split-Mouth Randomised Controlled Clinical Trial. *Eur. J. Oral Implant.* **2008**, *1*, 277–292.
41. Ottoni, J.M.P.; Oliveira, Z.F.L.; Mansini, R.; Cabral, A.M. Correlation between Placement Torque and Survival of Single-Tooth Implants. *Int. J. Oral Maxillofac. Implant.* **2005**, *20*, 769–776.
42. Karl, M.; Kelly, J.R. Influence of Loading Frequency on Implant Failure under Cyclic Fatigue Conditions. *Dent. Mater.* **2009**, *25*, 1426–1432. [CrossRef]
43. Lee, C.K.; Karl, M.; Kelly, J.R. Evaluation of Test Protocol Variables for Dental Implant Fatigue Research. *Dent. Mater.* **2009**, *25*, 1419–1425. [CrossRef]
44. Colpak, E.D.; Gumus, H.O. Effect of Surface Modifications of Abutment Screws on Reverse Torque Values: An In Vitro Study. *Int. J. Prosthodont.* **2020**, *33*, 401–409. [CrossRef]
45. Rosentritt, M.; Behr, M.; Gebhard, R.; Handel, G. Influence of Stress Simulation Parameters on the Fracture Strength of All-Ceramic Fixed-Partial Dentures. *Dent. Mater.* **2006**, *22*, 176–182. [CrossRef] [PubMed]
46. Corazza, P.H.; De Moura Silva, A.; Cavalcanti Queiroz, J.R.; Salazar Marocho, S.M.; Bottino, M.A.; Massi, M.; Rodrigo, R.O. Carbon Film Coating of Abutment Surfaces: Effect on the Abutment Screw Removal Torque. *Implant. Dent.* **2014**, *23*, 434–438. [CrossRef]
47. Gehrke, S.A.; Carvalho Serra, R.D. Load Fatigue Performance of Conical Implant-Abutment Connection: Effect of Torque Level and Interface Junction. *Minerva Stomatol.* **2015**, *64*, 1–7.
48. Thomé, G.; Gomes, A.; de Moura, M.; Beng, C.; Trojan, L. Evaluation of the Mechanical Reliability of Different Implant-Abutment Connections for Single Crowns According to the ISO 14801 Fatigue Test Protocol. *Int. J. Oral Maxillofac. Implant.* **2021**, *36*, 47–54. [CrossRef] [PubMed]
49. Park, I.S.; Won, S.Y.; Bae, T.S.; Song, K.Y.; Park, C.W.; Eom, T.G.; Jeong, C.M. Fatigue Characteristics of Five Types of Implant-Abutment Joint Designs. *Met. Mater. Int.* **2008**, *14*, 133–138. [CrossRef]
50. Lepesqueur, L.; Figueiredo, V.; Ferreira, L.; Sobrinho, A.; Massi, M.; Bottino, M.; Nogueira Junior, L. Coating Dental Implant Abutment Screws with Diamondlike Carbon Doped with Diamond Nanoparticles: The Effect on Maintaining Torque After Mechanical Cycling. *Int. J. Oral. Maxillofac. Implant.* **2015**, *30*, 1310–1316. [CrossRef]
51. Bacchi, A.; Regalin, A.; Brilhante Bhering, C.L.; Alessandretti, R.; Spazzin, A.O. Loosening Torque of Universal Abutment Screws after Cyclic Loading: Influence of Tightening Technique and Screw Coating. *J. Adv. Prosthodont.* **2015**, *7*, 375–379. [CrossRef]
52. Jiménez-Melendo, M.; Llena-Blasco, O.; Bruguera, A.; Llena-Blasco, J.; Yáñez-Vico, R.M.; García-Calderón, M.; Vaquero-Aguilar, C.; Velázquez-Cayón, R.; Gutiérrez-Pérez, J.L.; Torres-Lagares, D. Mechanical Behavior of Single-Layer Ceramized Zirconia Abutments for Dental Implant Prosthetic Rehabilitation. *J. Clin. Exp. Dent.* **2014**, *6*, e485–e490. [CrossRef] [PubMed]
53. Basílio, M.d.A.; Butignon, L.E.; Arioli Filho, J.N. Effectiveness of Screw Surface Coating on the Stability of Zirconia Abutments after Cyclic Loading. *Int. J. Oral Maxillofac. Implant.* **2012**, *27*, 1061–1067.

4. Kim, S.K.; Lee, J.B.; Koak, J.Y.; Heo, S.J.; Lee, K.R.; Cho, L.R.; Lee, S.S. An Abutment Screw Loosening Study of a Diamond Like Carbon-Coated CP Titanium Implant. *J. Oral Rehabil.* **2005**, *32*, 346–350. [CrossRef] [PubMed]
5. Silva, D.; Dziedzic, M.; Nhata, J.; Jamcoski, V.H.; Dziedzic, M. Assessment of Preload in Carbon Coated Prosthetic Screws. *RSBO* **2012**, *9*, 137–142.
6. Bulaqi, H.A.; Mousavi Mashhadi, M.; Safari, H.; Samandari, M.M.; Geramipanah, F. Dynamic nature of abutment screw retightening: Finite element study of the effect of retightening on the settling effect. *J. Prosthet. Dent.* **2015**, *113*, 412–419. [CrossRef]
7. Love, C.A.; Cook, R.B.; Harvey, T.J.; Dearnley, P.A.; Wood, R.J.K. Diamond like Carbon Coatings for Potential Application in Biological Implants—A Review. *Tribol. Int.* **2013**, *63*, 141–150. [CrossRef]
8. Fu, J.H.; Wang, H.L. Breaking the Wave of Peri-Implantitis. *Periodontology 2000* **2020**, *84*, 145–160. [CrossRef]
9. Berglundh, T.; Armitage, G.; Araujo, M.G.; Avila-Ortiz, G.; Blanco, J.; Camargo, P.M.; Chen, S.; Cochran, D.; Derks, J.; Figuero, E.; et al. Peri-Implant Diseases and Conditions: Consensus Report of Workgroup 4 of the 2017 World Workshop on the Classification of Periodontal and Peri-Implant Diseases and Conditions. *J. Periodontol.* **2018**, *89*, S313–S318. [CrossRef]

coatings

MDPI

Article

Corrosion Resistance and In Vitro Biological Properties of TiO_2 on MAO-Coated AZ31 Magnesium Alloy via ALD

Xiao Han [1,2], Yan Wang [1,2], Jianxiong Ma [1,2,*] and Xinlong Ma [1,2,*]

1 Tianjin Hospital, Tianjin University, Jie Fang Nan Road 406, Tianjin 300211, China; hx970120@126.com (X.H.); wangyan2519_@tju.edu.cn (Y.W.)

2 Tianjin Key Laboratory of Orthopedic Biomechanics and Medical Engineering, Tianjin Hospital, Tianjin 300050, China

* Correspondence: mjx969@163.com (J.M.); maxinlong86868@163.com (X.M.)

Abstract: The surface corrosion of magnesium alloys is effectively addressed currently by the creation of a micro-arc oxidation (MAO) ceramic layer. However, oxide film porousness restricts magnesium alloy use. Thus, this work used atomic layer deposition (ALD) to create a TiO_2 coating on MAO-coated AZ31B magnesium alloy to plug micropores and increase corrosion resistance and biological characteristics. The samples were analyzed using SEM, EDS, XPS, and XRD to determine their surface appearance, chemical content, and microstructure. Micro-arc oxidation produced a 20 μm oxide coating. The TiO_2 film reached 47.41 nm after 400 atomic layer deposition cycles. All corroded samples were tested for corrosion resistance using electrochemical and hydrogen evolution methods and examined for surface morphology. In vitro cell experiments examined biocompatibility. The results indicate that the TiO_2 layer sealed the MAO coating's micro-pores and micro-cracks, enhanced corrosion resistance, and preserved surface morphology following corrosion. The TiO_2/MAO composite coating is more biocompatible than the substrate and MAO coating. This research proposes coating AZ31B magnesium alloy for bio-remediation to increase corrosion resistance and biocompatibility.

Keywords: AZ31B magnesium alloy; micro-arc oxidation; atomic layer deposition; biocompatibility

1. Introduction

As a biodegradable biomedical metal material, magnesium alloy is comparable to human bone, meaning it satisfies the mechanical specifications needed for materials to be implanted into the body [1–5]. The magnesium ions generated by its degradation are easily absorbed by human tissues or excreted through body fluids and do not produce toxicity to harm the human body [6,7]. In addition, magnesium alloy also has good biocompatibility, osteoinductivity, and other biological properties [8–10]. On this basis, magnesium alloy has a broad variety of potential applications in the area of orthopedics, including the internal fixing of fractures and the repair of bone deformities. However, magnesium is a very active metal with poor corrosion resistance and a high degradation rate, which can affect cell viability and function [11,12]. Therefore, it is necessary to modify its surface in order to enhance its resistance to corrosion.

Surface modification is the easiest method of lowering the rate of corrosion of implanted materials since it forms a protective coating on the magnesium alloy itself, shielding the material substrate from the corrosive environment. Commonly used surface modification techniques include physical vapor deposition [1], chemical conversion [13], micro-arc oxidation (also known as plasma electrolytic oxidation (PEO)) [14], and the sol–gel method [15]. Among these, the micro-arc oxidation process is a straightforward, fast, and non-toxic technique that is often used for the surface modification of magnesium alloys. After MAO treatment, a layer of oxide ceramic coating with high adhesion and porous roughness will grow on the magnesium alloy's surface, which can enhance the material's resistance to corrosion [14]. Tang et al. performed micro-arc oxidation and solution treatment on AZ31 magnesium alloy, and the findings revealed that the MAO coating increased corrosion resistance while also forming apatite [16]. In addition, the rough pore surface stimulates cell proliferation and adhesion, resulting in fast local tissue repair. The results of a study by Ma et al. show that the MAO-treated magnesium samples can effectively improve the corrosion resistance of magnesium samples and promote bone formation [17]. AZ31 magnesium alloy coated with MAO, as proposed by Jian et al., has favorable cytocompatibility and corrosion resistance characteristics [18]. Therefore, this approach offers benefits for the development of novel implant materials and may be used in clinical settings.

However, the high-voltage spark discharges during the experimental process cause micropores and microcracks on the MAO film's surface, which let hostile ions travel through it and expose the samples to pitting corrosion [19]. Therefore, post-treatment of MAO film is required to improve the film's density and, hence, corrosion resistance. The main methods can be divided into two categories: conventional sealing (inorganic sealing [20] and organic sealing [21]) and composite film (chemical plating [22] and biofilm coating [23,24]). Among them, the atomic layer deposition (ALD) method is a popular choice for material modification of magnesium alloys because of its many benefits, including strong bonding, broad area homogeneity, and accurate thickness control. Li et al. prepared aluminium-doped zinc oxide (AZO) thin films on AZ31 magnesium alloy coated with a plasma electrolytic oxidation (PEO) layer by atomic layer deposition (ALD) method, which significantly increased the material's resistance to corrosion [25]. Li et al. discovered that the corrosion resistance of AZ31 magnesium alloy in Hanks' solution could be effectively improved by the dense, biocompatible tantalum oxide (Ta_2O_5) nanofilms deposited on MAO-coated AZ31 magnesium alloy by ALD, and the corrosion current density was reduced by three orders of magnitude [26]. Consequently, it is feasible to seal the surface micropores and cracks of MAO coatings via ALD.

Titanium dioxide is frequently utilized in medical implant materials since it is one of the most biocompatible inorganic materials. It is non-toxic to humans and has high mechanical qualities, such as corrosion resistance [27]. On alloys based on magnesium, titanium dioxide coatings have been employed in a number of investigations recently. Kania et al. evaluated the structural and corrosion behavior of TiO_2 coatings formed on MgCa4Zn1Gd1 alloys using magnetron sputtering and sol–gel techniques [28]. The results showed that the magnetron-sputtered TiO_2 surface had smaller and more uniform grains. Despite this, the samples' resistance to corrosion was greatly increased by both techniques. In addition, Kania et al. used ALD to deposit uniform TiO_2 films on MgCa2Zn1Gd3 alloy [29]. The magnesium base alloys under research saw a decrease in corrosion rate due to the presence of TiO_2 coatings, with thicker TiO_2 films exhibiting greater corrosion resistance. Nazarov et al. used ALD and in situ oxidation to prepare TiN/TiO_2 nanocomposite films [30]. The films adhered well to the substrate made of magnesium alloy. The film's strong adherence to the magnesium alloy substrate significantly increased the metal's resistance to corrosion. In conclusion, because of its mechanical qualities and biocompatibility, titanium dioxide is a good option for bone replacement and repair.

In this work, the ALD approach was used to create titanium oxide (TiO_2) films on MAO-coated AZ31B magnesium alloy in order to close the micropores and enhance corrosion resistance. Investigations were also conducted on the composite coatings' biocompatibility and resistance to corrosion.

2. Materials and Methods

2.1. Sample Preparation

2.1.1. Preparation of AZ31B Magnesium Alloy

Utilizing extruded AZ31B magnesium alloy (diameters: t = 2 mm, d = 10 mm, content: 3.000 wt% Al, 1.000 wt% Zn, 0.200 wt% Mn, Fe < 0.005 wt%, bal. Mg) as a substrate, 800

#, 1500 #, and 2000 # SiC sandpaper was employed in a sequential manner. After being sanded, the samples were kept in a dry environment after being ultrasonically cleaned many times in deionized water and acetone to eliminate surface contaminants.

2.1.2. Preparation of MAO

Magnesium alloy and stainless-steel plates were used as anode and cathode for the micro-arc oxidation reaction in an electrolyte containing 10 g/L Na_2SiO_4, 5 g/L NaOH, and 3 g/L $(NaPO_3)_6$, and the whole process was carried out in constant current mode using a 75-kW pulse power source (SAMMAO-100A-800SPM, Shangi Institute For Advanced Materials Co., Ltd., Shanghai, China). Table 1 displays the experimental parameters that were employed.

Table 1. Parameters of the experiment during microarc oxidation (MAO).

Experiment Groups	Current Density (A/dm^2)	Current Frequency (Hz)	Positive Duty Cycle (%)	Negative Duty Cycle (%)	Process Time (min)
MAO	5.5	500	40	20	15

2.1.3. Preparation of TiO_2/MAO

The PICOSUN™ R-200 standard atomic layer deposition system (ALD; Espoo, Finland) was used to deposit TiO_2 thin films on MAO coatings. The Ti source was Tetrakis (dimethylamino) titanium (IV) or TDMA-Ti, and the O source was H_2O. During the deposition process, the temperatures of TDMA-Ti, H_2O, and the feed line were 75 °C, 20 °C, and 120 °C, respectively.

Each cycle process included:

(1) TDMA-Ti precursor pulsed for 0.1 s and N_2 purged for 5 s.
(2) H_2O precursor pulsed for 0.015 s and N_2 purged for 5 s.

Both precursors were deposited at a flow rate of 100 sccm for a total of 400 cycles.

Figure 1 depicts the experimental flowchart used in this study to visually represent our design concepts.

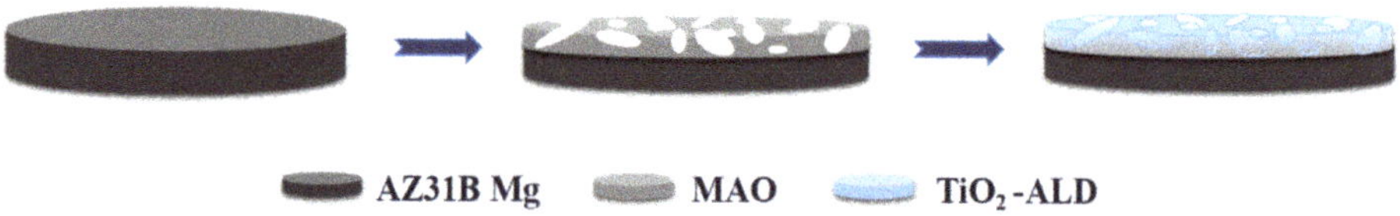

Figure 1. The experimental flowchart.

2.2. *Surface Structure and Chemistry Characterization*

Surface shape, chemical content, and film thickness of the micro-arc oxidation were examined by scanning electron microscopy (SEM; SU-8010, Hitachi, Tokyo, Japan). Spectrophotometric ellipsometry (ME-L, Wuhan Eoptics Technology Co., Ltd., Wuhan, China) was used to measure the thickness of the titanium dioxide film created via atomic layer deposition. The structure and phase distribution of the samples were investigated by X-ray diffraction (XRD, D8A, Bruker, Berlin, Germany) using Cu-Kα radiation (40 kV, 40 mA, λ = 1.5405 Å, step size 2°/min, test angle: 5°~80°) by Raman spectrometer (Witec alpha300, Ulm, Germany) and by X-ray photoelectron spectroscopy (XPS, Philips PHI-5300 ESCA, Philips, Amsterdam, The Netherlands) using an Al-Kα source, respectively. The water contact angle of the samples was measured using the contact angle equipment (CAMKSV021733, Gothenburg, Sweden). Three distinct surface areas were chosen, and the values were averaged.

2.3. Corrosion Resistance

Potentiodynamic polarization (PDP) measurements were performed in 0.01 M phosphate-buffered saline (PBS: NaCl, KCl, Na_2HPO_4 and KH_2PO_4) using an electrochemical workstation (CHI Instruments, Shanghai, China), with a sample exposure area of 0.28 cm^2. A calomel electrode served as the reference electrode, a graphite rod (Φ 6 * 65 mm) served as the counter electrode, and a coated sample served as the working electrode in this three-electrode setup. Finally, the potentiodynamic polarization curves were fitted.

Hydrogen evolution tests were carried out by immersion experiments over a period of 7 days using a homemade funnel apparatus. The AZ31B magnesium alloy and modified samples were submerged in 100 mL of Hanks' solution (composition: Nacl, Kcl, $Na_2HPO_4.12H_2O$, KH_2PO_4, $NaHCO_3$, and glucose, pH7.2–7.5) at 37 °C (three samples as a group). The relative height of the liquid level was recorded for each day.

2.4. In Vitro Cytocompatibility Evaluation

Protein electrophoresis using bovine serum albumin was used to assess the samples' ability to bind proteins. Electrophoretic blots were analyzed using Lane 1D L300 image analysis software.

MC3T3-E1 cells were grown in α-MEM medium with 10% fetal bovine serum and 1% penicillin-streptomycin. Culture conditions were 37 °C, 5% CO_2, and 100% humidity. The cells employed in the experiment were from the third to the fifth generation, and the culture medium was replaced every two days throughout the culture phase.

The samples that had been sterilized with UV were placed in α-MEM complete medium for a duration of 24 h at a temperature of 37 °C. The ratio of the sample's surface area to the extraction medium was 1.25 cm^2/mL, following the guidelines of ISO 10993-5:1999 [31]. Extracts were collected and filtered for subsequent cellular experiments.

MC3T3-E1 cells were inoculated into 96-well plates at a concentration of 1×10^4 cells/mL, and each well was filled with 100 μL of cell suspension. The cells were incubated for a duration of 12 h. Once the cells were fully attached to the wall, the original medium was removed and the culture was maintained by replacing the sample extracts every 2 days. The cells were tested for cck-8 at 1, 3, and 5 days, and cell viability was determined by measuring the absorbance at 450 nm in each well using an enzyme-labelled reader (Biotek ELX808, Richmond, VA, USA). Cells were inoculated into 24-well plates at a concentration of 2×10^4 cells/mL. The sample extract was replaced daily, and the cells were cultured until day 3. At this point, the cells were stained using Calcein/PI cell viability/cytotoxicity assay kit (Beyotime, Haimen, China) and examined using inverted fluorescence microscopy.

3. Results

3.1. Surface Characterization

Figure 2 depicts the surface topography of the three sets of samples. As illustrated in Figure 2, the magnesium alloy exhibits numerous scratches on its surface following the polishing treatment. Following the micro-arc oxidation treatment, the surface of the MAO coating exhibited the presence of micro-hole and micro-crack structures (shown by yellow circles). The holes are the result of the discharge channels of the micro-arc oxidation, while the fractures occur owing to the deformation of the magnesium alloy substrate caused by the significant amount of heat released throughout the experiment [32]. The morphology of the TiO_2 coating deposited by ALD was significantly altered, with a notable reduction in the number of microcracks on the surface of the MAO coating. Moreover, the size of the surface pores was considerably decreased and partly obscured. This demonstrates that TiO_2 has been successfully applied onto the surface of the samples, albeit it does not fully envelop the MAO layer. This might be attributed to the restricted number of ALD deposition cycles leading to the formation of a thin TiO_2 layer. Accordingly, we examined the cross-sectional structure of the MAO and found that the thickness of the magnesium oxide (MgO) layer was around 20 μm. The thickness of the ALD-deposited TiO_2 film, which is not easily visible in the SEM, was determined to be around 47.41 nm using a spectroscopic ellipsometer. Significant levels of magnesium were found on the sample surface, mostly from the substrate material and the MAO coatings, according to the EDS findings of the TiO_2/MAO coatings (Figure 3). On the other hand, the MAO coating and the presence of TiO_2 are responsible for elemental oxygen. Furthermore, elemental Ti is less common than Mg, O, but its distribution is uniform overall, supporting the effective deposition of the TiO_2 coating on the MAO samples' surface.

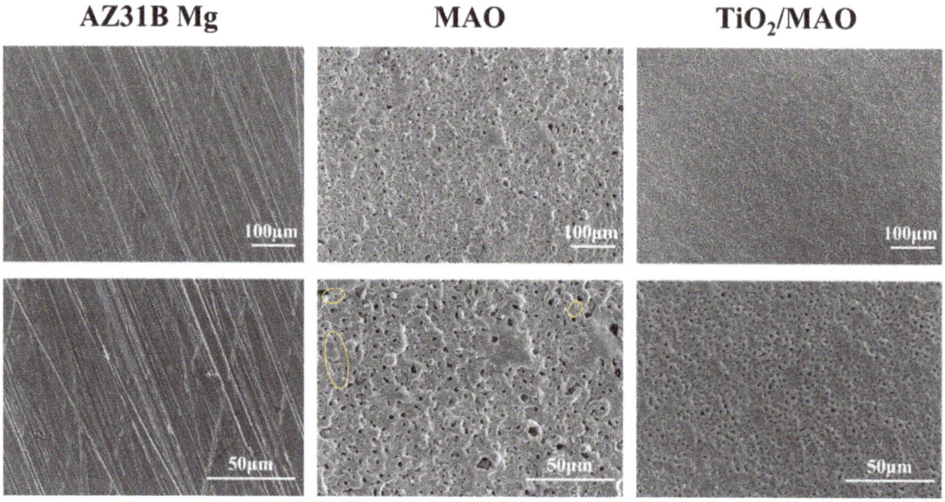

Figure 2. SEM images of AZ31B Mg, MAO, and TiO_2/MAO coatings.

Figure 3. (**a**) Cross-sectional morphology of MAO, elemental mapping images of (**b**) Mg, (**c**) O, (**d**) Ti, and (**e**) EDS surface analysis for TiO_2/MAO coating.

The X-ray diffraction (XRD) spectra were used to characterize the three sets of samples. As illustrated in Figure 4, diffraction peaks of Mg were observed for all samples ((100), (002), (101), (102), (110), (103), (200), (112), and (201)). In comparison to the other diffraction peaks, the (100) direction exhibited a notable increase in crystallinity due to the relatively low growth energy, which led to a large boost in the diffraction intensity of the crystalline surfaces. The MAO and TiO_2/MAO coatings exhibited diffraction peaks corresponding to magnesium oxide ((200) and (220)), which suggests that the coatings after the micro-arc oxidation are primarily composed of magnesium oxide. Furthermore, the intensity of the corresponding Mg diffraction peaks (100), (002), and (101) decreased significantly following ALD deposition. Unexpectedly, there were no discernible diffraction peaks seen for TiO_2.

This absence of peaks may be attributed to the very thin coating, which likely caused TiO_2 to reside in an amorphous state. Therefore, the TiO_2/MAO sample was subjected to Raman spectroscopy once again. The findings indicated that the Raman signal of the sample was not prominent, and there were no distinct features of anatase TiO_2, such as E_g, B_{1g}, A_{1g} + B_{1g}, and E_g at 143, 198, 395, 512, 518, and 639 cm^{-1} [33]. While peaks at 198 and 639 cm^{-1} were identified, they were not clearly discernible, as seen in Figure 4b. This may be due to the thinness of the surface TiO_2 layer and the weak Raman signal. The indication is that the surface of the TiO_2 is in an amorphous state and that there is no occurrence of phase separation in the TiO_2 film. This may be verified by the EDS analysis shown in Figure 3 and the subsequent XPS analysis shown in Figure 5.

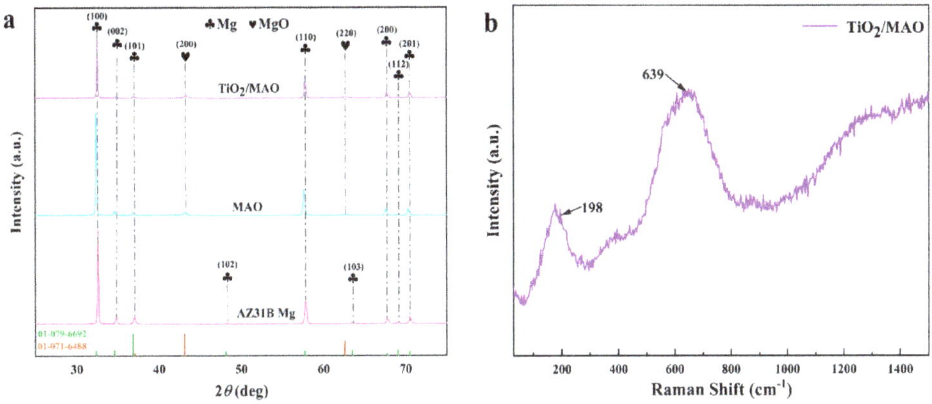

Figure 4. (**a**) XRD patterns of AZ31B Mg, MAO, and TiO_2/MAO, (**b**) Raman spectra of TiO_2/MAO samples.

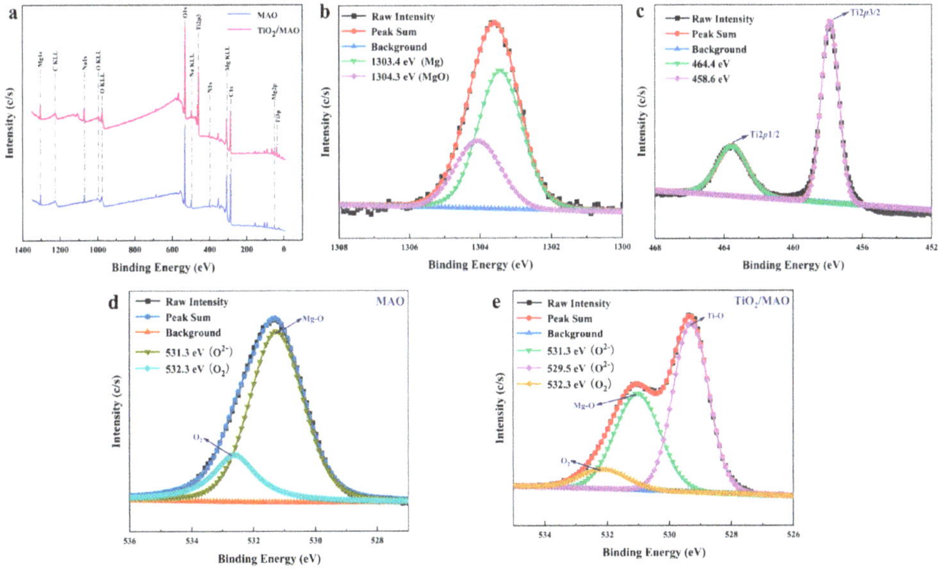

Figure 5. (**a**) The XPS spectra of MAO and TiO_2/MAO; high-resolution XPS spectra of implanted sample: (**b**) Mg 1s, (**c**) Ti 2*p*, (**d**) O 1s of MAO, and (**e**) O 1s of TiO_2/MAO.

In addition, the XPS technique was used to investigate the surface chemical composition of both materials. From the full spectra (Figure 5a), it can be seen that both sets of samples contain the elements magnesium, carbon, and oxygen. Figure 5b illustrates that the binding energy of magnesium on the sample surface is 1303.4 eV, corresponding to MgO, and that of magnesium metal is 1304.3 eV [34]. After the ALD deposition of TiO_2 on the MAO surface, significant Ti 2*p* and Ti 3*p* peaks appeared; peak O 1s also increased significantly (Figure 5a). The Ti 2p spectra, shown in Figure 5c, were comprised of two distinct bands, namely Ti $2p_{1/2}$ (464.4 eV) and Ti $2p_{3/2}$ (458.6 eV), which were predominantly in the form of TiO_2, indicating that TiO_2 was successfully deposited on the MAO surface [35]. In addition, the high-resolution O 1*s* spectra of both MAO and TiO_2/MAO samples were also fitted and analyzed (Figure 5d,e). According to the fitting results, the O 1*s* spectra of the MAO sample is divided into two distinct peaks at 531.3 eV and 532.3 eV [35,36]. The signal seen at 531.3 eV corresponds to O^{2-} ions associated with Mg-O bonds. This finding is consistent with the previous XRD data, suggesting that the surface of the sample consists of MgO after the process of micro-arc oxidation. The peak exhibiting a substantial binding energy of 532.3 eV can be ascribed to the presence of oxygen adsorbed on the surface. On the other hand, the O 1s spectra of TiO_2/MAO shows the presence of Ti-O bonds (529.5 eV), Mg-O bonds (531.3 eV), and adsorbed oxygen (532.3 eV) [37,38]. This result also indicates that the TiO_2 coating deposited during the experiment does not completely cover the MgO of the substrate, in agreement with the SEM results. Nevertheless, the deposition of TiO_2 did indeed have an impact on the Mg peak in Figure 4a, as can be clearly noticed. Both samples exhibit adsorbed oxygen on their surfaces. However, the TiO_2/MAO sample has a much lower fraction of adsorbed oxygen compared to the MAO sample. Following micro-arc oxidation, the sample's microporous surface shape has a higher probability of encountering oxygen upon exposure to air. The following ALD deposition partially fills the surface micropores, resulting in a decreased interaction with air and a corresponding decrease in the quantity of adsorbed oxygen. To summarize, ALD has effectively generated a TiO_2 coating, and the altered coating is present in the composition of MgO and TiO_2.

3.2. Electrochemical Measurements

In order to assess the samples' resistance to corrosion, dynamic potentiodynamic polarization experiments were conducted in a PBS solution. The results of these tests are shown in Figure 6a. The figure clearly demonstrates that the order of corrosion current density is as follows: TiO_2/MAO < MAO < AZ31B Mg. Conversely, the corrosion potential follows the exact opposite order: TiO_2/MAO > MAO > AZ31B Mg. The corrosion resistance of the samples is strongly correlated with the corrosion current density and corrosion potential. A lower corrosion current density indicates stronger corrosion resistance, whereas a larger corrosion potential suggests a lesser propensity for corrosion [39]. Hence, the TiO_2/MAO exhibits superior corrosion resistance and significantly mitigates corrosion tendencies. This may be attributed to the excellent sealing of pores and cracks on the original MAO coating surface by the TiO_2 coating, hence decreasing the corrosion of the coating. Table 2 also proves one of the results. Therefore, the TiO_2/MAO composite film can provide more effective protection for the magnesium alloy of the substrate and significantly improve the corrosion resistance of the material.

Table 2. Important data obtained by the Tafel polarization test.

	AZ31B Mg	MAO	TiO_2/MAO
I_{corr} (A/cm^2)	9.8230×10^{-6}	1.8732×10^{-6}	8.1325×10^{-7}
E_{corr} (V/SCE)	−1.4690	−1.3901	−1.3495

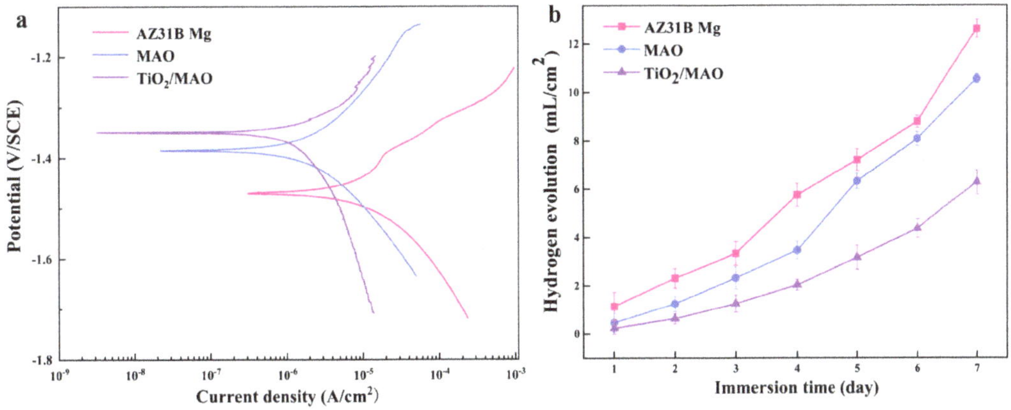

Figure 6. (**a**) Potentiodynamic polarization profiles of three samples; (**b**) HER curves of different samples immersed in Hank's solution for 7 days.

Following seven days of immersion in Hank's solution, the hydrogen evolution experimental findings for the three groups of samples are shown in Figure 6b. Among them, the potentiodynamic polarization experiment results are consistent with the significantly lower hydrogen evolution rate of TiO_2/MAO samples, suggesting that the TiO_2/MAO composite coating provides better substrate protection and can effectively stop the corrosion of AZ31B Mg. During the first three days of immersion, the AZ31B Mg exhibits a low rate of hydrogen evolution. This is attributed to the formation of corrosion products on the surface of the magnesium alloy, which effectively inhibits further corrosion [40]. With an increase in immersion duration, the corrosion products underwent progressive dissolution, leading to an increase in the surface corrosion rate. Consequently, the hydrogen production rate of the AZ31B magnesium alloy saw a substantial rise starting on the fourth day. While the MAO-coated samples showed a lower quantity of hydrogen evolution compared to the AZ31B magnesium alloy during immersion, the hydrogen evolution rate notably increased on the fifth day. This is caused by the deterioration of the permeable MAO coating and the interaction between the corrosive solution and the magnesium substrate. Furthermore, we used SEM to examine the surface structure of the samples after a 7-day immersion period. The results are shown in Figure 7. After 7 days of immersion, the sample's surface displayed many irregular fissures and a granular appearance due to the strong reactivity of the magnesium alloy. The presence of many cracks in the MAO coating may be attributed to corrosion resulting from the interaction between the solution and the substrate during the immersion process. This interaction leads to the generation of a significant quantity of hydrogen gas, which in turn disrupts and fractures the micro-arc oxidation coating. Conversely, the TiO_2/MAO surface exhibited no notable alterations, although there were minor fissures and surface indentations (shown by yellow circles). Hence, the TiO_2/MAO composite coating effectively safeguards the substrate and enhances its resistance to corrosion.

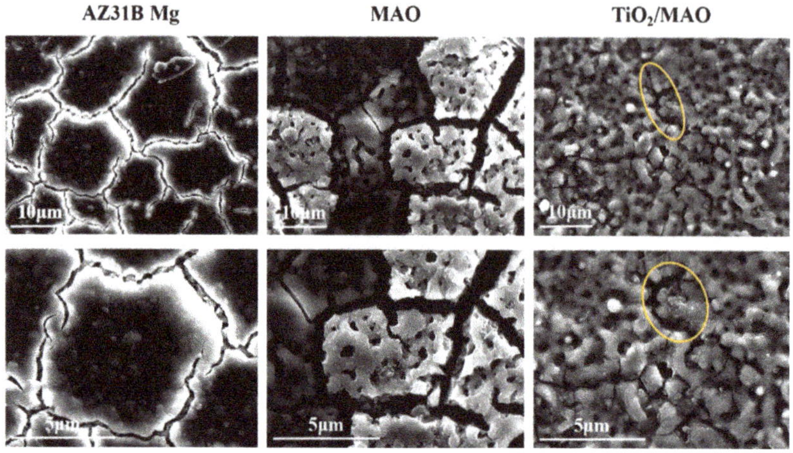

Figure 7. SEM images of AZ31B Mg, MAO, and TiO_2/MAO coatings immersed in Hank's solution for 7 days.

3.3. *In Vitro Cytocompatibility Studies*

Surface wettability, an essential surface property of biological implant materials, can be assessed using the water contact angle (WCA > 150°, classified as superhydrophobic; 150° > WCA > 90°, considered as hydrophobic; 90° > WCA > 10°, classified as hydrophilic; WCA < 10°, considered as superhydrophilic) [41]. Figure 8 demonstrates a steady reduction in the water contact angle for the three sample groups (101.02° ± 3.17° > 85.93° ± 2.46° > 77.40° ± 1.20°), indicating an increase in hydrophilicity. Furthermore, there is a strong correlation between the surface roughness and the surface wettability [42]. The wettability of the modified samples was enhanced as a result of the progressive augmentation in surface roughness after micro-arc oxidation and ALD deposition. The samples were graded based on their protein adsorption capability in the following order: TiO_2/MAO > MAO > AZ31B Mg. In contrast, a decrease in protein residue results in an increase in protein adsorption capability. The TiO_2/MAO picture exhibited the lowest protein residue and the highest brightness, confirming its better protein adsorption ability in comparison to the other two groups. Wettability also influences protein adsorption [42]. Surfaces that are superhydrophilic have a higher capacity to absorb water, resulting in a decrease in the adsorption of proteins [43]. Surfaces that are somewhat hydrophilic have the highest ability for protein adsorption [42], which aligns with our experimental findings: the protein adsorption capacity of TiO_2/MAO has been enhanced.

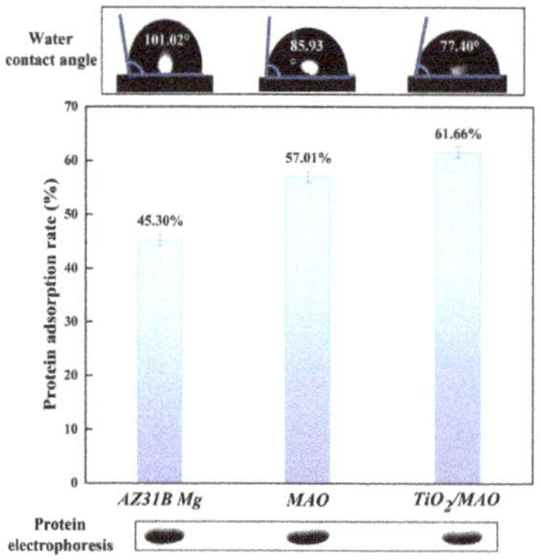

Figure 8. Protein adsorption of three groups and contact angle images (inset).

The proliferation results of MC3T3-E1 cells cultured in different sample extracts are shown in Figure 9a. After the first day of culture, the difference in proliferation status among the three groups of samples was not obvious, which might be due to the fact that the cells were sensitive to the growth environment at the early stage of culture and needed some time to adapt to the new culture environment [40]. A difference in cell proliferation between the MAO and TiO_2/MAO groups was less pronounced but superior to that of the AZ31B Mg group, suggesting that the MAO coatings and the composite coatings of TiO_2/MAO had a beneficial impact on enhancing cell development and proliferation. In order to assess the cytotoxicity of the various solutions, live/dead fluorescence imaging of MC3T3-E1 cells was conducted following a 2-day incubation period in the AZ31B Mg solution. Figure 9b displays the results of live/dead fluorescence imaging of MC3T3-E1 cells following a 2-day incubation in the AZ31B Mg, MAO, and TiO_2/MAO leachate samples, which were used to assess their cytotoxicity. The fluorescence images of all the sample groups showed obvious green fluorescence (living cells), while the AZ31B Mg and TiO_2/MAO groups had some red fluorescence (dead cells), indicating that the cells cultured in the extracts grew well and very few cells died. The cell growth status of the three sample groups was good; in particular, most of the cells in the TiO_2/MAO group were spread out, and the cells were tightly connected to each other. In addition, this outcome also indicates the cellular distribution state. The density of cells in the AZ31B Mg sample extract was marginally lower compared to the MAO and TiO_2/MAO sample groups. This finding aligns with the CCK-8 findings. The aforementioned findings suggest that the modified coated samples could promote cell proliferation and did not show cytotoxicity in the extracts of the samples.

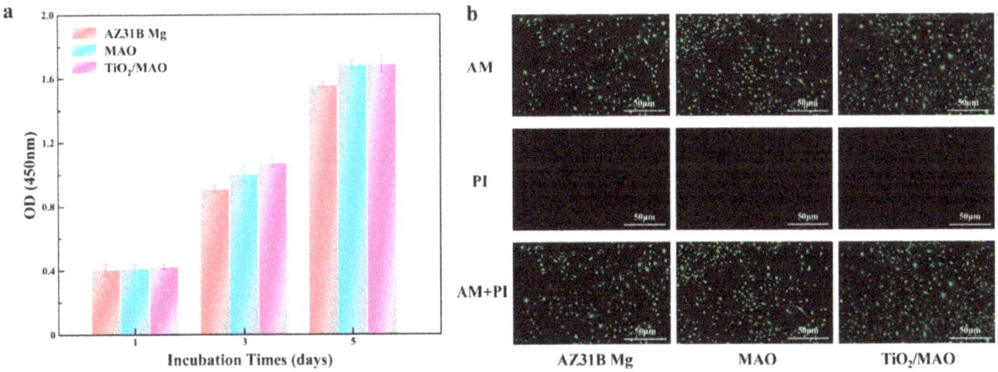

Figure 9. (**a**) The OD values of CCK-8 experiments for three groups. (**b**) Live/dead fluorescence staining results for three samples.

4. Discussion

Atomic layer deposition (ALD) is a precise method for depositing thin films on a substrate. It is based on chemical vapor deposition (CVD) and involves injecting two or more precursor molecules onto the substrate surface. These molecules then react chemically to form a single-atom film, which is deposited layer by layer. During the deposition process, the reactive precursors are sequentially deposited, with each reaction resulting in the deposition of a single layer of atoms. Hence, the quantity of deposition cycles may be regulated to attain meticulous management of film thickness. This work describes the production of titanium dioxide thin films utilizing atomic layer deposition (ALD) technology. The films were created by adsorbing two precursors, TDMA-Ti and H_2O, onto the surface of AZ31B magnesium alloy following micro-arc oxidation. This process resulted in the formation of a periodic atomic layer structure. The investigation resulted in the formation of a TiO_2 coating with a thickness of about 40 nm via the process of surface cycling, which was repeated 400 times. The samples coated with ALD-deposited TiO_2 on MAO exhibited a higher degree of completeness compared to the substrate and single layer MAO coatings. The reason for this is that the thick TiO_2 produced by ALD effectively plugs microcracks and holes. Nevertheless, the size of the holes formed during micro-arc oxidation differed, and the TiO_2 derived from 400 cycles did not fully envelop the MAO surface. This was mostly due to the excessively thin deposited film layer, which resulted in incomplete sealing of some of the bigger pores.

The TiO_2 coating functions as a physical barrier that hinders the entry of corrosive ions, successfully safeguarding the substrate from corrosion during the immersion test. Based on the PDP curve and immersion test, it is evident that the TiO_2/MAO coating enhances the corrosion resistance of the magnesium alloy, resulting in a lower corrosion rate. Nevertheless, when the immersion period rises, corrosive ions permeate through the thin TiO_2 layer, resulting in pitting corrosion. In this investigation, the TiO_2/MAO coating exhibited a lower degree of corrosion resistance compared to general titanium oxide, mostly due to its thin layer. Nevertheless, the presence of a TiO_2 film plays a crucial role in controlling the excessive corrosion rate of magnesium alloy. By adjusting the cycle duration, it is feasible to generate a thicker and more evenly distributed TiO_2 film, therefore further regulating the corrosion rate.

The biocompatibility of the material is intimately linked to the chemical characteristics and morphology of the implant surface. Upon first contact with the medium, such as mesenchyme or blood, the substance undergoes protein adsorption on its surface, subsequently impacting cell adhesion, proliferation, and migration. To provide the best possible attachment and growth of cells, the material must possess a specific capability for protein adsorption. Surface modification of the samples has been shown to boost the protein adsorption capability in in vitro investigations. This sets the groundwork for the future implantation of the samples in vivo. Furthermore, we specifically examined the impact of corrosion products on cell survival in the implant environment in relation to biocompatibility. In order to achieve this objective, we evaluated the feasibility of MC3T3-E1 cells in the sample extracts by using cck-8 and live/dead cell labeling techniques. The findings demonstrated that the application of ALD coating enhanced the cell survival of the untreated samples and did not exhibit any notable cytotoxic effects. Hence, the changed samples exhibited excellent biocompatibility.

5. Conclusions

1. Titanium dioxide (TiO_2) thin films were effectively coated upon MAO films utilizing the ALD method. The micropores and fractures that formed during the micro-arc oxidation process were effectively sealed.
2. Due to the thin TiO_2 coating formed by 400 cycles, the TiO_2 film exists mainly in an amorphous form.
3. The corrosion resistance of the TiO_2/MAO composite coating was shown to be superior by electrochemical and hydrogen evolution studies.
4. As shown by in vitro biological tests, the TiO_2/MAO composite coating exhibited superior biocompatibility compared to both the alloy matrix and MAO coating.

ALD has been increasingly used for surface modification of biomaterials. However, there has been limited research on the use of ALD coatings, such as TiO_2 coatings, on magnesium base alloys following micro-arc oxidation. Based on this, the current research may be considered one of the most efficient bio-coatings. Future studies may use ALD technology and other oxide coatings to enhance the corrosion resistance of magnesium-based alloys. Additionally, future studies will prioritize investigating the biocompatibility of these materials.

Author Contributions: X.H.: conception, methodology, investigation, analysis, writing—original manuscript preparation; Y.W.: investigation, methodology; J.M.: analysis, review and editing; X.M.: analysis, review and editing. All authors have read and agreed to the published version of the manuscript.

Funding: This work was supported by grants from the National Key Research and Development Program of China (2022YFC3601904) and the Tianjin Natural Science Foundation Key Program (22JCZDJC00340).

Institutional Review Board Statement: Not applicable.

Informed Consent Statement: Not applicable.

Data Availability Statement: The data presented in this study are available in the article.

Conflicts of Interest: The authors declare no conflicts of interest.

References

1. Liu, Y.; Zhang, Y.; Wang, Y.-L.; Tian, Y.-Q.; Chen, L.-S. Research progress on surface protective coatings of biomedical degradable magnesium alloys. *J. Alloys Compd.* **2021**, *885*, 161001. [CrossRef]
2. Lu, X.; Cai, H.; Li, Y.R.; Zheng, X.; Yun, J.; Li, W.; Geng, X.; Kwon, J.S.; Jiang, H.B. A Systematic Review and Network Meta-Analysis of Biomedical Mg Alloy and Surface Coatings in Orthopedic Application. *Bioinorg. Chem. Appl.* **2022**, *2022*, 4529520. [CrossRef]
3. Johari, N.A.; Alias, J.; Zanurin, A.; Mohamed, N.S.; Alang, N.A.; Zain, M.Z.M. Recent progress of self-healing coatings for magnesium alloys protection. *J. Coat. Technol. Res.* **2022**, *19*, 757–774. [CrossRef]
4. Peron, M.; Torgersen, J.; Berto, F. Mg and Its Alloys for Biomedical Applications: Exploring Corrosion and Its Interplay with Mechanical Failure. *Metals* **2017**, *7*, 252. [CrossRef]
5. Gawlik, M.M.; Wiese, B.; Desharnais, V.; Ebel, T.; Willumeit-Romer, R. The Effect of Surface Treatments on the Degradation of Biomedical Mg Alloys—A Review Paper. *Materials* **2018**, *11*, 2561. [CrossRef]
6. Hanzi, A.C.; Gerber, I.; Schinhammer, M.; Loffler, J.F.; Uggowitzer, P.J. On the in vitro and in vivo degradation performance and biological response of new biodegradable Mg-Y-Zn alloys. *Acta Biomater.* **2010**, *6*, 1824–1833. [CrossRef]
7. Reinhart, R.A. Magnesium metabolism. A review with special reference to the relationship between intracellular content and serum levels. *Arch. Intern. Med.* **1988**, *148*, 2415–2420. [CrossRef]
8. Kim, B.J.; Piao, Y.; Wufuer, M.; Son, W.C.; Choi, T.H. Biocompatibility and Efficiency of Biodegradable Magnesium-Based Plates and Screws in the Facial Fracture Model of Beagles. *J. Oral Maxillofac. Surg.* **2018**, *76*, 1055.e1–1055.e9. [CrossRef]
9. Dutta, S.; Gupta, S.; Roy, M. Recent Developments in Magnesium Metal-Matrix Composites for Biomedical Applications: A Review. *ACS Biomater. Sci. Eng.* **2020**, *6*, 4748–4773. [CrossRef]
10. Li, P.; Zhou, N.; Qiu, H.; Maitz, M.F.; Wang, J.; Huang, N. In vitro and in vivo cytocompatibility evaluation of biodegradable magnesium-based stents: A review. *Sci. China Mater.* **2018**, *61*, 501–515. [CrossRef]
11. Xia, D.; Liu, Y.; Wang, S.; Zeng, R.-C.; Liu, Y.; Zheng, Y.; Zhou, Y. In vitro and in vivo investigation on biodegradable Mg-Li-Ca alloys for bone implant application. *Sci. China Mater.* **2018**, *62*, 256–272. [CrossRef]
12. Atrens, A.; Song, G.L.; Liu, M.; Shi, Z.; Cao, F.; Dargusch, M.S. Review of Recent Developments in the Field of Magnesium Corrosion. *Adv. Eng. Mater.* **2015**, *17*, 400–453. [CrossRef]
13. Li, L.Y.; Cui, L.Y.; Zeng, R.C.; Li, S.Q.; Chen, X.B.; Zheng, Y.; Kannan, M.B. Advances in functionalized polymer coatings on biodegradable magnesium alloys—A review. *Acta Biomater.* **2018**, *79*, 23–36. [CrossRef]
14. Lin, Z.; Wang, T.; Yu, X.; Sun, X.; Yang, H. Functionalization treatment of micro-arc oxidation coatings on magnesium alloys: A review. *J. Alloys Compd.* **2021**, *879*, 160453. [CrossRef]
15. Murillo-Gutiérrez, N.V.; Ansart, F.; Bonino, J.P.; Menu, M.J.; Gressier, M. Protection against corrosion of magnesium alloys with both conversion layer and sol–gel coating. *Surf. Coat. Technol.* **2013**, *232*, 606–615. [CrossRef]
16. Tang, H.; Yu, D.; Luo, Y.; Wang, F. Preparation and characterization of HA microflowers coating on AZ31 magnesium alloy by micro-arc oxidation and a solution treatment. *Appl. Surf. Sci.* **2013**, *264*, 816–822. [CrossRef]
17. Ma, W.H.; Liu, Y.J.; Wang, W.; Zhang, Y.Z. Improved biological performance of magnesium by micro-arc oxidation. *Braz. J. Med. Biol. Res.* **2015**, *48*, 214–225. [CrossRef]
18. Jian, S.-Y.; Ho, M.-L.; Shih, B.-C.; Wang, Y.-J.; Weng, L.-W.; Wang, M.-W.; Tseng, C.-C. Evaluation of the Corrosion Resistance and Cytocompatibility of a Bioactive Micro-Arc Oxidation Coating on AZ31 Mg Alloy. *Coatings* **2019**, *9*, 396. [CrossRef]
19. Cui, L.-Y.; Zeng, R.-C.; Guan, S.-K.; Qi, W.-C.; Zhang, F.; Li, S.-Q.; Han, E.-H. Degradation mechanism of micro-arc oxidation coatings on biodegradable Mg-Ca alloys: The influence of porosity. *J. Alloys Compd.* **2017**, *695*, 2464–2476. [CrossRef]
20. Zhang, G.; Wu, L.; Tang, A.; Ma, Y.; Song, G.-L.; Zheng, D.; Jiang, B.; Atrens, A.; Pan, F. Active corrosion protection by a smart coating based on a MgAl-layered double hydroxide on a cerium-modified plasma electrolytic oxidation coating on Mg alloy AZ31. *Corros. Sci.* **2018**, *139*, 370–382. [CrossRef]

21. Toorani, M.; Aliofkhazraei, M. Review of electrochemical properties of hybrid coating systems on Mg with plasma electrolytic oxidation process as pretreatment. *Surf. Interfaces* **2019**, *14*, 262–295. [CrossRef]
22. Cui, X.-J.; Ping, J.; Zhang, Y.-J.; Jin, Y.-Z.; Zhang, G.-A. Structure and properties of newly designed MAO/TiN coating on AZ31B Mg alloy. *Surf. Coat. Technol.* **2017**, *328*, 319–325. [CrossRef]
23. Dai, X.-J.; Li, X.-C.; Wang, C.; Yu, S.; Yu, Z.-T.; Yang, X.-R. Effect of MAO/Ta_2O_5 composite coating on the corrosion behavior of Mg–Sr alloy and its in vitro biocompatibility. *J. Mater. Res. Technol.* **2022**, *20*, 4566–4575. [CrossRef]
24. Dou, J.; Yu, H.; Chen, C.; Lok-Wang Ma, R.; Ming-Fai Yuen, M. Preparation and microstructure of MAO/CS composite coatings on Mg alloy. *Mater. Lett.* **2020**, *271*, 127729. [CrossRef]
25. Li, Y.; Li, H.; Xiong, Q.; Wu, X.; Zhou, J.; Wu, J.; Wu, X.; Qin, W. Multipurpose surface functionalization on AZ31 magnesium alloys by atomic layer deposition: Tailoring the corrosion resistance and electrical performance. *Nanoscale* **2017**, *9*, 8591–8599. [CrossRef]
26. Li, C.Y.; Yu, C.; Zeng, R.C.; Zhang, B.C.; Cui, L.Y.; Wan, J.; Xia, Y. In vitro corrosion resistance of a Ta_2O_5 nanofilm on MAO coated magnesium alloy AZ31 by atomic layer deposition. *Bioact. Mater.* **2020**, *5*, 34–43. [CrossRef]
27. Hernández-Montes, V.; Betancur-Henao, C.P.; Santa-Marín, J.F. Titanium dioxide coatings on magnesium alloys for biomaterials: A review. *Dyna* **2017**, *84*, 261–270. [CrossRef]
28. Kania, A.; Pilarczyk, W.; Szindler, M.M. Structure and corrosion behavior of TiO_2 thin films deposited onto Mg-based alloy using magnetron sputtering and sol-gel. *Thin Solid Film* **2020**, *701*, 137945. [CrossRef]
29. Kania, A.; Szindler, M.M.; Szindler, M. Structure and Corrosion Behavior of TiO_2 Thin Films Deposited by ALD on a Biomedical Magnesium Alloy. *Coatings* **2021**, *11*, 70. [CrossRef]
30. Xu, Z.; Zhang, Q.; Luo, L.; Liu, Y.; Wan, J. Microstructure and corrosion resistance of TiN/TiO_2 nano-composite film on AZ31 magnesium alloy. *Surf. Coat. Technol.* **2021**, *406*, 126681. [CrossRef]
31. Li, W.; Tian, A.; Li, T.; Zhao, Y.; Chen, M. Ag/ZIF-8/Mg-Al LDH composite coating on MAO pretreated Mg alloy as a multi-ion-release platform to improve corrosion resistance, osteogenic activity, and photothermal antibacterial properties. *Surf. Coat. Technol.* **2023**, *464*, 129555. [CrossRef]
32. Ahmed, M.; Qi, Y.; Zhang, L.; Yang, Y.; Abas, A.; Liang, J.; Cao, B. Influence of Cu^{2+} Ions on the Corrosion Resistance of AZ31 Magnesium Alloy with Microarc Oxidation. *Materials* **2020**, *13*, 2647. [CrossRef] [PubMed]
33. Tanskanen, A.; Karttunen, A.J.; Karppinen, M. Substantially enhanced Raman signal for inorganic–organic nanocomposites by ALD-TiO_2 capping. *RSC Adv.* **2016**, *6*, 41087–41091. [CrossRef]
34. Zhu, J.; Jia, H.; Liao, K.; Li, X. Improvement on corrosion resistance of micro-arc oxidized AZ91D magnesium alloy by a pore-sealing coating. *J. Alloys Compd.* **2021**, *889*, 161460. [CrossRef]
35. Huang, L.; Su, K.; Zheng, Y.-F.; Yeung, K.W.-K.; Liu, X.-M. Construction of TiO_2/silane nanofilm on AZ31 magnesium alloy for controlled degradability and enhanced biocompatibility. *Rare Met.* **2019**, *38*, 588–600. [CrossRef]
36. Lv, L.; Lee, F.Y.; Zhou, J.; Su, F.; Zhao, X.S. XPS study on microporous titanosilicate ETS-10 upon acid treatment. *Microporous Mesoporous Mater.* **2006**, *96*, 270–275. [CrossRef]
37. Mohanta, R.R.; Medicherla, V.R.R.; Mohanta, K.L.; Nayak, N.C.; Majumder, S.; Solanki, V.; Varma, S.; Bapna, K.; Phase, D.M.; Sathe, V. Ion beam induced chemical and morphological changes in TiO_2 films deposited on Si(111) surface by pulsed laser deposition. *Appl. Surf. Sci.* **2015**, *325*, 185–191. [CrossRef]
38. Peron, M.; Bin Afif, A.; Dadlani, A.; Berto, F.; Torgersen, J. Comparing physiologically relevant corrosion performances of Mg AZ31 alloy protected by ALD and sputter coated TiO_2. *Surf. Coat. Technol.* **2020**, *395*, 125922. [CrossRef]
39. Peron, M.; Cogo, S.; Bjelland, M.; Bin Afif, A.; Dadlani, A.; Greggio, E.; Berto, F.; Torgersen, J. On the evaluation of ALD TiO_2, ZrO_2 and HfO_2 coatings on corrosion and cytotoxicity performances. *J. Magnes. Alloys* **2021**, *9*, 1806–1819. [CrossRef]
40. Peron, M.; Bertolini, R.; Cogo, S. On the corrosion, stress corrosion and cytocompatibility performances of ALD TiO_2 and ZrO_2 coated magnesium alloys. *J. Mech. Behav. Biomed. Mater.* **2022**, *125*, 104945. [CrossRef]
41. Song, W.; Mano, J.F. Interactions between cells or proteins and surfaces exhibiting extreme wettabilities. *Soft Matter* **2013**, *9*, 2985–2999. [CrossRef]
42. Oliveira, S.M.; Alves, N.M.; Mano, J.F. Cell interactions with superhydrophilic and superhydrophobic surfaces. *J. Adhes. Sci. Technol.* **2012**, *28*, 843–863. [CrossRef]
43. Zheng, J.; Li, L.; Tsao, H.K.; Sheng, Y.J.; Chen, S.; Jiang, S. Strong repulsive forces between protein and oligo (ethylene glycol) self-assembled monolayers: A molecular simulation study. *Biophys. J.* **2005**, *89*, 158–166. [CrossRef] [PubMed]

coatings

MDPI

Article

Enhancing Passivation Behaviors and Wear Resistance of Biomedical Ti-15Mo Alloy via {332} Twinning Pre-Tension and Aging

Shanshan Zhang [1,2], Fazhan Yang [1,2,*], Wei Sun [1], Ning Cui [1] and Tiewei Xu [1,*]

[1] School of Mechanical and Automotive Engineering, Qingdao University of Technology, Qingdao 266520, China; sszhangqd@163.com (S.Z.); sw290214281@163.com (W.S.); sdcuining@sina.com (N.C.)

[2] Key Laboratory of Industrial Fluid Energy Conservation and Pollution Control, Qingdao 266520, China

* Correspondence: fazhany@163.com (F.Y.); twxu@163.com (T.X.)

Abstract: The purpose of this investigation was to determine the extent to which a combination of pre-deformation and aging procedures could enhance the corrosion and wear resistance of the Ti-15Mo alloy for biomedical materials. The initial grains were refined with {332} twins and other defects after the pre-deformation, increasing the amount of precipitation nucleation and forming finer α phases during aging. The finer precipitates with numerous boundaries are beneficial in forming a thicker passivation film in phosphate-buffered saline (PBS) solution, and the corrosion resistance of the 20% pre-deform plus aged alloy is improved by 4.23 times. The result indicates that the corrosion passivation behavior and film structure of the biomedical Ti-15Mo alloy are significantly influenced by its microstructure. The worn track and debris of the alloy after pre-tension and aging gradually decreases with the increase in pre-deformation rates, caused by the increase in the hardness and wear resistance. The mechanism of corrosion and wear resistance in PBS solution were revealed, which showed the pre-tension and aging processes that were performed on the Ti-15Mo alloy. This study proposes that pre-tension with {332} twins contributed to precipitation refinement, which would enhance the passivation behaviors and wear resistance of the biomedical β titanium alloy.

Keywords: passivation behavior; corrosion resistance; biomedical Ti alloy; {332} twin; wear loss

1. Introduction

Titanium (Ti) alloy with excellent mechanical properties and biocompatibility has been widely used as a biomaterial in the past few decades [1–3]. The increase in human lifespan and a fast lifestyle has caused a higher requirement for biomedical implants that are more durable and reliable. However, traditional Ti alloys have a nonnegligible risk of toxicant elements due to corrosion behaviors, frictional wear and ion dissolution during long-term service. Thus, biomedical β-Ti alloys are designed to meet the non-toxicity requirements [4] and possess superior biological performance and compatibilities in the human body environment. On the other hand, the loss and failure of biomaterials are noteworthy issues due to long-term corrosion and friction during the service process. Excellent corrosion and wear resistance is also desired and favorable for extending the service life of biomedical implants, so that Ti implants should endure complex corrosion and wear in vivo [5].

As is well known, the crucial strategy for performance improvement in metals is microstructural refinement. In a previous study by the authors, the improved microstructure and strength were found in the Ti-15Mo-3Al-2.7Nb-0.2Si alloy after cold pre-rolling and aging [6]. The Ti-Nb alloy after pre-deformation and subsequent aging also presented the optimized β matrix texture and morphology of the α precipitates [7]. Similarly, the Ti29Nb13Ta4.6Zr alloy with nanoscale α precipitates after cold rolling and aging demonstrated improved mechanical properties and decreased the passivation current density [8].

Ji et al. reported that the microstructure of the Ti alloy significantly influenced the corrosion resistance via improving the nucleation and growth of the passive film [9]. This passive film acts as a barrier, preventing biological and chemical reactions between biomedical implants and the corrosion environment. The microstructural refinement facilitates the formation of a dense passive film and improves the corrosion resistance [10]. Enhanced passive films in PBS solution can be obtained by refining α precipitates in β-Ti alloys. The content of hypervalent oxides in the passive film also increases with the refinement of the microstructure [11]. A biomedical Ti-15Mo alloy does not have toxicant elements compared with other Ti alloys, and is easy to deform with a {332} <113> twinning mechanism [12–14]. The twins are different from slipping dislocations and have special effects on the precipitation behavior during the aging process after pre-deformation. However, pre-deformation with the twinning mechanism and the precipitation behaviors of β-Ti alloys are scarcely reported. Meanwhile, the passivation behaviors and corrosion resistance of the Ti-15Mo alloy after pre-deformation and aging should be investigated systematically.

In order to obtain excellent corrosion and wear resistance, this study introduced {332} twins and other defects via pre-tension, refining the coarse grains and providing assistant nucleation for the aging precipitation of the Ti-15Mo alloy. In contrast to conventional solution and aging treatments [15,16], the alloy after pre-deformation and aging treatments is desired for extending the service life of biomedical applications. Microstructural evolution and its effects on the corrosion resistance were also investigated using electron backscatter diffraction (EBSD), X-ray diffraction (XRD), transmission electron microscopy (TEM), electrochemical impedance spectroscopy (EIS) and X-ray photoelectron spectroscopy (XPS). This study would more deeply reveal the mechanism of corrosion and wear resistance of the alloy after pre-tension and aging.

2. Experimental Materials and Methods

2.1. Specimen Preparation

The Ti-15Mo alloy used in this study was melted as an industrial ingot by vacuum arc melting three times. The chemical composition is listed as follows: 14.7 mass% Mo, 0.08 mass% O and Ti balanced. The transformation temperature of the α/β phase is approximately 755 °C via the metallography method. The billet of the alloy was hot-rolled to a ϕ24 mm bar followed by homogenization and forging. After the solution treatment (ST) at 790 °C for 1 h and quenching, the pre-tension of the alloy was conducted on an MTS 370.02 universal testing machine (Eden City, MN, USA) at room temperature under 0.001 s^{-1}. The samples after the 3%, 9%, 15% and 20% deformation rate were marked as Tens-3, Tens-9, Tens-15 and Tens-20, respectively. According to our previous studies [17,18], the aging treatment following ST/pre-deformation was conducted at 550 °C for 8 h. The Tens-3, Tens-9, Tens-15 and Tens-20 specimens after aging treatment were marked as Tens-3-A, Tens-9-A, Tens-15-A and Tens-20-A, respectively. After pre-tension and aging, the alloys were, respectively, subjected to various characteristic tests, including microstructure, hardness, corrosion and wear resistance, as shown in Figure 1. The specimens for the tests were cut by DF7745 electrical discharge machining (Taizhou, China), and paralleled to the tensile axis.

2.2. Electrochemical Measurements

The corrosion test was conducted in electrolyte using the ivium electrochemical workstation (Eindhoven, Holland). The commercial PBS solution with 7.4 PH was chosen in this study with the following composition (unit: mM): 139.7 Cl^-, 153 Na^+, 8 HPO_4^{2-}, 4.7 K^+ and 2 $H_2PO_4^-$, which was manufactured by Labcoms™ LIFE SCIENCES (Shanghai, China). The electrochemical measurement was performed with the conventional three-electrode cell system, including a working electrode, a platinum sheet as the counter electrode and a reference electrode of saturated AgCl. The EIS, potential dynamic (PD) polarization and Mott–Schottky (M-S) method were employed in this study. The open-circuit potential (OCP) was recorded for 3600 s to ensure stability before other electrochemical tests. The negative voltage of −1.5 V was loaded for 5 min before the OCP tests to dissolve the oxide layer formed in the atmospheric environment during the sample preparation. The EIS experiment was performed with a scan frequency range from 10^{-2} Hz to 10^5 Hz. And the EIS data were fitted to an equivalent electric circuit (EEC) model by the ZSimpWin software, which is integrated in the electrochemical workstation. The M-S test was conducted in a potential range from −1000 mV to 500 mV. The PD polarization curve was recorded from −1000 mV to 2000 mV at a scanning rate of 1 mV/s. For each condition, at least two specimens were tested to ensure reproducibility.

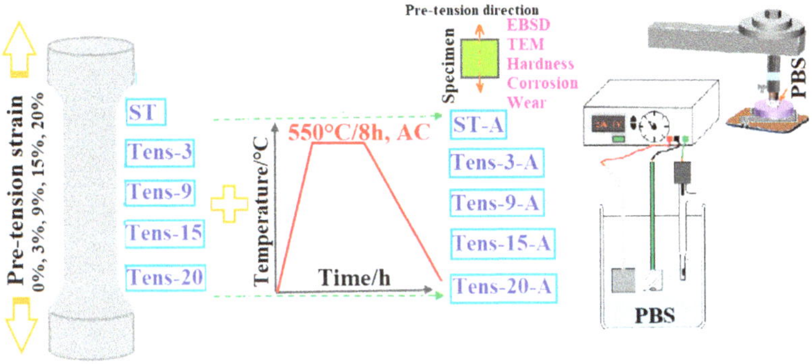

Figure 1. The schematic diagram of the experimental process.

2.3. Friction Tests and Characterization Methods

The friction test was conducted at an ambient temperature using Bruker's universal mechanical tester (UMT, Billerica, MA, USA). Prior to testing, the specimen for the friction tests underwent mechanical grinding and polishing to achieve a mirror-like finish surface. The specimens were immersed in PBS solution as the friction medium during testing. The applied load was 1 N, the frequency was set at 1 Hz and a stroke length of 2 mm was used. The duration of the test was 30 min. An aluminum oxide ball with a diameter of 6 mm was employed as the friction pair. The worn tracks of tribological experiments were observed using a DCM8 confocal microscope (Wetzlar, Germany). To ensure reliability of results, three replicate wear tests were conducted for each specimen under identical conditions.

The hardness of the alloy was measured by a micro-Vickers hardness tester (HV-1000, Xi'an, China), with 500 g for 5 s and measured 10 times to obtain an average value. The JSM6460 scanning electron microscope (SEM, Tokyo, Japan) was utilized to observe the morphology of the microstructure, worn tracks and debris. Electron backscatter diffraction (EBSD) analysis was conducted using an Oxford Instruments AZtecSynergy operated at 20 kV and a step of 0.3 μm. The EBSD data were post-processed by the channel 5 software. The morphology of deformed structures and α precipitates were observed using a JEM-3010 transmission electron microscope (TEM, Tokyo, Japan) operated at 100–300 kV.

3. Results

3.1. Microstructure of ST and Its Evolution during Pre-Tension

Figure 2 illustrates the orientation imaging map (OIM) and pole figures (PFs) of the Ti-15Mo alloy before pre-tension. The size of the grains in the ST alloy is about 50 μm and presents a regular shape (Figure 2a). It suggests that recrystallization had completed after the ST at 790 °C for 1 h. The ST specimen consists of a single β phase, which has a body-centered cubic (BCC) crystal structure. The initial texture of the alloy investigated by EBSD (Figure 2) and the OIM of the rolling direction (RD) illustrated that no dominant texture had been formed after the rolling and ST processes. The distribution of the main orientations in the Ti-15Mo alloy after the ST is relatively random. However, a slight strong {100} orientation could be observed from the PF analysis in Figure 2b. This recrystallized texture was liable to form in bcc metals [19–21], and also indicated that the recrystallization of the ST alloy had completed. No twin was observed in Figure 2a and it indicated that the stacking fault energy (SFE) of the Ti-15Mo alloy was insufficiently low to form an annealing twin during the ST and water quenching processes. Differences in the ω and α″ phase were also not observed in the alloy due to the low resolution of EBSD. However, previous works of the authors had investigated the ST Ti-15Mo alloy with TEM and revealed that these metastable phase transformations had taken place slightly during quenching [17,18].

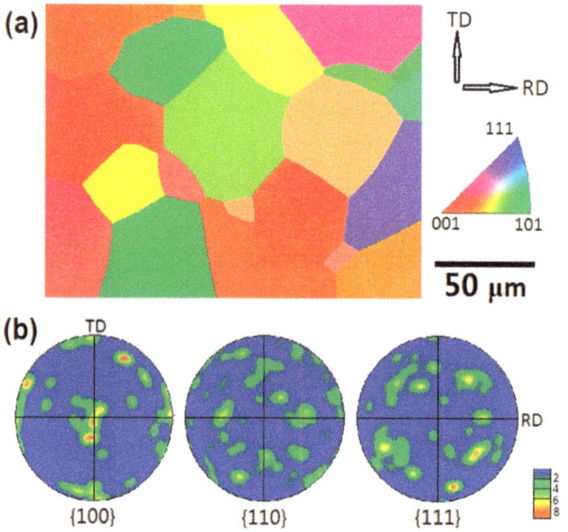

Figure 2. OIM and PFs of the Ti-15Mo alloy treated at 790 °C for 1 h. (**a**) OIM, (**b**) PFs. The scanning step size was 2 μm.

{332} <113> twins of β-Ti alloys could be formed during cold deformation and had 50.57° misorientation [22–25]. The pre-tension in this work was performed in the Ti-15Mo alloy at the 3%, 9%, 15% and 20% rate at room temperature, respectively. The OIMs and boundary character distribution (BCD) of the EBSD after pre-tension are shown in Figure 3. The boundaries with 10~49° misorientation are marked with green color. Black color marks the boundaries where the misorientation was over 53°. The 49~53° misorientation boundaries are marked with red color, and most of them should be {332} <113> twins formed during the pre-tension deformation of the Ti-15Mo alloy. Figure 3a shows the OIM and BCD of the Ti-15Mo alloy, which was performed by 3% pre-tension deformation. The density difference in the {332} <113> twins is significantly observed in different recrystallized grains due to the original texture. {332} <113> twins are apt to forming in some certain texture matrices, which was decided with the difference in the critical resolved shear stress (CRSS) during deformation [13]. The grain with a higher Schmidt factor (SF) would be apt to form {332} <113> twins and deform with the twinning mechanism [26]. It could be observed that a few {332} <113> twins appeared in the interior of recrystallization grains. This phenomenon was also reported by Wu et al. while {112} twins were investigated [27]. The Ti-15Mo alloy with 3% pre-tension is considered in an earlier stage for tension deformation, so some {332} <113> twins terminated in the interior of grains. With an increase in pre-tension deformation, some twins propagated to the neighboring grains across the low-angle boundary (LAB), but some ended at the high-angle boundary (HAB) and induced a new twinning formation in the neighboring grain due to the large stress concentration. Figure 3b shows the OIM and BCD of the alloy with 9% pre-tension and presents a significant difference from Figure 3a. Firstly, the number of twins increased obviously with the increase in the pre-tension deformation. Boundaries crossing the black line in Figure 3b were still about 50.57° misorientation and suggested that {332}

<113> twins were the main deformation mechanism while the alloy was conducted by 9% pre-tension. Secondly, most of the twins have propagated to the whole grain, indicating that the length of twins increased with the development of pre-tension deformation. However, the width of the twins in the Ti-15Mo alloy with 9% pre-tension never has an obvious change and indicates that the amount and length of twins increased with the increase in pre-tension deformation but the width did not.

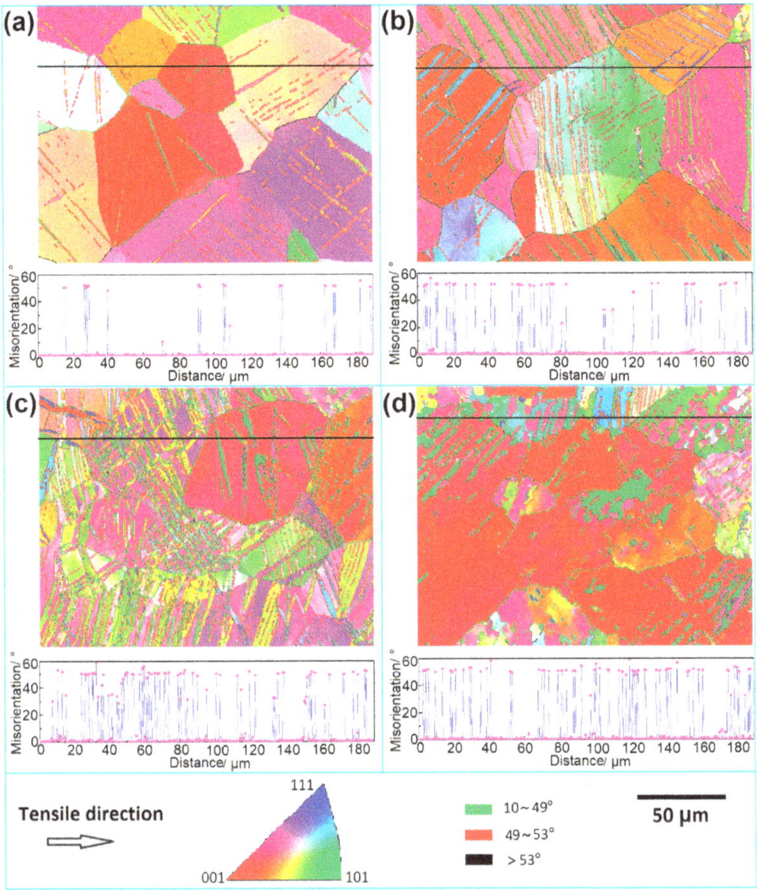

Figure 3. OIMs and BCD of the Ti-15Mo alloy after pre-tension with 3%, 9%, 15% and 20% strain. (**a**) 3%, (**b**) 6%, (**c**) 15%, (**d**) 20%. The scanning step size was 0.1 μm.

The OIM and misorientation of boundaries in the alloy with 15% pre-tension are shown in Figure 3c. In contrast to the 9% pre-tension, Figure 3c presents a higher density of twins in the Ti-15Mo alloy after 15% pre-tension deformation. Furthermore, lots of {332} <113> twins had different orientations and were distributed crosswise in the initial grain. The sufficient stress caused by the increased pre-tension deformation is considered to induce these crisscrossed twins due to the requirement for a higher CRSS. The crisscrossed twins also refined initial grains and formed lots of sub-structures with different orientations. Some misorientation of twin boundaries at the black line in Figure 3c is lower or higher than 50.57° obviously and indicates that the sub-structures might be rotated and form deviation from the {332} <113> twinning misorientation during 15% pre-tension deformation. Even so, the interface of the twin and matrix still presents a rough straight shape due to the sub-structures maintaining the morphology of twinning boundaries. The alloy performed by 20% pre-tension presents a different microstructure in contrast to the 3%~15% deformation. Twinning boundaries formed in high SF grains conducted sufficient deformation and rotation, which induced some grains to be refined due to there being lots of irregular sub-structures (in Figure 3d). Although boundaries of these sub-structures at the black line in Figure 3d are no longer straight, most of them are around 50.57° misorientation and show as the HABs. It is different from the Ti-15Mo-3Al-2.7Nb-0.2Si alloy, which is dominated by dislocation motion during cold-rolled deformation and showed low-angle grain boundaries [6]. Another interesting result could be obtained from the fact that twins are rarely found in the <001> orientation grains (red color in Figure 3) compared with other orientation grains. According to the previous literature [28,29], these <001> orientation grains are not easy to form {332} <113> twins during pre-tension deformation due to their low SF and requirement for a higher CRSS [28,30].

Figure 4 shows the boundary size distribution (BSD) of the alloy after different pre-tensions. As the deformation increased, the initial grains were continuously refined by various twin variants. Compared with the ST specimen, the alloy after pre-deformation has more refined sub-structures. The size of sub-grains in the Tens-3 specimen is 0.5~4 μm, even over a 46.3% proportion of them are approximately 1.0 μm, which is caused by twinning formation during the pre-tension. Over a 53.1% proportion of the sub-grains have a size of approximately 1.0 μm, exhibiting more refined sub-structures in the alloy after 9% pre-tension. The number of sub-grains in the Tens-15 specimen is slightly higher than that of Tens-9. The number of grain sizes of about 0.5 μm became maximum in the Tens-20 specimen, due to the formation of nanoscale sub-structures at the twin and twin-crossing positions. These deformed sub-grains in the alloy after pre-tension would have an influence on precipitation behaviors during the subsequent aging.

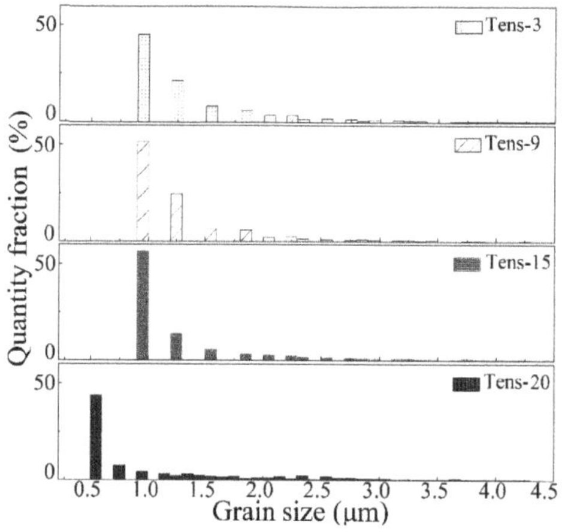

Figure 4. Grain size distribution of the Ti-15Mo alloy after different pre-tensions.

As seen in Figure 5 in the TEM images, the microstructure of the alloys subjected to 9% and 20% pre-tension is shown. The bright-field (BF) image in Figure 5a was used to analyze the morphology of the twins and boundaries of the Ti-15Mo alloy deformed with 9% pre-tension. The width of the twins is less than 1 μm and the boundaries present as a straight shape. The TEM of the 9% pre-tension alloy also shows that the orientations of the twins and the matrix have a significant difference with the white and black contrast. Figure 5b presents the selected area electron diffraction (SAED) pattern of the twins (a red circle in Figure 5a) and indicates their characteristic of {332} <113> twins. Furthermore, in the alloy after the Tens-9 process, the athermal ω phase has been identified by the diffraction spots at 1/3 and 2/3 [211] M (in Figure 5b). A different morphology is shown in

Figure 5c and indicates that evolution of the twinning should take place during pre-tension from 9% to 20%. The BF image of TEM (in Figure 5c) illustrates that dense and crossed twins have cut the original grain into many small sub-structures. Furthermore, the density of dislocations in these twins and sub-structures is significantly higher in contrast to that of the 9% pre-tension alloy (in Figure 5a). The SEAD pattern in the bottom-left corner in Figure 5c also indicates that many fine ω phases formed in the alloy after 20% pre-tension. These ω phases would serve to reduce the energy barrier and play a crucial role in assisting the nucleation of α precipitates during aging.

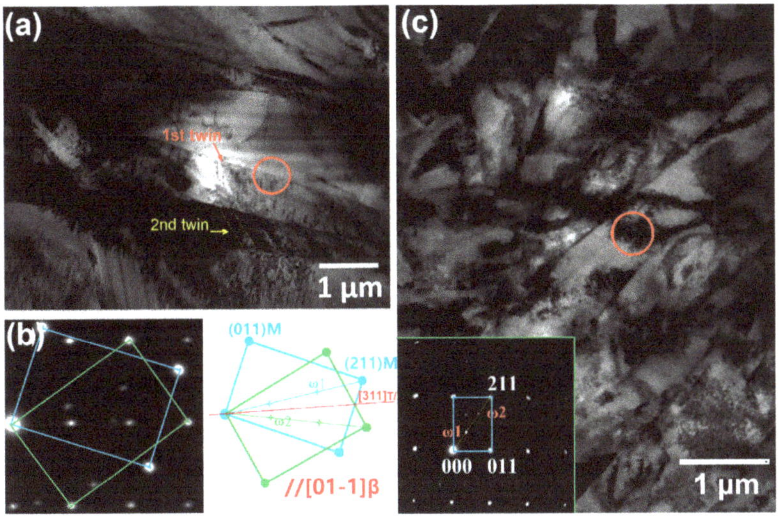

Figure 5. BF and DF TEM microstructure of the Ti-15Mo alloy after pre-tension with 9% and 20% strain. (**a**) BF with 9% pre-tension, (**b**) SAED with 9% pre-tension, (**c**) BF and SAED with 20% pre-tension.

Figure 6 shows the synthetic images of the {332} twin boundary and Kernel average misorientation (KAM) of the alloy after 3%, 9%, 15% and 20% pre-tension, respectively. The boundaries of the twins are exhibited as brown lines, which are automatically generated by channel 5 software with the corresponding misorientation between the adjacent regions. In this study, the 49–53° misorientation is identified as {332} twin boundaries, because the misorientation of {332} twins induced by pre-tension could have a small deviation from the standard of 50.57°. Figure 6a shows that a few twin boundaries appear in some initial grains of the alloy after 3% pre-tension deformation. The alloy after 9% pre-tension formed more twin boundaries, which could be observed in all grains (in Figure 6b). With the increase in pre-tensive deformation, these twinning variants with different orientations exhibited a cross-division observation, resulting in the refinement of the initial grains. While the pre-tensive deformation was over 9%, misorientation of some boundaries was beyond the range of 49–53°. This deviation from the twinning characteristic might be caused by the torsion of {332} twins. The further microstructural refinement of the Ti-15Mo alloy would occur with the increase in the cross-division and torsion of the twins. These sub-structures became smaller and more boundaries deviated from the standard misorientation of {332} twins (shown in Figure 6c,d). The more deviated boundaries were observed with the increase in the pre-tension (in Figure 6a–d), indicating that more twin structures occur due to torsion with the further deformation. These boundaries still play an important role in the microstructural refinement, even if they no longer have a perfect {332} twin boundary.

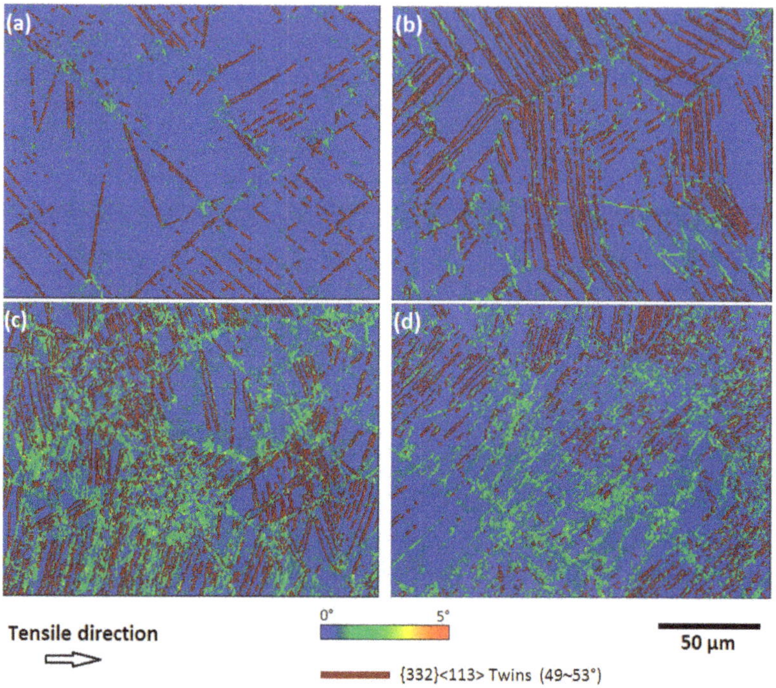

Figure 6. KAM and {332} boundaries' synthetic images of the Ti-15Mo alloy after pre-tension: (**a**) Tens-3, (**b**) Tens-9, (**c**) Tens-15 and (**d**) Tens-20 sample. The scanning step of EBSD was 0.1 μm.

The KAM maps were used to analyze the corresponding lattice rotation after pre-tension. In this study, the red color in the KAM map denotes the region with a 5° rotation, signifying the highest lattice rotation, while the blue color indicates a negligible rotation with 0° value. The KAM maps of the Tens-3, Tens-9, Tens-15 and Tens-20 specimens are presented in Figure 6. The KAM of the twins displays a higher value compared to the surrounding matrix, indicating a greater rotation and more defects due to deformation. Notably, the KAM value in the joint region of the matrix and twins is higher than that of the surrounding regions, indicating that the twin boundaries are prone to acting as dislocation sources during pre-tensive deformation. The difference in the KAM from Figure 6 also illustrated that more dislocations and residual stresses were formed in the alloy with the increase in the pre-tensive deformation.

3.2. Microstructure and Precipitation Behaviors during Aging

Generally, the α phase with a close-packed hexagonal (CPH) crystal structure will precipitate from the β phase during aging in the STed Ti-15Mo alloy. In this study, the alloy after the solution and aging treatments is marked as ST-A for short. Similarly, the 3%, 9%, 15% and 20% pre-deformed alloy after aging is labeled as Tens-3-A, Tens-9-A, Tens-15-A and Tens-20-A, respectively. Figure 7 presents the SEM and TEM morphology of α precipitates in the ST-A sample. The ST over T_β causes the β matrix phase to be metastable, and more α precipitates would form during aging to change the stability of the alloy. However, the coarse grains in the ST alloy provide sufficient spaces for the growth of the α precipitates, resulting in a relatively large size of the α phase (in Figure 7). The TEM image of the ST-A sample is shown in Figure 7b and illustrates that the size of α precipitates in the ST-A sample is approximately 50 nm in width and 5 μm in length. The result of SAED, shown in the top-right corner of Figure 7b, exhibits that there are only α precipitates after the 550 °C aging treatment.

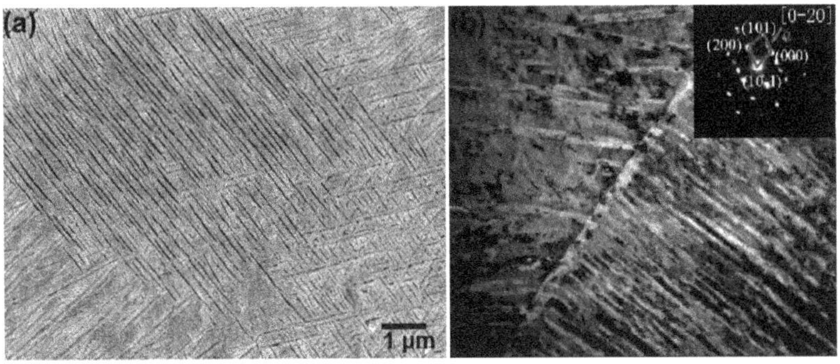

Figure 7. (**a**) SEM and (**b**) TEM images of ST-A specimen.

Figure 8a–d show the SEM microstructure of the Tens-3-A, Tens-9-A, Tens-15-A and Tens-20-A specimens, respectively. In contrast to the morphology of the α precipitates in the ST-A specimen (in Figure 7), the α precipitates formed during aging in the pre-deformed alloy are significantly finer. The pre-tension influenced the precipitation behaviors due to two main factors. Firstly, the dislocations and ω metastable phases formed during pre-tension facilitate the nucleation of the β→α phase transformation in a non-equilibrium condition. Secondly, the {332} twins induced by pre-tension refined the initial grains, restricting the growth of α precipitates and preventing the formation of coarse precipitation. It can be observed from Figure 8 that the size of the α precipitates formed during aging becomes finer and distributes more randomly with the increase in the pre-deformation. In the pre-tensive deformation over 15%, the length of precipitation is less than 1 μm (in Figure 8c,d), which is significantly refined compared with the size of the α precipitates in the ST-A specimen. The nanoscale α precipitates formed during aging in the pre-tensive alloy illustrate short-needle morphology, and distribute in the β matrix grains homogeneously. Figure 8e–h show the TEM microstructure of the alloy after 3%, 9%, 15% and 20% pre-deformation and aging. Different from the ST-A specimen, two changes could be observed in the alloy after pre-deformation and aging. Firstly, numerous dislocations and ω metastable phases formed during pre-tension provide sufficient non-equilibrium nucleation sites for the precipitation phase transformation. Secondly, the alloy after pre-tensive deformation formed numerous twins and sub-structures, which restricted the growth of α precipitates due to insufficient space, resulting in microstructural refinement. The amount of α precipitates increased, the size of them decreased and the distribution became more random with the increase in the pre-deformation rate (as seen in Figure 8).

3.3. *Passivation Behaviors and Corrosion Resistance of Biomedical Ti-15Mo Alloy*

Figure 9a shows the open-circuit potential (OCP) curves of the ST-A, Tens-3-A, Tens-9-A, Tens-15-A and Tens-20-A specimens in PBS solution. It was observed that as corrosion proceeded, the corrosion potential in all the specimens increased, indicating that the surface of the Ti-15Mo alloy spontaneously formed an oxide film, which played a role in inhibiting corrosion. The formation of the oxide film on the surface of titanium metal affects the electrochemical reaction at the corrosion electrode, causing the open-circuit potential to shift positively, which is a gradual passivation process. As the metal is immersed in PBS solution for a longer period, the thickness of the passivation film increases, enabling protection against the continuous dissolution of the metal. The initial stage of the OCP curve shows a rapid increase in the positive potential, and as corrosion proceeded, the change in the OCP becomes smaller, eventually reaching a dynamic equilibrium. The result indicates that the stabilized structure of the passivation film (chemical composition, thickness, defect concentration, etc.) has been formed completely. The required time for reaching a stable OCP reflects the difference in the passivation speed of the Ti-15Mo alloy after different pre-tension and aging. Compared with the ST-A sample, in general, the alloy after pre-tension and aging needs a shorter time to be a stable OCP. Furthermore, this trend is more obvious with the increase in the pre-deformation rates of the Ti-15Mo alloy. Furthermore, the open-circuit potential after 3600 s became stable in all specimens, and the value increased with the increase in the pre-tensive deformation (in Figure 9b), indicating an enhanced corrosion resistance of the alloy.

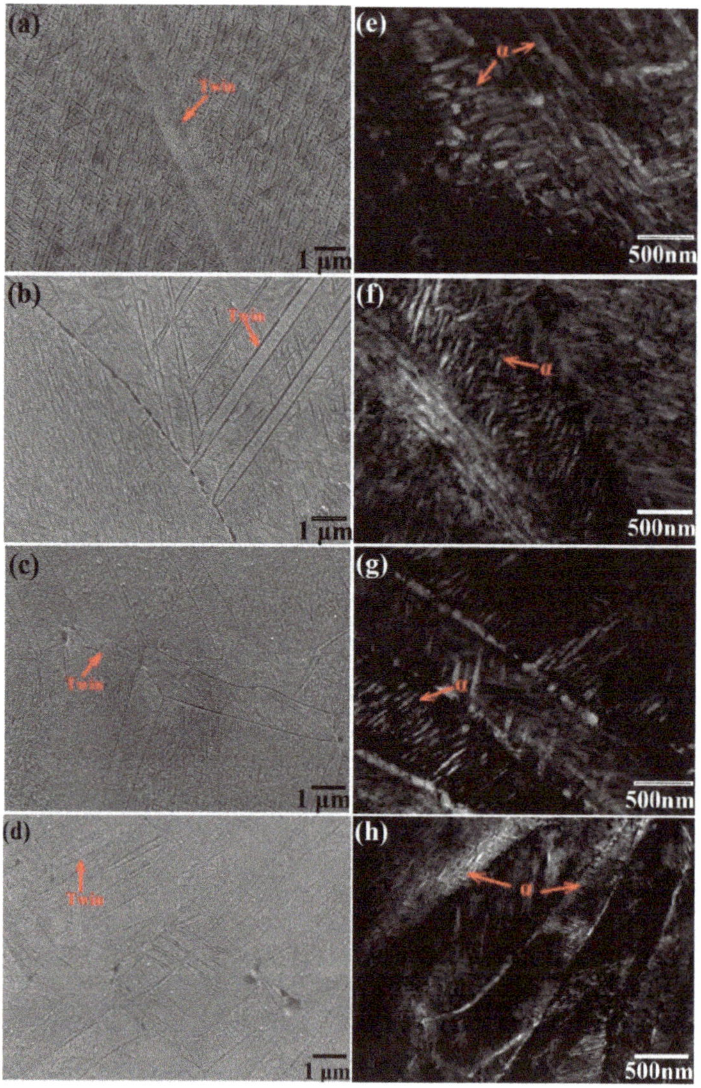

Figure 8. SEM and TEM microstructure of the Ti-15Mo alloy after pre-tension and aging: (**a**,**e**) Tens-3-A specimen, (**b**,**f**) Tens-9-A specimen, (**c**,**g**) Tens-15-A specimen, (**d**,**h**) Tens-20-A specimen, (**a**–**d**) SEM microstructure, (**e**–**h**) TEM microstructure.

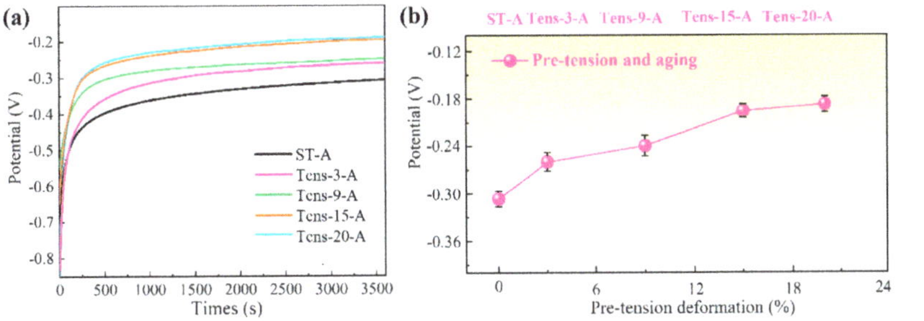

Figure 9. (**a**) OCP curves, (**b**) 3600 s potential of ST-A, Tens-3-A, Tens-9-A, Tens-15-A and Tens-20-A specimens in PBS solution.

Figure 10a,b present the Nyquist plots and the Bode plots of the alloy after pre-tension and aging. The results indicate the noticeable difference in the semi-circle radius of the Nyquist plots due to different pre-deformations conducted in the ST-A, Tens-3-A, Tens-9-A, Tens-15-A and Tens-20-A specimens. In principle, a larger circle radius in the Nyquist plots suggests a higher impedance of the passivation film. Notably, the passivation film on the ST-A specimen with a smaller radius was attributed to its initial coarse grains and rod-like precipitates. The Nyquist plots of the passivation films on all the specimens shown in Figure 10a illustrate an incomplete single capacitor arc, indicating their typical capacitance characteristics. The size of the impedance arc reflects the electrochemical reaction rate of the passivation film of the alloy. Generally, a larger diameter of the impedance arc suggests a slower charge transfer process and a lower corrosive rate. Compared to the ST-A alloy, the passivation film on the pre-deformed and aged alloy exhibits a significantly larger impedance arc diameter, slowing down the corrosion reaction effectively. This observation is caused by pre-tension, which significantly refined the initial grains and facilitated the formation of finer precipitates. The more ultra-fine microstructures of the pre-deformed and aged alloys possess massive boundaries and promote the formation of more stable passivation films. The impedance arc of the Tens-3-A specimen was slightly larger compared with other pre-tension and aged specimens, due to the difference in the vacancy concentration in the passivation film.

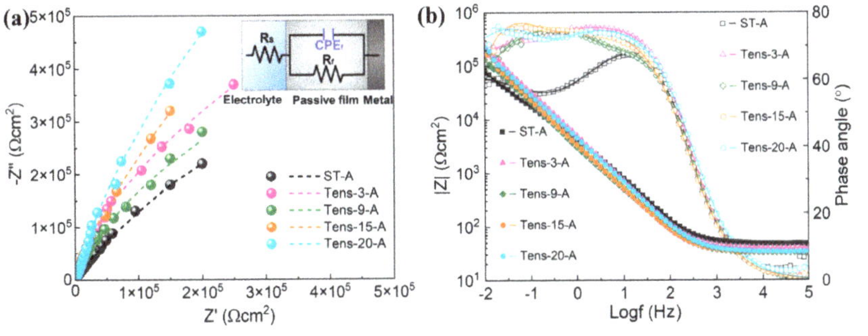

Figure 10. (**a**) Nyquist plots and (**b**) Bode and impedance plots of ST-A, Tens-3-A, Tens-9-A, Tens-15-A and Tens-20-A specimens in PBS solution.

The Bode plots of the ST-A, Tens-3-A, Tens-9-A, Tens-15-A and Tens-20-A specimens are shown in Figure 10b. The results reveal a single peak in the Bode-phase plots for each specimen with a range of 50~75°. The observation indicates that the passivation film formed during OCP has a single time constant, resulting in relatively stable passivation behaviors in PBS solution. In both the Bode and impedance plots, two frequency bands can be observed. In a high-frequency region (10^3~10^5 Hz), the phase angle values dropped to

0°, and the impedance almost remains constant, which is a reflection of the resistance of the electrolyte between the reference and working electrodes. However, in the low-frequency region (10^{-2}~10^{2} Hz), the slope of impedance/frequency is close to −1, suggesting the capacitive behavior of passivation films on all specimens. This observation also suggests that a stable oxide film has formed on the alloy's surface, demonstrating the stability of the surface–electrolyte interaction. Additionally, the phase angle decreases with the increase in frequency within the range of 10 to 10^{3} Hz. The results of different frequency regions suggest that there are different electrochemical mechanisms due to the characteristics of the passivation film. In the high-frequency region, the phase angle was close to 0, reflecting that the passivation film effectively inhibited the impedance change.

The stable passivation film improves the corrosion resistance of the Ti-15Mo alloy in PBS solution and improves its biomedical performance. The result of Figure 10b reveals that all specimens exhibit a single time constant. To evaluate the resistance of the passivation film on the ST-A, Tens-3-A, Tens-9-A, Tens-15-A and Tens-20-A specimens in PBS solution, an electrical equivalent circuit (EEC) was used to analyze the EIS data. The values of the EEC elements are calculated via the experimental EIS fitting results using ZSimpWin software and are shown in the top-right corner of Figure 10a and Table 1. The elements of the EEC have specific meanings: R_s signifies the solution resistance, R_f is the passivation film resistance and CPE stands for the constant phase element. Noteworthy, the CPE is used instead of a traditional capacitor due to the phase angle values being less than 90° in the low-frequency range (10^{-2}~10^{2} Hz), suggesting a nonideal capacitor at the electrode/electrolyte interface [31]. This deviation from ideal capacitive behavior may be caused by varying relaxation times due to microscopic inconsistencies in the passivation film. The impedance of a CPE is expressed as follows: [32]

$$z_{CPE} = \frac{1}{(j\omega)^n CPE_f} \quad (1)$$

where Z_{CPE} signifies the constant of CPE, j as the imaginary number and ω as the angular frequency. The value of n determines whether the constant phase element behaves like a capacitor, resistor or inductor. For n = 1, it functions like an ideal capacitor, for n = 0, as a pure resistor and for n = −1, as an inductor. The n value of all specimens in this study is above 0.82, according to Table 1, indicating that the passivation film on the Ti-15Mo alloy's surface behaves more like a capacitor. The R_f and CPE_f, as the elements of the EEC, are listed in Table 1, representing the resistive and capacitive responses of the passivation film. The effective capacitance (C_{eff}) of the passivation film can be calculated using the following equation [33–35]:

$$C_{eff} = \left(R_f^{\,1-n} CPE_f\right)^{\frac{1}{n}} \quad (2)$$

Table 1. The fitted values of R_s, R_f, CPE, n and calculated values of C_{eff}, L_{ss} of Ti-15Mo alloy performed with different processes.

Specimen	R_s (Ω·cm²)	R_f (kΩ·cm²)	CPE_f (μΩ⁻¹·cm⁻²·sⁿ)	n	C_{eff} (μF·cm⁻²)	L_{SS} (nm)
ST-A	52.4	843	6.53	0.837	9.10	6.32
Tens-3-A	38.3	243	5.75	0.851	6.10	9.43
Tens-9-A	32.0	228	3.94	0.816	3.85	15.00
Tens-15-A	32.8	123	2.93	0.834	2.39	24.12
Tens-20-A	33.2	187	2.43	0.834	2.07	27.71

Since the passivation film shows similar capacitor characteristics, its thickness can be likened to the space between parallel plates in a capacitor. Consequently, the passivation film's thickness is calculated as the following equation [33]:

$$L_{SS} = (\varepsilon_r \varepsilon_0)/C \quad (3)$$

where ε_0 is the vacuum permittivity and 8.854×10^{-14} F/cm, as is commonly known [34]. ε_r represents the permittivity of the Ti-based oxide film, which has been identified in the previous literature as 65 [33].

Table 1 presents the EEC fitted parameters including the calculated C_{eff}, L_{ss} values for the specimens. It can be observed that the R_s values of all specimens in PBS solution are quite similar. R_f shifts from 123 to 843 kΩ·cm^2 due to variations in the passivation films. In PBS solution, the *CPE* values were 6.53 μΩ^{-1}·cm^{-2}·s^n for the ST-A specimen and 2.43 μΩ^{-1}·cm^{-2}·s^n for the Tens-20-A specimen, suggesting a significant difference in capacitor characteristics due to the microstructural refinement. The thickness (L_{ss}) of the passivation film in the alloy after pre-tension and aging is approximately 1.5~4.4 times more than that of the ST-A alloy. The L_{ss} values calculated using Equation (3) show that the ST-A specimen has a film thickness of 6.32 nm, Tens-3-A has 9.43 nm, Tens-9-A has 15.0 nm, Tens-15-A has 24.12 nm and Tens-20-A has 27.71 nm. These results suggest that the thickness of the passivation film increases with the increase in the pre-deformation. Specifically, the L_{ss} value of the Tens-20-A specimen is 4.4 times that of the ST-A specimen. The variation in the EEC parameters suggests that the microstructure is significantly influenced by the formation of passivation films on the specimens. The twin boundaries, ω phase and dislocations formed during the pre-deformation refined the initial grain and deduced finer precipitates during aging, which also increases the nucleation sites for forming a thicker passivation film, which would decrease the electrode reaction rate and enhance the corrosion resistance of the metal. The refined grains and finer precipitates are considered as dominant factors, which promoted the formation of Ti- and Mo-based oxides and improved the corrosion resistance.

Figure 11a presents the Mott–Schottky (M-S) plots and fitted curves for the passivation film on the ST-A, Tens-3-A, Tens-9-A, Tens-15-A and Tens-20-A specimens in PBS solution. The M-S curves illustrate linearity within the potential range of 0 mV to 500 mV. In contrast, the nonlinearity of the Mott–Schottky curves was caused by the interference of the Faraday reaction on space charge capacitors like TiO_2 and Ti_2O_3, which was reported in the other metal and solution [36]. In this study, the potential interval of 0 mV to 500 mV (vs. AgCl) was chosen for calculating the donor concentration. In general, M-S measurements reveal that the passivation films on all alloys act as n-type semiconductors similar to other Ti alloys, with oxygen vacancies as the dominant charge carriers [36]. These vacancies introduce additional active sites, influencing the chemical reactions and the surface energy of the passivation film. The oxygen vacancies also facilitate the formation of the passivation film, since the oxygen vacancy formation energy (2.7 eV) is lower than that of a titanium interstitial (4.7 eV). Hence, oxygen vacancies are the primary donors in the passivation film, and their densities can be calculated from the slope of the Mott–Schottky fitted lines, as shown in Table 2. According to the M-S relation, the space charge capacitance of N-type semiconductors can be calculated using the following equation [36,37]:

$$\frac{1}{C^2} = \frac{2}{\varepsilon \varepsilon_0 e N_D}\left(E - E_{FB} - \frac{kT}{e}\right) \quad (4)$$

$$N_D = \frac{2}{e \varepsilon_r \varepsilon_0 b} \quad (5)$$

where b represents the slope of the fitted lines and N_D denotes the donor concentration (cm^{-3}). Table 2 displays the values of the slope of the M-S fitted lines and the corresponding ND values calculated using Equations (4) and (5). The calculated N_D values for all specimens fall within the range of 6.28×10^{20} to 11.4×10^{20} cm^{-3}, which is in a narrow range. The concentration of oxygen vacancies in the passivation films remains relatively constant regardless of pre-deformation. However, a slight variation is still observed in the Tens-3-A specimen, indicating a lower N_D value than in the Tens-9-A and Tens-15-A specimens. This result explains a larger impedance arc in the Nyquist plots observed in the Tens-3-A specimen, though there is a thinner passivation film compared to the Tens-9-A specimen.

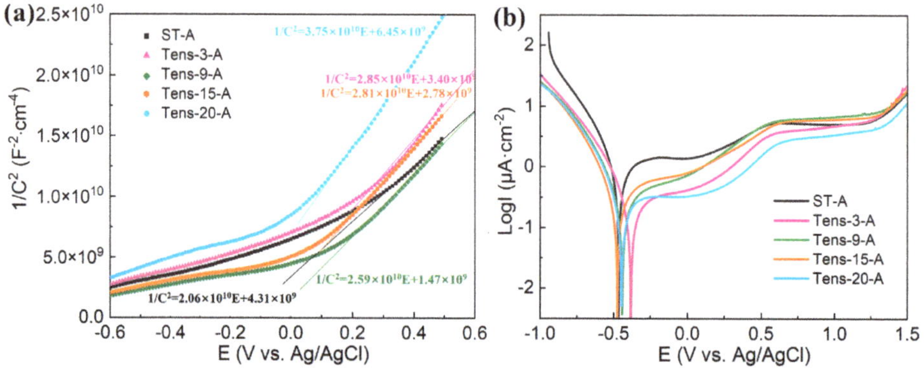

Figure 11. (**a**) M-S data and fitting curves and (**b**) potential dynamic polarization curves of ST-A, Tens-3-A, Tens-9-A, Tens-15-A and Tens-20-A specimens in PBS solution.

Table 2. The slope values of fitted M-S curves and calculated N_D values of passive film on the Ti-15Mo alloy after different pre-deformation and aging.

Specimen	b (10^{10})	N_D (10^{20} cm^{-3})
ST-A	2.06	11.4
Tens-3-A	2.85	8.26
Tens-9-A	2.59	9.09
Tens-15-A	2.81	8.38
Tens-20-A	3.75	6.28

Potential dynamic (PD) polarization measurements were conducted to analyze the performance of the passivation films in PBS solution. Figure 11b displays the PD polarization curves of the ST-A, Tens-3-A, Tens-9-A, Tens-15-A and Tens-20-A specimens in PBS solution. Despite the surfaces of the Ti alloy being covered with typical barrier-type passivation films, electronic and ionic conductivity can still generate the current. Figure 11b illustrates two distinct passivation platforms for the current density in the polarization ranges of −300 mV~300 mV and 500 mV~1500 mV (vs. Ag/AgCl), indicating the stability of the passivation films of the Ti-15Mo alloy within these voltage ranges. These films are crucial in preventing direct contact between the electrolyte and fresh metal surface. The values of the corrosion voltage (E_{corr}), corrosion current (I_{corr}) and passivation current (I_{pass}) derived from fitting and analyzing the PD polarization curves are summarized in Table 3. The E_{corr} values of all specimens are similar. However, in contrast to the ST-A alloy, the I_{corr} values are significantly lower due to the enhancement in the oxide film for the alloy after pre-tension and aging.

Passivation currents in the first platforms (1st I_{pass}) decrease with the increase in pre-tension, indicating that the corrosion resistance of the alloy after pre-rolling and aging is significantly improved. The values of 1st I_{pass} at 0 mV in the ST-A, Tens-3-A, Tens-9-A, Tens-15-A and Tens-20-A specimens are 1.38, 0.42, 0.73, 0.80 and 0.33 μA/cm^2, respectively. However, the values of 2nd I_{pass} at 1000 mV do not exhibit this trend. As the potential in the human body is generally lower than 500 mV, giving more attention to the 1st I_{pass}

value is valuable. The results from Table 3 indicate that the corrosion resistance of the Tens-20-A specimen is 4.2 times better than that of the ST-A specimen. The alloys deformed by pre-rolling form lots of {332} twins, ω phases and dislocations, causing the refinement of grains. Subsequently, finer α precipitates and their boundaries provide more sites and channels to form a thicker oxide film, which is a critical factor for the corrosion resistance of the alloy.

Table 3. The values of E_{corr}, I_{corr} and I_{pass} of the Ti-15Mo alloy performed by different processes in PBS solution.

Specimen	E_{corr} (V)	I_{corr} ($\mu A/cm^2$)	1st I_{pass} ($\mu A/cm^2$) @0 mV	2nd I_{pass} ($\mu A/cm^2$) @1000 mV
ST-A	−0.49	0.89	1.38	5.04
Tens-3-A	−0.40	0.15	0.42	4.99
Tens-9-A	−0.47	0.19	0.73	6.90
Tens-15-A	−0.51	0.28	0.80	6.55
Tens-20-A	−0.48	0.20	0.33	3.63

3.4. *Friction Performance of the Alloy after Pre-Tension and Aging*

Figure 12a shows the friction coefficient of the Ti-15Mo alloy in PBS solution, suggesting that a significant fluctuation is observed in all specimens with the duration of friction. This phenomenon was caused by the friction characteristics and friction type of the Ti alloy. The exfoliation of abrasive particles in the PBS solution leads to the observed fluctuation in the friction coefficient. This is consistent with the previous literature [38], which had reported instability of the friction coefficient in Ti alloys. Figure 12b presents the microhardness and wear loss of the ST-A, Tens-3-A, Tens-9-A, Tens-15-A and Tens-20-A specimens after friction in PBS solution. By analyzing the mechanical properties of the aged alloys following different pre-tensions, a relationship between the friction coefficient, wear and microstructure is established. The results affirmed that the increased hardness of the pre-deformed and aged alloy provided a positive impact on the wear resistance. It is an effective method for improving the hardness and the wear resistance of the alloy via microstructural refinement, such as the pre-tension and aging processes.

Figure 12b shows that the hardness values of the aged alloy increase with the increase in the pre-tensive deformation. Compared with the ST-A specimen, the hardness value of the Tens-20-A specimen increases to 335.4 from 272.0, indicating a 23.3% increase. The result is caused by two factors. Firstly, the {332} twins contribute to the strengthening effect of refinement by segmentizing the initial grains. Secondly, the finer nanoscale α precipitates formed during pre-tension cause a severe dispersion in the strengthening effect. Figure 13a,c,e,g,i show the three-dimensional morphology of the worn tracks of the ST-A, Tens-3-A, Tens-9-A, Tens-15-A and Tens-20-A specimens, respectively. The wear loss in PBS solution decreases with the increase in the hardness. Although the passivation film on the specimen influences the friction behaviors, the decisive factor is the hardness. A higher hardness would result in a lower wear loss. The result of the wear loss indicates that a significant enhancement in the wear resistance occurred in the alloy after pre-tension and aging. The wear loss of the ST-A specimen is 3.32×10^7 μm^3, but that of the Tens-20-A specimen is 2.24×10^7 μm^3 in the same friction condition, exhibiting a 33% increase in wear resistance. Additionally, the Tens-20-A specimen had the smallest friction coefficient (in Figure 12a); thus, the best wear resistance could be concluded as the lowest wear loss.

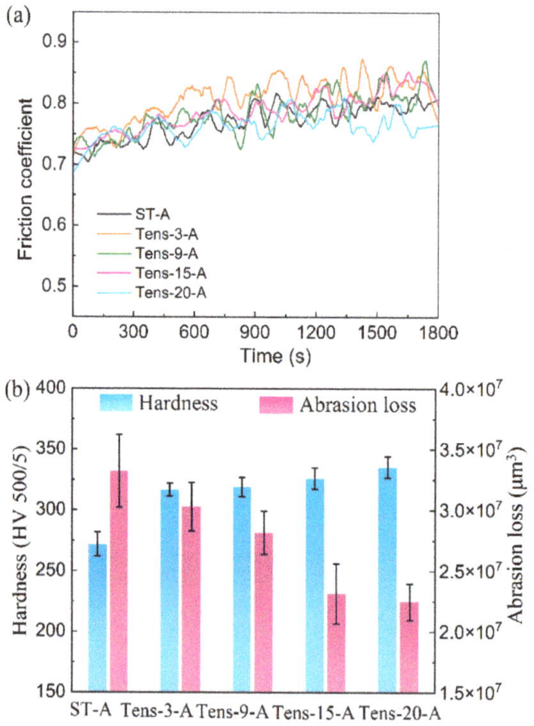

Figure 12. (**a**) Friction coefficient, (**b**) hardness and wear loss of Ti-15Mo alloy performed by different processes.

The SEM images in Figure 13b,d,f,h,j present the micro-morphology of the worn tracks on the ST-A, Tens-3-A, Tens-9-A, Tens-15-A and Tens-20-A specimens, respectively. The results illustrate ridges and grooves along the sliding direction with randomly distributed metallic debris, suggesting abrasive wear as the primary wear mechanism. The worn tracks on the specimens mostly have a flat appearance, suggesting surface damage with micro-cutting and flaky stripping. The ST-A specimen displays deep grooves and larger particles adhering to the surface, suggesting adhesive wear resulting from the continuous formation and attachment of particles during friction (in Figure 13b). Numerous grooves can be seen on the surface along the sliding direction, with a layer of worn debris covering the wear grooves. Plastic deformation of the alloy surface during wear results in work-hardening behavior, consistent with previous reports on the sliding wear of Ti alloys [39]. In contrast, the higher hardness of the Tens-20-A specimen leads to smaller grooves, less debris and reduced adhesion of particles to the surface. Figure 13i,j demonstrate the distribution of smaller particles on the surface of the wear patterns. This observation suggests the presence of an abrasive wear mechanism during the friction process of the Ti-15Mo alloy. Higher hardness results in worn tracks with shallower grooves. However, in a reciprocating sliding process, the surface undergoes repeated work hardening, leading to the development of layered cracks that eventually propagate and contribute to layered cracking of the material. The motion of the counter ball induces plastic deformation, generating abrasion particles that contribute to wear debris or form a transfer layer. Accumulated wear debris forms significant grooves on the surface, with a varying width of the wear tracks based on the alloy strength, exhibiting undulation and differences in width. The worn surfaces of all specimens display shallow grooves, confirming the presence of an abrasive wear mechanism [40].

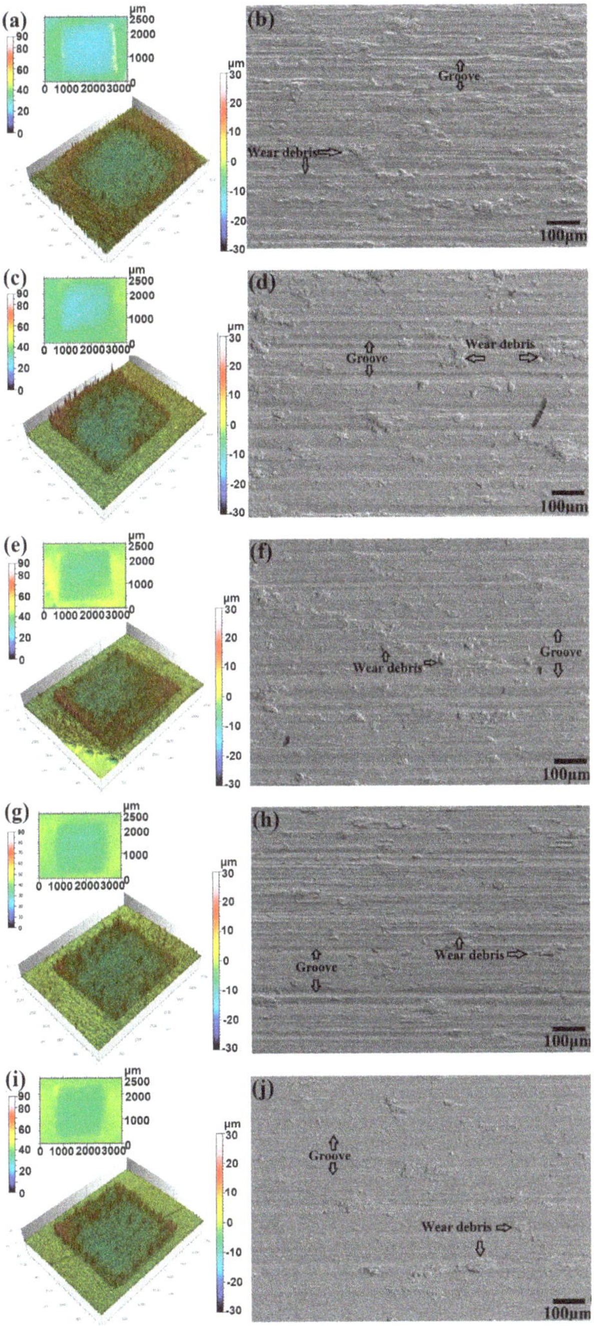

Figure 13. Three-dimensional morphology of the worn tracks on (**a**,**b**) ST-A specimen, (**c**,**d**) Tens-3-A specimen, (**e**) and SEM images of the worn tracks on (**f**) Tens-9-A specimen, (**g**,**h**) Tens-15-A specimen, (**i**,**j**) Tens-15-A specimen.

4. Discussion

Different from traditional solution and aging treatment methods [41], based on the twinning-induced plasticity (TWIP) effect, {332} twin structures were introduced into the Ti-15Mo alloy through pre-tension deformation. Numerous HABs via twinning in the matrix have refined the initial grain formed during the ST. Based on these refined sub-structures and induced ω phases, the finer α phases at the nanoscale were precipitated during the aging treatment and distributed homogeneously. The microstructure of the alloy after pre-tension and aging enhanced the passivation effect and corrosion resistance of the alloy in the simulated human environment. The microstructural refinement via {332} twins and finer precipitates improved the hardness of the alloy and enhanced the wear resistance during friction in PBS solution. A schematic diagram (in Figure 14) has been drawn to depict the passivation behaviors of the ST-A and Tens-20-A alloy based on the relationship between the microstructure, passivation film and corrosion resistance. In PBS solution, the passivation film on the biomedical Ti-15Mo alloy mainly comprises the oxidation of Ti and Mo, initiated at surface flaws. In contrast to the limited element diffusion in the ST-A alloy (as Figure 14a), the extensive boundaries of the ultra-fine alloy offer numerous sites for the nucleation and growth of the passivation film (as Figure 14). The ultra-fine alloy is capable of forming a thicker film containing Ti and Mo oxide, enhancing the corrosion resistance of the alloy. Initially, the passivation film enables the creation of a more stable and efficient oxide layer to isolate electron and ion exchange, despite the existence of donor concentrations (including oxygen vacancy ($O_{vacancy}$) and metallic vacancy ($M_{vacancy}$)). Furthermore, a thicker oxide layer can prolong the corrosive and dissolution processes of the metal surface through the passivation film. Ultimately, the presence of a superior oxide film can impede the corrosion reaction, thereby enhancing the corrosion resistance of the ultra-fine alloy.

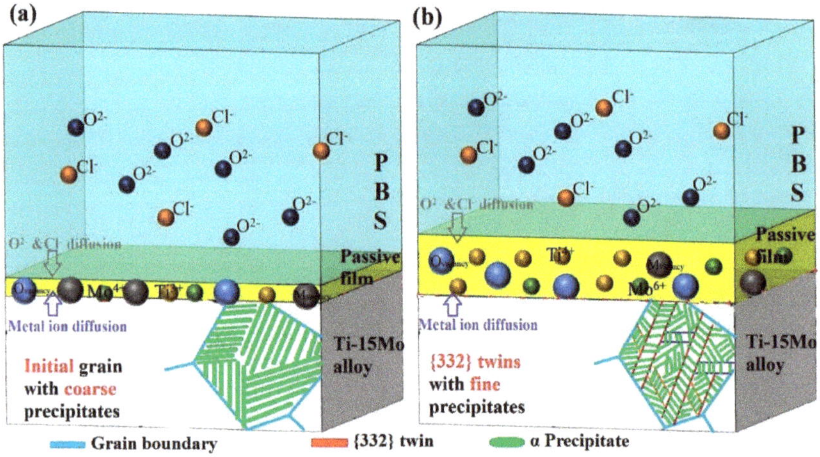

Figure 14. The schematic diagram of corrosion behavior and passive film of the Ti-15Mo alloy in PBS solution: (**a**) after solution treatment and aging, (**b**) after pre-deformation and aging.

5. Conclusions

In this study, the effects of finer α phases within {332} twin structures via pre-tension and aging on the corrosion and wear resistance of a Ti-15Mo alloy were investigated systematically. The conclusions are as follows:

1. The refined sub-structures of the Ti-15Mo alloy with {332} twins via pre-tension played the role of restricting the growth of α phases during aging, resulting in forming finer needle-shaped α precipitates. With the increase in pre-tension, finer {332} twins of the Ti-15Mo alloy after pre-deformation contributed to decreasing the size of the α

precipitates formed during the aging process. The hardness of the aged alloy after pre-tension also increased from 272.0 HV in the ST-A specimen to 335.4 HV in the Tens-20-A specimen.

2. The microstructure of the alloy after pre-tension and aging is finer and has more boundaries, which enhanced the passivation film structure and corrosion resistance of the alloy. The thickness of the passivation film on the surface of the aged alloy after pre-tension increases with the increase in the pre-deformation. The improved passivation film structure on the surface of the alloy after pre-deformation and aging enhanced the corrosion resistance. The I_{pass} value of the Tens-20-A specimen was 0.327 $\mu A/cm^2$, less than 1.382 $\mu A/cm^2$ of the ST-A specimen, increasing the corrosion resistance by 4.23 times.
3. The worn track depth and debris size of the aged alloy after pre-tension decreased with the increase in pre-deformation. Compared with the ST-A specimen, which had a wear loss of 3.32×10^7 μm^3, the alloy after pre-tension and aging shows a better wear resistance. The wear loss of the Tens-20-A specimen in the same conditions was 2.24×10^7 μm^3, and the wear resistance was enhanced by 33%.

Author Contributions: Methodology—writing and original draft, S.Z.; conceptualization—review and editing, F.Y.; methodology, W.S.; investigation, N.C. and T.X. All authors have read and agreed to the published version of the manuscript.

Funding: This study was co-supported by the Shandong Provincial Natural Science Foundation (No. ZR2024ME184, ZR2020ME006) and the Key Research and Development Program of Shandong Province of China (No. 2018GGX102027).

Institutional Review Board Statement: Not applicable.

Informed Consent Statement: Not applicable.

Data Availability Statement: Data are contained within the article.

Conflicts of Interest: The authors declare no conflicts of interest.

References

1. Xu, T.; Sun, B.; Zhang, S.; Liu, Y.; Sun, W.; Cui, N.; Lv, B. The {332} Twinning Behavior of a Ti-15Mo Medical Alloy during Cyclic Deformation and Its Effect on Microstructure and Performance. *Materials* **2024**, *17*, 1462. [CrossRef] [PubMed]
2. Macías, R.; Olmos, L.; Garnica, P.; Alanis, I.; Bouvard, D.; Chávez, J.; Jiménez, O.; Márquez-Beltrán, C.; Cabezas-Vila, J.L. Processing of Porous-Core Materials for Bone Implant Applications: A Permeability and Mechanical Strength Analysis. *Coatings* **2024**, *14*, 65. [CrossRef]
3. Jáquez-Muñoz, J.M.; Gaona-Tiburcio, C.; Mendez-Ramirez, C.T.; Carrera-Ramirez, M.G.; Baltazar-Zamora, M.A.; Santiago-Hurtado, G.; Lara-Banda, M.; Estupiñan-Lopez, F.; Nieves-Mendoza, D.; Almeraya-Calderon, F. Corrosion of Anodized Titanium Alloys. *Coatings* **2024**, *14*, 809. [CrossRef]
4. Zhang, W.; Ren, J.; Liu, B.; Liu, Y.; Wu, Z.; Qiu, J. Microstructure and Mechanical Properties of Cold Drawn Ti-Nb-Ta-Zr-O Wires for Orthodontic Applications. *Met. Mater. Int.* **2020**, *26*, 973–978. [CrossRef]
5. Lan, C.; Chen, F.; Chen, H.; Wu, Y.; Wu, X. Influence of oxygen content on the microstructure and mechanical properties of cold rolled Ti-32.5Nb-6.8Zr-2.7Sn-xO alloys after aging treatment. *J. Mater. Sci. Technol.* **2018**, *34*, 2100–2106. [CrossRef]
6. Xu, T.W.; Li, J.S.; Zhang, S.S.; Zhang, F.S.; Liu, X.H. Cold deformation behavior of the Ti-15Mo-3Al-2.7Nb-0.2Si alloy and its effect on a precipitation and tensile properties in aging treatment. *J. Alloys Compd.* **2016**, *682*, 404–411. [CrossRef]
7. Talbot, C.E.P.; Church, N.L.; Jones, N.G. Effect of prior processing on the decomposition of the β phase in Ti–Nb alloys during intermediate heat treatments. *Scr. Mater.* **2024**, *238*, 115729. [CrossRef]
8. Silva, J.; Viana, C.C.; Silva, R.G.; Vacchi, S.; Filho, A.A.M.; Rovere, C.A.D.; Plaine, A.H. Electrochemical Corrosion Behavior in Simulated Body Fluid of a Metastable β-Type Ti29Nb13Ta4.6Zr Biomedical Alloy with Improved Mechanical Properties. *Adv. Eng. Mater.* **2023**, *25*, 2301175. [CrossRef]
9. Ji, P.F.; Li, B.; Liu, S.G.; Zhang, X.; Chen, B.H.; Zhang, X.Y.; Ma, M.Z.; Liu, R.P. Controlling the corrosion behavior of Ti-Zr alloy by tuning the α/β phase volume fraction and morphology of β phase. *J. Alloys Compd.* **2020**, *825*, 154153. [CrossRef]
10. Toptana, F.; Alvesa, A.C.; Carvalho, Ó.; Bartolomeu, F.; Pinto, A.M.P.; Silva, F.; Miranda, G. Corrosion and tribocorrosion behaviour of Ti6Al4V produced by selective laser melting and hot pressing in comparison with the commercial alloy. *J. Mater. Process. Technol.* **2019**, *266*, 239–245. [CrossRef]

11. Ji, P.; Liu, S.; Shi, C.; Xu, K.; Wang, Z.; Chen, B.; Li, B.; Yang, Y.; Liu, R. Synergistic effect of Zr addition and grain refinemen on corrosion resistance and pitting corrosion behavior of single α-phase Ti-Zr-based alloys. *J. Alloys Compd.* **2022**, *896*, 16301 [CrossRef]
12. Min, X.; Yao, K.; Zhou, X.; Emura, S.; Tsuchiya, K. Continuous microstructure evolution of ball rolled TWIP Ti-15Mo alloy and it effect on phase precipitation behavior. *Mater. Charact.* **2018**, *145*, 116–125. [CrossRef]
13. Im, Y.D.; Lee, Y.K.; Song, K.H. Effect of Grain Misorientation Angle on Twinning Propagation in Ti-15Mo Alloy. *Met. Mater. In* **2018**, *24*, 913–917. [CrossRef]
14. Xiang, L.; Min, X.H.; Ji, X.; Emura, S.; Cheng, C.Q.; Tsuchiya, K. Effect of Pre-cold Rolling-Induced Twins and Subsequen Precipitated ω-Phase on Mechanical Properties in a β-Type Ti–Mo Alloy. *Acta Metall. Sin.* **2018**, *31*, 604–614. [CrossRef]
15. Wu, X.; Yang, K.; Cheng, J.; Lin, J. Evolution of the Microstructure and Mechanical Properties of a Biomedical Ti-20Zr-40Ta Allo during Aging Treatment. *Coatings* **2024**, *14*, 590. [CrossRef]
16. Xu, T.; Zhang, S.; Liang, S.; Cui, N.; Cao, L.; Wan, Y. Precipitation behaviour during the $\beta \rightarrow \alpha/\omega$ phase transformation and it effect on the mechanical performance of a Ti-15Mo-2.7Nb-3Al-0.2Si alloy. *Sci. Rep.* **2019**, *9*, 17628. [CrossRef]
17. Xu, T.W.; Zhang, S.S.; Cui, N.; Cao, L.; Wan, Y. Microstructural and Textural Evolutions during $\beta \rightarrow \alpha/\omega$ Transformation i Biomedical Ti-15Mo Alloy and Its Effects on Mechanical Performance. *J. Mater. Eng. Perform.* **2019**, *28*, 6969–6975. [CrossRef]
18. Xu, T.; Zhang, S.; Cui, N.; Cao, L.; Wan, Y. Precipitation Behavior of Ti15Mo Alloy and Effects on Microstructure and Mechanica Performance. *J. Mater. Eng. Perform.* **2019**, *28*, 7188–7194. [CrossRef]
19. Zhang, S.S.; Xu, T.W.; Sun, M.X.; Lv, B.J.; Ma, X.H. Effects of microstructure and texture evolution during the industrial ECAE an recrystallization on tensile properties of pure niobium. *Mater. Sci. Eng. A* **2021**, *807*, 140896. [CrossRef]
20. Xu, T.W.; Zhang, S.S.; Cui, N.; Cao, L.; Wan, Y. Effects of Strain Rate and Annealing Temperature on the Evolution of Microstructur and Texture in Pure Niobium. *Metall. Mater. Trans. A* **2019**, *50*, 5297–5313. [CrossRef]
21. Liu, X.; Wang, T.; Ren, X.; Fu, J.; Cheng, L.; Zhu, B.; Wang, K. New Insights into the Ingot Breakdown Mechanism of Near- Titanium Alloy: An Orientation-Driven Perspective. *Metals* **2024**, *14*, 792. [CrossRef]
22. Castany, P.; Yang, Y.; Bertrand, E.; Gloriant, T. Reversion of a Parent {130}<310>α″ Martensitic Twinning System at the Origin o {332}<113>β Twins Observed in Metastable β Titanium Alloys. *Phys. Rev. Lett.* **2016**, *117*, 245501. [PubMed]
23. Shin, S.; Zhu, C.; Vecchio, K.S. Effect of twinned-structure on deformation behavior and correlated mechanical properties in metastable b-Ti alloy. *J. Alloys Compd.* **2019**, *811*, 152054. [CrossRef]
24. Zhang, J.Y.; Fu, Y.Y.; Wu, Y.J.; Qian, B.; Chen, Z.; Inoue, A.; Wu, Y.; Yang, Y.; Sun, F.; Li, J.; et al. Hierarchical {332}<113> twinnin in a metastable β Ti-alloy showing tolerance to strain localization. *Mater. Res. Lett.* **2020**, *8*, 247.
25. Im, Y.D.; Kim, H.G.; Lee, Y.K.; Song, K.H. Effect of Deformation Twinning in Metastable β-Type Ti-15Mo Alloys. *JOM* **2018**, *7* 2582–2589. [CrossRef]
26. Gao, J.; Huang, Y.; Guan, D.; Knowles, A.J.; Ma, L.; Dye, D.; Rainforth, W.M. Deformation mechanisms in a metastable bet titanium twinning induced plasticity alloy with high yield strength and high strain hardening rate. *Acta Mater.* **2018**, *152*, 301–30 [CrossRef]
27. Wu, S.Q.; Yang, Y.; Li, G.P.; Ping, D.H.; Hu, Q.M.; Yang, R. ω Lattice mechanism of {112}<111> twinning nucleation and growt and termination. *Acta Metall. Sin.* **2016**, *52*, 249–256.
28. Zhou, X.; Min, X.; Emura, S.; Tsuchiya, K. Accommodative {332}<113> primary and secondary twinning in a slightly deforme β-type Ti-Mo titanium alloy. *Mater. Sci. Eng. A* **2017**, *684*, 456.
29. Zhou, X.Y.; Min, X.H. Effect of grain boundary angle on {332}<113> twinning transfer behavior in β-type Ti-15Mo-5Zr allo *J. Mater. Sci.* **2018**, *53*, 8604.
30. Zhang, J.Y.; Li, J.S.; Chen, Z.; Meng, Q.K.; Sun, F.; Shen, B.L. Microstructural evolution of a ductile metastable β titanium allo with combined TRIP/TWIP effects. *J. Alloys Compd.* **2017**, *699*, 775. [CrossRef]
31. Lu, Z.; Macdonald, D.D. Transient growth and thinning of the barrier oxide layer on iron measured by real-time spectroscopi ellipsometry. *Electrochim. Acta* **2008**, *53*, 7696–7702. [CrossRef]
32. Sun, C.; Xiao, R.; Li, H.; Ruan, Y. Effects of phase selection and microsegregation on corrosion behaviors of Ti-Al-Mo alloys *Corros. Sci.* **2022**, *200*, 110232. [CrossRef]
33. Su, B.; Wang, B.; Luo, L.; Wang, L.; Su, Y.; Wang, F.; Xu, Y.; Han, B.; Huang, H.; Guo, J.; et al. The corrosion behavior o Ti-6Al-3Nb-2Zr-1Mo alloy: Effects of HCl concentration and temperature. *J. Mater. Sci. Technol.* **2021**, *74*, 143–154. [CrossRef]
34. Yan, X.; Xu, X.; Zhou, Y.; Wu, Z.; Wei, L.; Zhang, D. Constructing a novel bi-lamellar microstructure in selective laser melte Ti-6Al-4V alloy via electropulsing for improvement of strength and corrosion resistance. *J. Mater. Sci. Technol.* **2024**, *193*, 37–50 [CrossRef]
35. Zhang, S.; Xu, T.; Sun, B.; Sun, W.; Lv, B.; Yang, F. Enhancing the mechanical properties and corrosion resistance of biomedica Ti15Mo alloy with ultra-finer {332} twins via cyclic deformation. *J. Alloys Compd.* **2024**, *1008*, 176578. [CrossRef]
36. Li, J.; Li, S.J.; HAO, Y.L.; Huang, H.H.; Bai, Y.; Hao, Y.Q.; Guo, Z.; Xue, J.Q.; Yang, R. Electrochemical and surface analyses o nanostructured Ti-24Nb-4Zr-8Sn alloys in simulated body solution. *Acta Biomater.* **2014**, *10*, 2866–2875. [CrossRef]
37. Qin, P.; Chen, L.Y.; Liu, Y.J.; Liang, S.X.; Sun, H.; Zhang, L.C. Corrosion and passivation behavior of laser powder bed fusio produced Ti-6Al-4V under various prior plastic deformation strains. *Corros. Sci.* **2024**, *230*, 111919. [CrossRef]
38. Weng, Z.; Gua, K.; Cui, C.; Cai, H.; Liu, X.; Wang, J. Microstructure evolution and wear behavior of titanium alloy under cryogeni dry sliding wear condition. *Mater. Charact.* **2020**, *265*, 110385. [CrossRef]

39. Dai, Y.; Jiang, X.; Ou, M.; Li, K.; Xiang, Q.; Yang, F.; Liu, J. Tribocorrosion Behaviour of a Ti–25Nb–3Zr–2Sn–3Mo Alloys Induction Nitride Layer in a Simulated Body Fluid Solution. *Coatings* **2023**, *13*, 231. [CrossRef]
40. Liu, G.; Su, Y.; Xin, S.; Li, C.; Deng, Z.; Liu, D.; Lin, Y.C. Enhanced wear resistance of Ti reinforced Inconel 718 superalloy manufactured by laser directed energy deposition. *Mater. Charact.* **2024**, *209*, 113736. [CrossRef]
41. Liu, Y.C.; Xu, T.W.; Zhang, S.S.; Ji, H.; Lv, B.J.; Cui, N. Effect of Microstructural and Textural Anisotropy on the Tensile Properties of Selective Laser Melted and Annealed Ti6Al4VE Alloy. *Adv. Eng. Mater.* **2022**, *25*, 2201552. [CrossRef]

coatings

MDPI

Article

Potential of Cranberry to Stimulate Osteogenesis: An In Vitro Study

Yasmim Guterres Bauer [1,2], Eduarda Blasi Magini [1,2], Ingrid Vicente Farias [3], José Della Pasqua Neto [1], Gislaine Fongaro [2,4], Flávio Henrique Reginatto [3], Izabella Thais Silva [2,3,4] and Ariadne Cristiane Cabral Cruz [1,2,*]

1 Post-Graduation Program of Dentistry, Center for Education and Research on Dental Implants, Federal University of Santa Catarina, Florianopolis 88040-900, Brazil; yasinha@gmail.com (Y.G.B.); eduardablasimagini@gmail.com (E.B.M.); dellapasquaneto@hotmail.com (J.D.P.N.)

2 Laboratory of Applied Virology, Federal University of Santa Catarina, Florianopolis 88040-900, Brazil; gislaine.fongaro@ufsc.br (G.F.); izabella.thais@ufsc.br (I.T.S.)

3 Post-Graduation Program in Pharmacy, Federal University of Santa Catarina, Florianopolis 88040-900, Brazil; iridvf@gmail.com (I.V.F.); flavio.reginatto@ufsc.br (F.H.R.)

4 Post-Graduation Program in Biotechnology and Biosciences, Federal University of Santa Catarina, Florianopolis 88040-900, Brazil

* Correspondence: ariadne.cruz@ufsc.br; Tel.: +55-48-3721-3407

Abstract: This study investigated the potential of Cranberry extract to stimulate osteogenesis in vitro. The total phenolic and monomeric anthocyanin contents in the Cranberry were determined. Liquid chromatography coupled with mass spectrometry was used to identify the Cranberry's constituents. To assess the Cranberry's cytotoxicity, a thiazolyl blue tetrazolium bromide assay was employed. Concerning the osteogenesis potential of Cranberry, alkaline phosphatase (ALP) activity, bone morphogenetic protein 2 (BMP-2) expression, and extracellular matrix mineralization were evaluated. The total phenolic content was 522.72 $\pm$ 9.80 mg GAE g-1 ES and 364.95 $\pm$ 12.49 mg GAE/g detected by the Fast Blue BB and Folin–Ciocalteu method, respectively. For monomeric anthocyanin, the content was 460 $\pm$ 30 mg ECG g-1 ES. Moreover, Cranberry concentrations ranged from 62.5 to 500 mg/mL and were found to be biocompatible with osteoblasts and mesenchymal stromal cells. Regarding osteogenesis, 20 mg/mL of Cranberry promoted 2-fold more ALP activity and almost 1.5-fold more BMP-2 than compared to the positive control group. Additionally, 200 mg/mL of Cranberry stimulated a 1.7-fold increase in extracellular matrix mineralization compared to the positive control group. In conclusion, Cranberry displayed potential in stimulating early and late markers of osteogenesis. Its ability to promote osteogenesis and its biocompatibility at higher concentrations hold promise for future application into biomaterials for bone regeneration.

Keywords: cranberry; osteogenesis; mesenchymal stromal cells; osteoblast; cell culture; in vitro

1. Introduction

Moderate or severe jaw atrophy, stemming from tooth loss, trauma, or surgical resection, often necessitates vertical bone augmentation to increase bone height, thereby facilitating the subsequent placement of dental implants followed by implant-supported prosthesis [1,2]. Nevertheless, vertical bone augmentation procedures present biological and technical challenges. Current biomaterials as bone graft exhibit several limitations, including associated morbidity, inadequate resorption rates leading to compromised dimensional stability, suboptimal structural integrity, and challenging handling properties [2,3]. In terms of biological considerations, bone formation entails a complex process reliant on osteogenesis and osteoclastogenesis. Furthermore, biomaterials for vertical bone augmentation within the oral cavity introduce an additional requirement for antimicrobial and anti-inflammatory properties, owing to the presence of a contaminated environment [3,4].

Cranberry components have garnered increasing attention within the sphere of health, including dentistry, owing to their antibacterial efficacy against cariogenic, mycotic, endodontic, and periodontal pathogens [5,6], along with their anti-inflammatory attributes [7]. Cranberry extract, derived from *Vaccinium macrocarpon, Vaccinium oxycoccus*, or *Vaccinium microcarpum*, abounds in numerous bioactive compounds, with polyphenols, especially phenolic acids, flavonoids (anthocyanins and flavonols), and tannins constituting major polyphenolic entities. Moreover, diverse fractions of Cranberry have undergone investigation, encompassing Cranberry extract [8], high-molecular-weight constituents [9,10], total polyphenol [11], phenolic acids [11], flavonols, proanthocyanidins, and specifically, A-type Cranberry proanthocyanidins [7].

High-molecular-weight constituents extracted from *Cranberry* have demonstrated the capability to inhibit nuclear factor kappa B and matrix metalloproteinase 3, thereby impacting fibroblast functionality during inflammatory responses [10]. Additionally, proanthocyanidins, present in high-molecular-weight non-dialyzable material extracted from Cranberry [9] , are associated with noteworthy attributes such as anti-adhesive and antibacterial properties, as well as the ability to hinder collagenase and proteinase activity [10]. Furthermore, proanthocyanidins, primarily composed of epicatechin subunits, featuring at least one type A bond and known as A-type proanthocyanidins from grape seeds, appear to disrupt the maturation and physiology of osteoclastic cells, as well as the bone matrix itself [12]. This effect is achieved by inhibiting the functional development of osteoclasts through the NF-kB pathway and mitogen-activated protein kinase signaling pathways [13]. Polyphenolic compounds, such as flavonoids sourced from materials other than Cranberries, have been explored for their potential in preserving bone. These polyphenolic compounds appear to regulate cell signaling pathways that impact the differentiation of osteoblasts and osteoclasts, thereby promoting bone formation and suppressing bone resorption [14–16]. Given the absence of research evaluating the role of any Cranberry fraction in osteogenesis, this study aimed to investigate the potential of Cranberry extract to stimulate osteogenesis in an in vitro setting. In the event that Cranberry extract exhibits osteogenic potential, it can be embedded into biomaterials applied to promote bone regeneration. Therefore, the amalgamation of antimicrobial, anti-inflammatory, and osteogenic properties of Cranberry could booster the vertical bone augmentation within the oral cavity.

2. Materials and Methods

2.1. Cranberry Extract Solution

Dry Cranberry extract (*Vaccinium macrocarpon*, Shaanxi Honghao, Xi'an, China) was solubilized in Dulbecco's Modified Eagle Medium (DMEM). For cytotoxicity, Cranberry solutions ranged from 62.5 to 500 mg/mL. Additionally, concentrations of 20 mg/mL, 2 mg/mL, and 0.2 mg/mL were prepared for alkaline phosphatase (ALP) activity and bone morphogenetic protein type 2 (BMP-2) expression, while concentrations of 250 mg/mL, 200 mg/mL, 150 mg/mL, 100 mg/mL, 50 mg/mL, 20 mg/mL, 2 mg/mL, and 0.2 mg/mL were employed for extracellular matrix mineralization.

2.2. Total Phenolic Content by Folin–Ciocalteu and Fast Blue BB

The total phenolic content of the Cranberry extract was ascertained employing the Folin–Ciocalteu [17] and Fast Blue BB [18] techniques. The findings were reported in mg of gallic acid equivalents per g of dry extract (mg GAE g-1 ES). The analytical curve generated using gallic acid spanned concentrations from 500 to 5 μg/mL, yielding an $R^2 = 0.999$ for both analyses.

2.3. Total Monomeric Anthocyanin Content

To quantify the total monomeric anthocyanin content in the Cranberry extract, the differential pH method was employed using wavelengths of 520 and 700 nm. An absorptivity coefficient (ε) of 26,600 L/M/cm was applied for cyanidin-3-*O*-glycoside in methanol.

The outcomes were reported as mg equivalent of cyanidin 3-O-glycoside per 100 g of dry extract (mg ECG 100 g-1 ES) [18,19].

2.4. Liquid Chromatography Coupled with Mass Spectrometry (LC/MS)

For anthocyanin identification and separation, a high-performance liquid chromatography system coupled with a photodiode array and a high-resolution mass spectrometry unit equipped with electrospray ionization and operating in positive mode was employed. The Cranberry extract was solubilized in methanol and water, reaching a concentration of 250 µg/mL. A SynergiTM—Phenomenex® hydro-RP 80 Å, (Phenomenex, Torrance, CA, USA) LC column polar end-capped C18, with 150 mm × 2 mm × 4 µm was maintained at 40 °C, with 5 µL of injection volume. The mobile phases comprised 2% formic acid (solvent A) and acetonitrile (solvent B). The gradient elution scheme consisted of 0 min at 5% B, 10 min at 12% B, and 16 min at 15% B. A flow rate of 0.4 mL/min was employed, and chromatograms were monitored at 520 nm. Mass spectrometry detection involved a full sweep from 50 to 1500 m/z, with a sweep time of 1 s. Data acquisition and processing were executed using MassLynx v4.1 software. For MS/MS analyses, argon served as the collision agent with energy varying from 10 to 20 to 30 eV depending on the analyzed compound. The capillary voltage was 3.5 kV with a temperature of 90 °C, while the desolvation temperature was 400 °C. The gas flow (N2) was 900 L/h, and the cone voltage was 30 V [20].

2.5. Cell Lineage and Culture Conditions

To evaluate the cytotoxicity and osteogenic potential of the Cranberry extract, cell lineages described below were incubated at 37 °C in a humidified atmosphere with 5% CO_2. For experiments, 2.0×10^4 cells per well were seeded in a 96-well plate. After seeding, the plates were maintained at room temperature for 1h for reducing edge effect, where cell density is different in the inner and edge of the well. The culture media were changed every 2–3 days. To determine the sample size (n) for independent groups, data from a pilot study [21] were used and the minimum n required was 1.83.

Osteoblasts MC3T3-E1 subclone 4 (American Type Culture Collection, CRL 2593, Manassas, VA, USA) were cultured with modified α-Minimum Eagles' Medium (α-MEM), supplemented with 10% fetal bovine serum (FBS), 50 µM L-ascorbic acid, and 10 mM β-glycerophosphate [21]. Mesenchymal stromal cells from human exfoliated deciduous teeth (SHED, Curityba Biotech™ Cell Processing Center, Curitiba, Brazil) were cultured in a standard medium (DMEM with 10% FBS) [22]. SHED were used in passage 4 to 6.

2.6. Cranberry Extract Cytotoxicity

The cytotoxicity of Cranberry extract was determined using the thiazolyl blue tetrazolium bromide (MTT) colorimetric test, assessing MC3T3-E1 and SHED on day 3. Upon removing the culture media, cells were exposed to the MTT solubilized in culture media (at a ratio of 1:5 *v:v*) and incubated for 4 h at 37 °C. Subsequently, formazan crystals were solubilized using dimethyl sulfoxide. Absorbances were measured at 540/630 nm. Cellular viability percentages were calculated relative to the cellular control group (cells treated with standard medium) [23].

2.7. Osteogenic Differentiation

SHED were utilized for osteogenic differentiation analysis. In the positive control group, SHED were exposed to the osteogenic medium (DMEM supplemented with FBS, 50 µM L-ascorbic acid, and 10 mM β-glycerophosphate).

Based on cytotoxicity findings, Cranberry extract concentrations up to 500 mg/mL exhibited biocompatibility with MC3T3-E1 and SHED. Additionally, concentrations below 250 mg/mL maintained nearly 100% cellular viability. Therefore, doses below the cytotoxic threshold were considered safe for further osteogenic experiments, which guided the selection of 0.2, 2, and 20 mg/mL for the ALP and BMP-2 assays. This, along with the maintained cell viability at concentrations bellow 250 mg/mL, motivated the use of higher concentrations on extracellular matrix mineralization. As a result, concentrations ranging from 0.2 to 200 mg/mL were employed in the mineralization assays.

The subsequent assays were carried out as follows:

2.8. Alkaline Phosphatase Activity

The ALP activity in the cell supernatant was assessed by the liberation of ρ-nitrophenol (ρ-NP) from ρ-nitrophenol phosphate (ρNPP) on days 3, 7, and 10. Briefly, a mixture of 0.5 mL of diethanolamine buffer (1.02 M, pH 9.8, 0.6 mM magnesium chloride) and 0.5 mL of 60.8 mM ρNPP was incubated for 2 min at 37 °C. Following this, 0.1 mL of the supernatant was added. After a 30 min incubation, the reaction was stopped by the addition of 1 mL of 3 N NaOH. Absorbances were measured at 405 nm, and ALP activity was calculated based on a ρNP standard curve [24].

2.9. Bone Morphogenetic Protein Type 2 Expression

To determine the BMP-2 expression on day 3, the Quantikine™ Human BMP-2 Immunoassay was employed. Briefly, 100 µL of the RD1–19 Assay Diluent and 50 µL of each supernatant were introduced to a well of the kit plate, sealed with adhesive tape, and subjected to 2 h incubation on a horizontal orbital microplate shaker set to 500 rpm. Subsequent to aspiration, each well received 400 µL of wash buffer, with this step being repeated four times. Following this, 200 µL of BMP-2 conjugate was added to each well, and the mixture was incubated for 2 h. After another aspiration/wash cycle, 200 µL of substrate solution was introduced and incubated for 30 min. Subsequently, 50 µL of stop solution was added, and the optical density was read at 450 nm with wavelength correction set to 540 nm or 570 nm [25].

2.10. Extracellular Matrix Mineralization

The extracellular matrix mineralization was assessed via Alizarin Red S staining on day 15. Cells were fixed with 4% paraformaldehyde for 15 min, followed by treatment with 3% Alizarin Red S for 30 min. Analysis was conducted using a stereo microscope, and images were captured with the acquisition software. For quantification of mineralization, each well was washed three times with 200 µL of PBS to remove the excess of stain. Additionally, to ensure unbiased quantification and account for the edge effect, the Alizarin Red S staining was solubilized using 0.5 N hydrochloric acid for 30 min, and the absorbance was subsequently measured at 490 nm [21]. This procedure ensures that all the stain is quantified, avoiding analysis limited to specific regions of the well.

2.11. Statistical Analysis

Statistical analyses were conducted using GraphPad Prism version 8. Data normality was assessed initially. For normally distributed data, one-way ANOVA was conducted, while the Kruskal–Wallis test was applied to non-parametric data. When statistically significant differences were detected, post hoc analysis was performed using the Tukey's test for ANOVA or Dunn's test for Kruskal–Wallis test. A significance threshold of $p < 0.05$ was applied.

3. Results

3.1. Cranberry with Heightened Total Monomeric Anthocyanin and Phenolic Content

We determined the total phenolic content of the Cranberry extract to be 522.72 ± 9.80 mg GAE/g and 364.95 ± 12.49 mg GAE/g using the Fast Blue BB (FBBB) and Folin–Ciocalteu (FC-B) methods, respectively. Moreover, the FBBB/FC-B ratio was 1.43, indicating the presence of phenolics compounds [18]. Additionally, the Cranberry extract contained 460 ± 30 mg ECG g-1 ES of monomeric anthocyanin. This finding signifies that the Cranberry used in our study displayed a 140-fold increase in monomeric anthocyanin compared to the minimum to indicate the authenticity of the Cranberry extract [26].

Regarding the phytoconstituents, authentic Cranberry extract may encompass delphini din-galactoside, cyanidin-3-*O*-glycoside, cyanidin arabinoside, peonidin-3-*O*-galactoside, peonidin-3-*O*-glycoside, and peonidin-arabinoside. Also, cyanidin-3-*O*-glucoside and peonidin-3-*O*-glucoside should be present in low concentrations. As depicted in Figure 1a,b, the analysis identified cyanidin-3,5-glycoside, cyanidin-3-glycoside, and peonidin-3-*O* glycoside. In contrast, it was also identified the presence of cyanidin-rutinoside in the samples, suggesting a possible contamination by *Morus nigra* (mulberry) [26].

a)

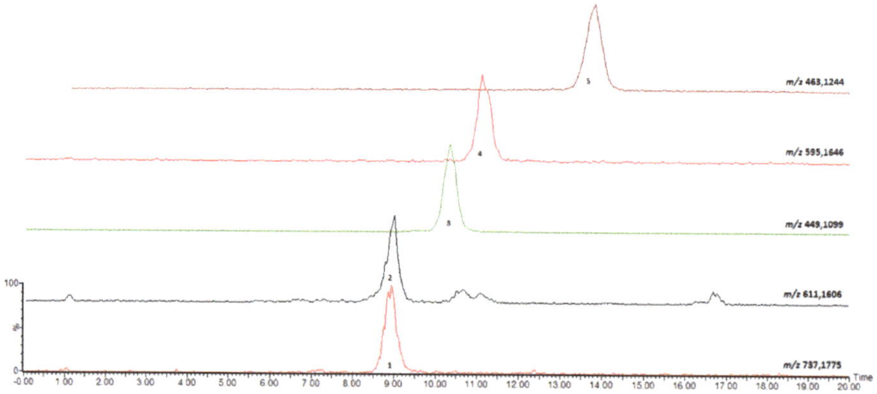

b)

Cranberry extract	TR (min)	Exp. *m/z*	Fragments m/z	Molecular formula	Error (ppm)	Compound suggestion
1	8.90	737.17	287.05 437.07 575.11	$C_{29}H_{37}O_{22}^{+}$	-0.1	Cyanidin derivative
2	9.01	611.16	287.05 449.11	$C_{27}H_{31}O_{16}^{+}$	-1.0	Cyanidina-3,5-glycoside
3	10.35	449.10	287.05	$C_{21}H_{21}O_{11}^{+}$	-3.3	Cyanidin-3- glycoside [a]
4	11.06	595.16	287.05 449.11	$C_{27}H_{31}O_{15}^{+}$	1.5	Cyanidin-rutinoside[a]
5	14.32	463.12	301.06	$C_{22}H_{23}O_{11}^{+}$	0.9	Peonidin-3-o-glycosideo

Figure 1. (**a**) Chromatographic profile of Cranberry extract depicting the ions extracted under the positive ionization mode. (**b**) Phytoconstituents characterization of the of Cranberry extract. [a] injection verified with the assistance of an authentic sample. The experiments were carried out in biological triplicate and experimental duplicate ($n = 6$).

3.2. Heightened Concentrations of Cranberry Extract Were Biocompatible

As depicted in Figure 2A,B, concentrations of Cranberry extract ranging from 62.5 mg/mL to 500 mg/mL exhibited biocompatibility with mesenchymal stromal cells and osteoblasts, since all concentrations promoted a cellular viability higher than 70% [27]. Additionally, all concentrations promoted a similar cellular viability compared to the positive control (cell without Cranberry extract, $p < 0.05$). Also, Cranberry concentrations below 250 mg/mL maintained a SHED viability around 100%.

3.3. Cranberry Extract Increased the Alkaline Phosphatase Activity

As indicated in Figure 3, the presence of 0.2 or 2 mg/mL Cranberry extract promoted similar ALP activity compared to the positive control group (SHED treated with osteogenic medium) on days 3, 7, and 10. Additionally, 20 mg/mL Cranberry extract resulted in an almost 2.5-fold increase in ALP activity compared to the positive control group (SHED treated with osteogenic medium) on day 3 ($p < 0.0001$) and an almost 3-fold increase on days 7 ($p < 0.0001$) and 10 ($p < 0.0001$). Furthermore, a correlation between the concentration of the Cranberry extract and ALP activity was identified (r = 1).

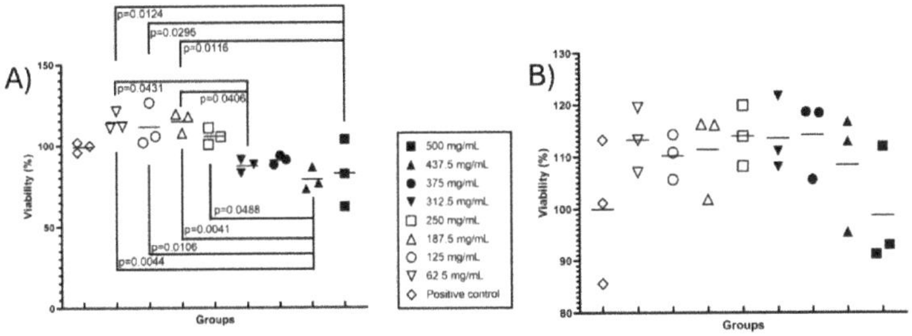

Figure 2. Assessment of Cranberry extract biocompatibility in (**A**) stromal cells from human exfoliated deciduous teeth (SHED) and (**B**) Osteoblasts MC3T3-E1 through thiazolyl blue tetrazolium bromide (MTT) on day 3. Cell viability percentages were determined in reference to the positive control (cell treated without Cranberry extract). The experiments were carried out in biological triplicate and experimental triplicate ($n = 9$ per group). Each geometric shape represents mean (SHED) or median (MC3T3-E1) from one experimental repetition. The bars within the geometric shapes represent the median or mean of all the repetitions, encompassing all samples. Statistical analyses were performed by comparing the mean or median from all experimental repetitions. Statistical analysis for data from SHED employed a one-way ANOVA test ($p = 0.0004$), followed by Tukey's multiple comparisons test. The post-test p values are presented in the graph. No significant difference was found among groups (Kruskal–Wallis test, $p = 0.4880$) concerning MC3T3-E1 findings.

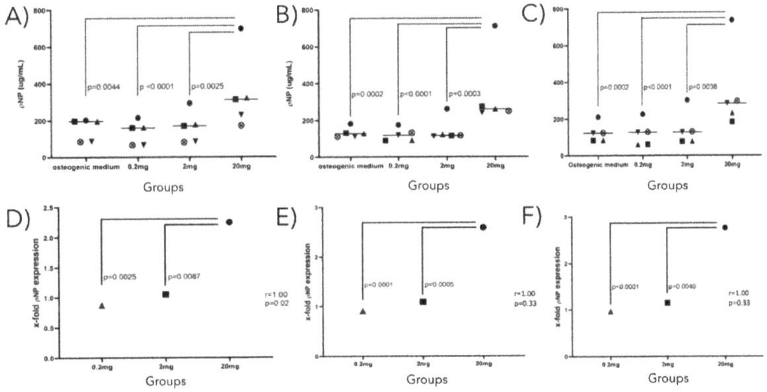

Figure 3. Alkaline phosphatase (ALP) activity assessed by measuring ρ-nitrophenyl (ρPNP) produced from ρ-nitrophenyl phosphate (ρPNPP) on days (**A**) 3, (**B**) 7, and (**C**) 10 in stromal cells from human exfoliated deciduous teeth (SHED). The experiments were carried out in biological triplicate and experimental quadruplicate ($n = 12$ per group). Each geometric shape represents median from one experimental repetition. The bars within the geometric shapes represent the median or mean of all the repetitions, encompassing all samples. Statistical analyses were performed by comparing the mean or median from all experimental repetitions. Cells were exposed to Cranberry extract (0.2 mg/mL, 2 mg/mL, or 20 mg/mL) or an osteogenic medium (Dulbecco's Modified Eagles Medium supplemented with 10% fetal bovine serum, 50 μM ascorbic acid, and 10 mM β-glycerophosphate). Statistical analysis was based on the Kruskal–Wallis test ($p < 0.0001$ for all experimental time) followed by Dunn's multiple comparisons test. The post-test p values are presented in the graph. The normalized data presents the x-fold increase caused by the presence of Cranberry extract compared to the osteogenic medium on days (**D**) 3, (**E**) 7, and (**F**) 10.

3.4. Cranberry Extract Increased the Bone Morphogenetic Protein-2 Expression

On day 3, the presence of 20 mg/mL Cranberry extract resulted in an approximate 1.5-fold increase in BMP-2 expression compared to the positive control (SHED treated with osteogenic medium), as illustrated in Figure 4. Also, a modest correlation (r = 0.6, p = 0.58) between the Cranberry concentration and the BMP-2 expression was discerned.

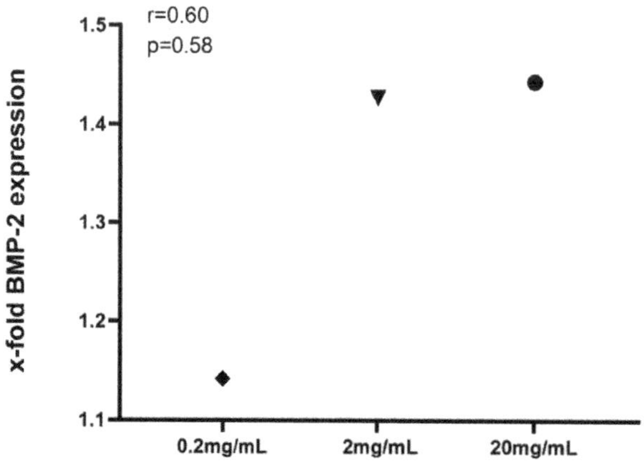

Figure 4. Bone morphogenetic protein type 2 (BMP-2) expression assessed in stromal cells from human exfoliated deciduous teeth (SHED) on day 3. The experiments were carried out in biological triplicate and experimental triplicate (n = 9 per group). Cells were exposed to Cranberry extract (0.2 mg/mL, 2 mg/mL, or 20 mg/mL) or an osteogenic medium (Dulbecco's Modified Eagles Medium supplemented with 10% fetal bovine serum, 50 μM ascorbic acid, and 10 mM β-glycerophosphate). The normalized data presents the x-fold increase caused by the presence of Cranberry extract compared to the osteogenic medium.

3.5. Cranberry Extract Increased the Extracellular Matrix Mineralization

As depicted in Figure 5A–C, the treatment of cells with Cranberry extract resulted in elevated extracellular matrix mineralization, as evidenced by Alizarin Red S staining on day 15. Furthermore, the presence of 0.2, 2, 20, 50, 100, 150, and 250 mg/mL of Cranberry extract promoted a similar extracellular matrix mineralization compared to the positive control (SHED treated with osteogenic medium). Additionally, 200 mg/mL of Cranberry extract was the most efficient and prompted a 1.7-fold increase in extracellular matrix mineralization relative to the positive control group, as evidenced in Figure 5C. Additionally, a correlation between the concentrations of the Cranberry extract and the levels of mineralization was identified (r = 0.911, p = 0.001).

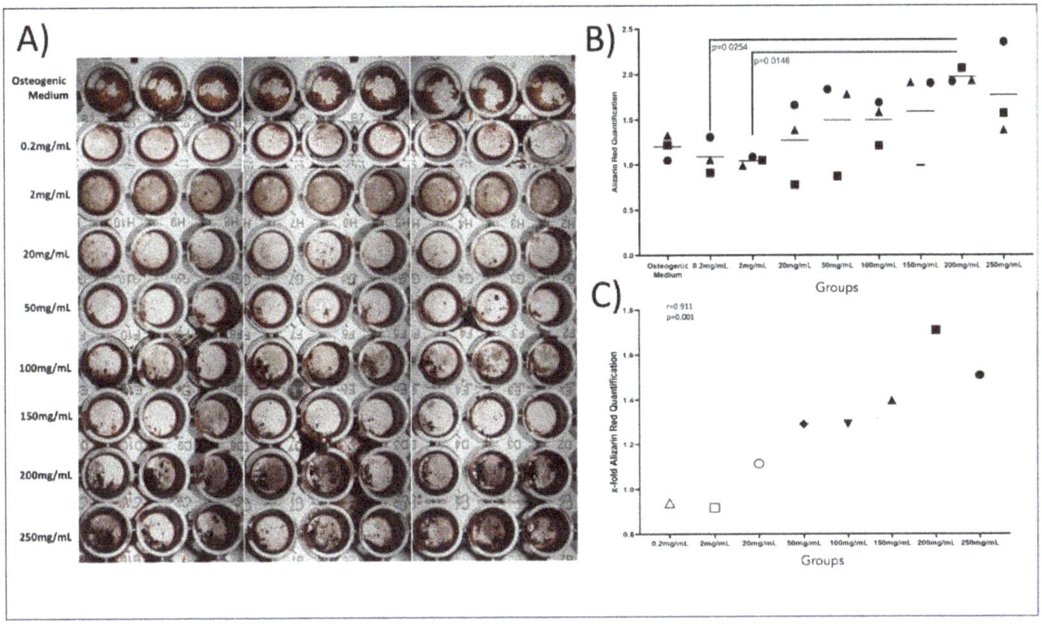

Figure 5. (**A**) This panel displays a representative image of Alizarin Red S staining conducted on day 15. Stromal cells from human exfoliated deciduous teeth (SHED) were subjected to various concentration of Cranberry extract (0.2 mg/mL, 2 mg/mL, 20 mg/mL, 50 mg/mL, 100 mg/mL, 150 mg/mL, 200 mg/mL, and 250 mg/mL) as well as an osteogenic medium (Dulbecco's Modified Eagles' Medium supplemented with 10% fetal bovine serum, 50 μM ascorbic acid, and 10 mM β-glycerophosphate). The experiments were carried out in biological triplicate and experimental triplicate ($n = 9$ per group). (**B**) Quantification of the Alizarin Red S. Each geometric shape represents mean from one experimental repetition. Statistical analyses were performed by comparing the mean or median from all experimental repetitions. The bars within the geometric shapes represent the median or mean of all the repetitions, encompassing all samples. Statistical analysis was performed using the one-way ANOVA test ($p = 0.0060$) following by Tukey's multiple comparisons test. The post-test p values are presented in the graph. (**C**) The normalized data present the x-fold increase caused by the presence of Cranberry extract compared to the osteogenic medium.

4. Discussion

Given the biological and technical challenges inherent to vertical bone augmentation procedures, the investigation of substances to be incorporated into scaffolds to enhance osteogenesis becomes imperative. Furthermore, considering the oral cavity's status as a contaminated environment, the incorporation or coating with additional antimicrobial and anti-inflammatory properties would confer distinct advantages. In this context, our aim was to assess the potential of Cranberry extract to stimulate osteogenesis in vitro. Initially, this study demonstrated that heightened concentrations of Cranberry extract (ranging from 62.5 to 500 mg/mL) exhibited biocompatibility with osteoblasts and mesenchymal stromal cells. In terms of osteogenic potential, a concentration of 20 mg/mL of Cranberry extract prompted an almost 3-fold increase in ALP activity and a 1.5-fold increase in BMP-2 expression relative to the positive control group consisting of SHED treated with an osteogenic medium. Additionally, concentrations of 200 mg/mL of Cranberry extract induced an approximately 1.7-fold increase in extracellular matrix mineralization. Furthermore, a correlation between the concentrations of the Cranberry extract and the levels of the early and late osteogenic markers was established.

As mentioned earlier, Cranberry extract concentrations ranging from 62.5 to 500 mg/mL were not considered cytotoxicity for either SHED or MC3T3-E1. However, concentrations above 250 mg/mL reduced SHED viability, an effect not observed in MC3T3-E1. It is important to highlight some inherent biological differences between these cell types. MC3T3E1 is a murine cell line, which presents limitations for translational research into human biology. Mouse and human cells differ significantly in their genetics, physiology, and cellular responses, including immune reactions, signaling pathways, and gene expression. Therefore, findings in murine cells often do not fully translate to human biology, reducing their predictive value for human disease research and drug development. Additionally, MC3T3E1 is an immortalized cell line, meaning it bypasses normal cellular aging (senescence). This results in altered metabolism, growth behavior, and chromosomal stability, which can make these cells more resilient but less representative of normal cell physiology. Furthermore, MC3T3-E1 are already in a late stage of osteoblast differentiation, limiting their plasticity and responsiveness to external stimuli. In contrast, SHED are mesenchymal stem cells, known for their high regenerative capacity and multipotent differentiation potential. Their stemness and higher proliferation rate likely explain their more robust response to stimuli compared to the more lineage-committed MC3T3-E1. SHED allow for the exploration of early molecular events and signaling pathways involved in the commitment to the osteogenic lineage. Since bone regeneration in vivo depends on the differentiation of undifferentiated progenitor cells, SHED better reflect the physiological context for studying this process. In summary, the divergent responses observed between SHED and MC3T3-E1 are likely due to differences in their origin and developmental state. SHED exhibit greater plasticity and responsiveness due to their stem cell nature, while MC3T3-E1 cells, as pre-committed osteoblasts, display more specific and potentially limited reactions to stimuli. Therefore, SHED provide aa more suitable model for human studies, as they are human-derived and have not undergone any immortalized process.

It is imperative to underscore that, based on the parameters used to assess Cranberry extract quality in this study, the extract demonstrated a high quality. Therefore, the analyzed Cranberry extract exhibited the presence of phenolic compounds devoid of interferents [18]. Additionally, this extract showcases a 140-fold surge in monomeric anthocyanin content in comparison to the minimum threshold required for indicating the authenticity of the Cranberry extract [26]. However, it is not possible to affirm that the effect of Cranberry observed was promoted by any of these compounds. Nevertheless, the specific compound within Cranberry responsible for mediating these osteogenic effects remains unidentified.

Hence, the findings revealed that the Cranberry extract led to heightened BMP-2 expression in comparison to cells treated solely with an osteogenic medium. This observation aligns with prior research that demonstrated the stimulation of ALP through anthocyanidins extracted from *Hibiscus syriacus* (Malvaceae) petals [28]. Additionally, proanthocyanidins were found to enhance ALP activity, contributing to osteogenic differentiation in human cells derived from the periodontal ligament [29]. Likewise, grape-extracted proanthocyanidins were shown to stimulate BMP-2 gene expression [30]. These cumulative findings underscore the potential of the compounds within Cranberry extract to augment ALP and BMP-2 levels, in concordance with this study.

It is crucial to note that elevate concentrations of Cranberry extract exhibited biocompatibility with osteoblast and stromal cells. This, coupled with the observed correlation between the Cranberry extract concentrations and osteogenic markers, motivated us to explore the impact of elevated concentrations of Cranberry extract in promoting extracellular matrix mineralization. Therefore, concentrations of 200 mg/mL of Cranberry extract induced an approximately 1.7-fold increase in extracellular matrix mineralization compared to the positive control group. It is important to note that lower Cranberry concentrations were able to stimulate ALP activity, whereas higher concentrations were necessary to potentiate extracellular matrix mineralization.

This study presents the inaugural investigation into the osteogenic potential of Cranberry extract. Given that the primary objective did not encompass unraveling the precise mechanism underlying Cranberry-induced osteogenic differentiation, it remains unclear. Nonetheless, as previously alluded to, the effects of polyphenols from other sources than Cranberry have been investigated, particularly regarding their impact on osteoblasts mediated through diverse signaling pathways, including Wingless-INT/β-catenin [31], insulin-like growth factor [32], bone morphogenetic proteins [33], Runt-related transcription factor 2 (Runx2) [32], and Osterix (OSX) [32]. Chen et al. [31] proposed that the mechanisms by which blueberry polyphenols promote bone formation involve β-catenin activation and the nuclear accumulation and transactivation of TCF/LEF gene transcription in bone and in osteoblasts. These authors suggested that the stimulation of molecular events leading to osteoblast differentiation triggered by P38 MAP kinase (MAPK)/β-catenin canonical Wnt signaling resulted in significant increases in bone growth in young rats consuming blueberry-supplemented diets. Moreover, previous studies have suggested that polyphenols may also modulate oxidative stress and inflammatory responses via pathways such as nuclear factor erythroid-2 related factor (Nrf2), a key regulator of the antioxidant response, and NF-κB, a major regulator of inflammation, both of which are known to affect osteoblast differentiation [34]. Additionally, Bu et al. [32] studied how dried plum polyphenols influence osteoblast activity and mineralized nodule formation under normal and inflammatory conditions. The Runx2 promoter contains a homologous sequence of NF-κB and activated protein-1 (AP-1) binding sites, which confer responsiveness to TNF-α, leading to the suppression of osteoblast activity. Consequently, anti-inflammatory drugs or natural compounds that inhibit NF-κB or AP-1 signaling could potentially counter the negative effects of TNF-α on osteoblast Runx2 expression and enhance mineralized nodule formation. In our study, no dose-dependent relationship was observed between BMP-2 expression and Cranberry extract concentrations. Additionally, BMP-2 levels were mildly expressed compared to ALP activity, which may suggest the involvement of pathways other than BMP signaling. For instance, the Wnt/β-catenin signaling pathway plays a critical role in osteoblast differentiation and bone formation by regulating the expression of key osteogenic markers like Runx2 and OSX [34]. Considering these findings, it is plausible that Cranberry extract could exert its osteogenic effects via one or more of these alternative pathways. Further research is required to explore whether Cranberry extract directly influences Wnt/β-catenin, Runx2, or other signaling pathways involved in osteogenic differentiation. Expanding on this point in future investigations could provide greater insight into the precise mechanisms through which Cranberry extract promotes osteogenesis.

Beyond its role in osteogenesis, it is pertinent to acknowledge that proanthocyanidins modulate the differentiation and physiology of osteoclastic cells, as well as the bone matrix itself [12]. This influence stems from their ability to hinder the functional development of osteoclasts via the NF-kB pathway and signaling pathways of mitogen-activated protein kinases (MAPKs), including c-Jun N-terminal kinase (JNK), which is responsible for activating the nuclear factor of activated T cells c1 (NFATc1), the primary gene regulator in osteoclast [13].

The potential of Cranberry extract to foster osteogenesis, coupled with the biocompatibility of the heightened Cranberry extract concentrations, offers promising prospects for integrating Cranberry into scaffolds intended for bone regeneration. Furthermore, the precise concentrations and amalgamations of Cranberry compounds that wield maximal efficacy, as well as the potential interactions with other factors, warrant further scrutiny. In summation, investigations into the effects of Cranberry on osteogenesis constitute an arena of ongoing scientific exploration. While the polyphenols found in Cranberries could indeed influence signaling pathways and enzyme activities related to bone formation, additional research is indispensable to comprehend and validate these effects. Concerning the limitations of this manuscript, it is worth mentioning that it is a preliminary in vitro study. As previously mentioned, further studies are warranted.

5. Conclusions

Heightened concentrations of Cranberry extract exhibited biocompatibility with both osteoblasts and mesenchymal stromal cells. Additionally, 20 mg/mL of Cranberry extract induced a nearly 3-fold increase in ALP activity and 1.5-fold increase in BMP-2 expression compared to the osteogenic control group. Furthermore, 200 mg/mL of Cranberry extract led to an approximately 1.7-fold increase in extracellular matrix mineralization. In conclusion, Cranberry extract stimulated both early and late markers of osteogenesis, with a correlation between the extract concentration and the levels of these osteogenic markers. Given its biocompatibility, along with the previously established antibacterial and anti-inflammatory properties, these findings suggest that Cranberry extract holds promise for future incorporation into scaffolds designed for bone regeneration.

Author Contributions: Conceptualization: Y.G.B., E.B.M., I.V.F., F.H.R., I.T.S. and A.C.C.C.; Methodology: Y.G.B., E.B.M., I.V.F., J.D.P.N., G.F., F.H.R., I.T.S. and A.C.C.C.; Formal analysis: Y.G.B. and A.C.C.C.; Investigation: Y.G.B., E.B.M., I.V.F. and J.D.P.N.; Writing—original draft: Y.G.B., I.V.F. and A.C.C.C.; Writing—review and editing: Y.G.B., E.B.M., I.V.F., J.D.P.N., G.F., F.H.R., I.T.S. and A.C.C.C.; Supervision: A.C.C.C. All authors have read and agreed to the published version of the manuscript.

Funding: No funding was obtained for this study.

Institutional Review Board Statement: Not applicable.

Informed Consent Statement: Not applicable.

Data Availability Statement: Data are contained within the article.

Acknowledgments: The authors are grateful to the Laboratório Multiusuário de Estudos em Biologia at the Universidade Federal de Santa Catarina for providing their equipment and infrastructures to perform some experimental analyses. The authors thank Curityba Biotech™ for providing stromal cells from human exfoliated deciduous teeth. The authors would like to thank Lenin Israel Proano Minaca for preliminary data. Yasmim Guterres Bauer and Eduarda Blasi Magini are supported by a scholarship from Conselho Nacional de Desenvolvimento Científico e Tecnológico (CNPq) and Fundação de Amparo à Pesquisa e Inovação do Estado de Santa Catarina—Brasil (FAPESC), respectively.

Conflicts of Interest: All authors declare they have no conflicts of interest.

References

1. Chiapasco, M.; Casentini, P.; Zaniboni, M. Bone Augmentation Procedures in Implant Dentistry. *Int. J. Oral Maxillofac. Implants* **2009**, *24*, 237–259. [PubMed]
2. Elnayef, B.; Monje, A.; Gargallo-Albiol, J.; Galindo-Moreno, P.; Wang, H.-L.; Hernández-Alfaro, F. Vertical Ridge Augmentation in the Atrophic Mandible: A Systematic Review and Meta-Analysis. *Int. J. Oral Maxillofac. Implants* **2017**, *32*, 291–312. [CrossRef] [PubMed]
3. Retzepi, M.; Donos, N. Guided Bone Regeneration: Biological Principle and Therapeutic Applications. *Clin. Oral Implants Res.* **2010**, *21*, 567–576. [CrossRef] [PubMed]
4. Chen, Y.; Gan, W.; Cheng, Z.; Zhang, A.; Shi, P.; Zhang, Y. Plant Molecules Reinforce Bone Repair: Novel Insights into Phenol-Modified Bone Tissue Engineering Scaffolds for the Treatment of Bone Defects. *Mater. Today Bio* **2024**, *24*, 100920. [CrossRef] [PubMed]
5. de Medeiros, A.K.B.; de Melo, L.A.; Alves, R.A.H.; Barbosa, G.A.S.; de Lima, K.C.; Porto Carreiro, A. da F. Inhibitory Effect of Cranberry Extract on Periodontopathogenic Biofilm: An Integrative Review. *J. Indian Soc. Periodontol.* **2016**, *20*, 503–508. [CrossRef]
6. Kumar, V.; Ankola, A.V.; Sankeshwari, R.; Jalihal, S.; Deepak, V.; Jois, H.S. Assessment of the Antimicrobial Efficacy of Hydroalcoholic Fruit Extract of Cranberry against Socransky Complexes and Predominant Cariogenic, Mycotic and Endodontic Climax Communities of the Oral Cavity: An Extensive in-vitro Study. *J. Oral Maxillofac. Pathol.* **2019**, *23*, 407–411. [CrossRef]
7. Galarraga-Vinueza, M.E.; Dohle, E.; Ramanauskaite, A.; Al-Maawi, S.; Obreja, K.; Magini, R.; Sader, R.; Ghanaati, S.; Schwarz, F. Anti-Inflammatory and Macrophage Polarization Effects of Cranberry Proanthocyanidins (PACs) for Periodontal and Peri-Implant Disease Therapy. *J. Periodontal Res.* **2020**, *55*, 821–829. [CrossRef]
8. Vaillancourt, K.; Ben Lagha, A.; Grenier, D. Effects of a Berry Polyphenolic Fraction on the Pathogenic Properties of Porphyromonas Gingivalis. *Front. Oral Health* **2022**, *3*, 923663. [CrossRef]

9. Neto, C.C.; Penndorf, K.A.; Feldman, M.; Meron-Sudai, S.; Zakay-Rones, Z.; Steinberg, D.; Fridman, M.; Kashman, Y.; Ginsburg, I.; Ofek, I.; et al. Characterization of Non-Dialyzable Constituents from Cranberry Juice That Inhibit Adhesion, Co-Aggregation and Biofilm Formation by Oral Bacteria. *Food Funct.* **2017**, *8*, 1955–1965. [CrossRef]
10. Tipton, D.A.; Hatten, A.A.; Babu, J.P.; Dabbous, M.K. Effect of Glycated Albumin and Cranberry Components on Interleukin-6 and Matrix Metalloproteinase-3 Production by Human Gingival Fibroblasts. *J. Periodontal Res.* **2016**, *51*, 228–236. [CrossRef]
11. Philip, N.; Leishman, S.J.; Bandara, H.M.H.N.; Healey, D.L.; Walsh, L.J. Randomized Controlled Study to Evaluate Microbial Ecological Effects of CPP-ACP and Cranberry on Dental Plaque. *JDR Clin. Trans. Res.* **2020**, *5*, 118–126. [CrossRef] [PubMed]
12. Tanabe, S.; Santos, J.; La, V.D.; Howell, A.B.; Grenier, D. A-Type Cranberry Proanthocyanidins Inhibit the RANKL-Dependent Differentiation and Function of Human Osteoclasts. *Molecules* **2011**, *16*, 2365–2374. [CrossRef] [PubMed]
13. Zhu, W.; Yin, Z.; Zhang, Q.; Guo, S.; Shen, Y.; Liu, T.; Liu, B.; Wan, L.; Li, S.; Chen, X.; et al. Proanthocyanidins Inhibit Osteoclast Formation and Function by Inhibiting the NF-KB and JNK Signaling Pathways during Osteoporosis Treatment. *Biochem. Biophys. Res. Commun.* **2019**, *509*, 294–300. [CrossRef]
14. Ramesh, P.; Jagadeesan, R.; Sekaran, S.; Dhanasekaran, A.; Vimalraj, S. Flavonoids: Classification, Function, and Molecular Mechanisms Involved in Bone Remodelling. *Front. Endocrinol.* **2021**, *12*, 779638. [CrossRef]
15. Inchingolo, A.D.; Inchingolo, A.M.; Malcangi, G.; Avantario, P.; Azzollini, D.; Buongiorno, S.; Viapiano, F.; Campanelli, M.; Ciocia, A.M.; De Leonardis, N.; et al. Effects of Resveratrol, Curcumin and Quercetin Supplementation on Bone Metabolism—A Systematic Review. *Nutrients* **2022**, *14*, 3519. [CrossRef]
16. Zhang, Z.; Li, B.; Wu, S.; Yang, Y.; Wu, B.; Lai, Q.; Lai, F.; Mo, F.; Zhong, Y.; Wang, S.; et al. Bergenin Protects against Osteoarthritis by Inhibiting STAT3, NF-KB and Jun Pathways and Suppressing Osteoclastogenesis. *Sci. Rep.* **2024**, *14*, 20292. [CrossRef]
17. Amarowicz, R.; Cwalina-Ambroziak, B.; Janiak, M.A.; Damszel, M.; Stępień, A.; Sulewska, K.; Karamać, M.; Penkacik, K. Effect of Fertilization on Phenolics of Rapeseeds and Their Antioxidant Potential. *Foods* **2024**, *13*, 561. [CrossRef]
18. Medina, M.B. Simple and Rapid Method for the Analysis of Phenolic Compounds in Beverages and Grains. *J. Agric. Food Chem.* **2011**, *59*, 1565–1571. [CrossRef]
19. Singleton, V.L.; Orthofer, R.; Lamuela-Raventós, R.M. Reagent. In *[14] Analysis of Total Phenols and Other Oxidation Substrates and Antioxidants by Means of Folin-Ciocalteu Methods in Enzymology*; Elsevier: Amsterdam, The Netherlands, 1999; pp. 152–178.
20. da Silva Oliveira, J.P.; de Oliveira, R.T.; Guedes, A.L.; da Costa Oliveira, M.; Macedo, A.F. Metabolomic Studies of Anthocyanins in Fruits by Means of a Liquid Chromatography Coupled to Mass Spectrometry Workflow. *Curr. Plant Biol.* **2022**, *32*, 100260. [CrossRef]
21. Sordi, M.B.; Curtarelli, R.B.; da Silva, I.T.; Fongaro, G.; Benfatti, C.A.M.; de Souza Magini, R.; Cabral da Cruz, A.C. Effect of Dexamethasone as Osteogenic Supplementation in in Vitro Osteogenic Differentiation of Stem Cells from Human Exfoliated Deciduous Teeth. *J. Mater. Sci. Mater. Med.* **2021**, *32*, 1. [CrossRef]
22. Littuma, G.J.S.; Sordi, M.B.; Borges Curtarelli, R.; Aragonês, Á.; da Cruz, A.C.C.; Magini, R.S. Titanium Coated with Poly(Lactic-co-glycolic) Acid Incorporating Simvastatin: Biofunctionalization of Dental Prosthetic Abutments. *J. Periodontal Res.* **2020**, *55*, 116–124. [CrossRef] [PubMed]
23. Sordi, M.B.; Fredel, M.C.; da Cruz, A.C.C.; Sharpe, P.T.; de Souza Magini, R. Enhanced Bone Tissue Regeneration with Hydrogel-Based Scaffolds by Embedding Parathyroid Hormone in Mesoporous Bioactive Glass. *Clin. Oral Investig.* **2022**, *27*, 125–137. [CrossRef]
24. Cruz, A.C.C.; Silva, M.L.; Caon, T.; Simões, C.M.O. Addition of Bone Morphogenetic Protein Type 2 to Ascorbate and β-Glycerophosphate Supplementation Did Not Enhance Osteogenic Differentiation of Human Adipose-Derived Stem Cells. *J. Appl. Oral Sci.* **2012**, *20*, 628–635. [CrossRef]
25. Yasuda, Y.; Ogawa, M.; Arakawa, T.; Kadowaki, T.; Saito, T. The Effect of Mineral Trioxide Aggregate on the Mineralization Ability of Rat Dental Pulp Cells: An In Vitro Study. *J. Endod.* **2008**, *34*, 1057–1060. [CrossRef]
26. Gardana, C.; Scialpi, A.; Fachechi, C.; Simonetti, P. Identification of Markers for the Authentication of Cranberry Extract and Cranberry-Based Food Supplements. *Heliyon* **2020**, *6*, e03863. [CrossRef]
27. Technical Committees. *ISO/TC 194 (Biological and Clinical Evaluation of Medical Devices) Biological Evaluation of Medical Devices—Part 5: Tests for in Vitro Cytotoxicity*, 3rd ed.; (Report ISO 10993-5: 2009); International Organization for Standardization: Geneva, Switzerland, 2009; pp. 1–34.
28. Karunarathne, W.A.H.M.; Molagoda, I.M.N.; Lee, K.T.; Choi, Y.H.; Jin, C.-Y.; Kim, G.-Y. Anthocyanin-Enriched Polyphenols from *Hibiscus syriacus* L. (Malvaceae) Exert Anti-Osteoporosis Effects by Inhibiting GSK-3β and Subsequently Activating β-Catenin. *Phytomedicine* **2021**, *91*, 153721. [CrossRef]
29. Huang, J.; Liu, L.; Jin, S.; Zhang, Y.; Zhang, L.; Li, S.; Song, A.; Yang, P. Proanthocyanidins Promote Osteogenic Differentiation of Human Periodontal Ligament Fibroblasts in Inflammatory Environment Via Suppressing NF-KB Signal Pathway. *Inflammation* **2020**, *43*, 892–902. [CrossRef]
30. Torre, E.; Iviglia, G.; Cassinelli, C.; Morra, M.; Russo, N. Polyphenols from Grape Pomace Induce Osteogenic Differentiation in Mesenchymal Stem Cells. *Int. J. Mol. Med.* **2020**, *45*, 1721–1734. [CrossRef]
31. Chen, J.; Lazarenko, O.P.; Wu, X.; Kang, J.; Blackburn, M.L.; Shankar, K.; Badger, T.M.; Ronis, M.J. Dietary-induced Serum Phenolic Acids Promote Bone Growth via P38 MAPK/Β-catenin Canonical Wnt Signaling. *J. Bone Miner. Res.* **2010**, *25*, 2399–2411. [CrossRef]

32. Bu, S.Y.; Hunt, T.S.; Smith, B.J. Dried Plum Polyphenols Attenuate the Detrimental Effects of TNF-α on Osteoblast Function Coincident with up-Regulation of Runx2, Osterix and IGF-I. *J. Nutr. Biochem.* **2009**, *20*, 35–44. [CrossRef]
33. Wang, Y.; Xia, C.; Chen, Y.; Jiang, T.; Hu, Y.; Gao, Y. Resveratrol Synergistically Promotes BMP9-Induced Osteogenic Differentiation of Mesenchymal Stem Cells. *Stem Cells Int.* **2022**, *2022*, 8124085. [CrossRef] [PubMed]
34. Iantomasi, T.; Palmini, G.; Romagnoli, C.; Donati, S.; Miglietta, F.; Aurilia, C.; Falsetti, I.; Marini, F.; Giusti, F.; Brandi, M.L. Dietary Polyphenols and Osteoporosis: Molecular Mechanisms Involved. *Int. J. Bone Fragil.* **2022**, *2*, 97–101. [CrossRef]

coatings

MDPI

Article

Silver Nanoparticles–Chitosan Nanocomposites as Protective Coatings for Dental Remineralization Treatment: An In Vitro Study

Katleen A. Aguirre de Rodríguez [1,2], Wendy Y. Escobar de González [1,2], Vianney Castañeda Monroy [2], Sean Murphy [3], Gabriel-Alejandro Martínez-Castañón [4,*], Horacio Bach [5] and Nereyda Niño-Martínez [6,*]

1 Facultad de Estomatología, Zona Universitaria, Av. Dr. Manuel Nava No. 2, San Luis Potosí CP 78290, Mexico; katleen.aguirre@ues.edu.sv (K.A.A.d.R.); wendy.escobar@ues.edu.sv (W.Y.E.d.G.)
2 Laboratorio de Biología Molecular, CENSALUD, Universidad de El Salvador (UES), San Salvador CP 1101, El Salvador; vianney.monroy@ues.edu.sv
3 Faculty of Dentistry, Department of Oral Biological and Medical Sciences, University of British Columbia, Vancouver, BC V6H 3Z6, Canada; sean.murphy@dentistry.ubc.ca
4 Facultad de Estomatología, Universidad Autónoma de San Luis Potosi, Zona Universitaria, Av. Dr. Manuel Nava No. 2, San Luis Potosí CP 78290, Mexico
5 Faculty of Medicine, Division of Infectious Diseases, University of British Columbia, Vancouver, BC V6H 3Z6, Canada; horacio.bach@ubc.ca
6 Facultad de Ciencias, Universidad Autónoma de San Luis Potosi, Av. Parque Chapultepec No. 1570, Privadas del Pedregal, San Luis Potosí CP 78295, Mexico
* Correspondence: mtzcastanon@fciencias.uaslp.mx (G.-A.M.-C.); nereyda.nino@uaslp.mx (N.N.-M.)

Abstract: Research with nanoparticles for the treatment and prevention of dental caries is of special interest given the high prevalence of the disease worldwide. Several studies support the use of nanoparticles associated with materials given their antimicrobial properties and potential demineralization reduction. This study aimed to evaluate the impact of the application of silver nanoparticles (AgNPs) and chitosan gel in combination with commercial fluoride varnish on the remineralization of dental enamel. Ninety-six tooth blocks were macroscopically evaluated via stereomicroscopy, ICDAS II, and laser fluorescence. Enamel blocks were subjected to artificial demineralization and divided into four exposure groups (24, 48, 120, and 168 h), and five different remineralizing agents were applied, namely, FV (fluoride varnish), FV + CG (fluoride varnish + chitosan gel), FV + AgNPs (fluoride varnish + AgNPs), FV + AgNPs + CG (fluoride varnish + AgNPs + chitosan gel), and AgNPs + CG (AgNPs + chitosan gel). Enamel surface changes were evaluated via laser fluorescence, X-ray diffraction, scanning electron microscopy, and energy-dispersive X-ray spectroscopy. Laser fluorescence results obtained from demineralized blocks and subsequently exposed to remineralizing treatment indicate significant differences. After exposure to remineralizing agents, hydroxyapatite and modified apatite phases were identified mainly in the samples treated with FV + AgNPs + CG in the groups exposed for 24, 48, and 120 h. The FV + AgNPs + CG and AgNPs + CG indicate good performance in terms of the Ca/P ratio in in vitro demineralization compared to the group treated with fluorine varnish.

Keywords: enamel remineralization; demineralization; chitosan; silver nanoparticles

1. Introduction

Dental caries is a highly prevalent disease in the world population. Its progression from the enamel surface to the innermost structures of the tooth depends on multiple factors such as cariogenic diet, hygiene habits, time, and salivary pH variations [1]. The onset of the lesion is related to the acidification of the environment, causing the demineralization of the teeth, which implies the loss of minerals in the initial stages. This process can be reversed naturally by repair through remineralization, which has the purpose of restoring the lost minerals. This process occurs at near-neutral physiological pH [2,3].

The timely identification of incipient lesions in enamel is of great importance in establishing preventive treatments and immediate intervention to stop their progression. The use of materials such as varnish fluoride for the management of the remineralization of dental tissues has been widely studied. Kim et al. (2021) compared the demineralization resistance and remineralization effects of various fluoride-releasing restorative commercial materials under clinical conditions and found that alkasite restorative material is a superior fluoride-releasing restorative material in stimulating demineralization and remineralization in an enamel [3]. However, in a different study, Chhatwani et al. (2021) found that the effectiveness of fluoride varnishes is compromised and limited due to its permanence in the oral environment, which stresses the materials under different temperature and pH conditions [4]. It has also been reported that the use of fluoride varnish (FV) could produce sensitivity and a burning sensation in some patients and, in regions where the levels of fluoride in water are high, the use of this varnish is limited; Parisay et al. (2024) explain that excessive fluoride consumption can result in symptoms like nausea, vomiting, diarrhea, escalated salivation, abdominal discomfort, muscle frailty, and spasms. Fluoride varnish is not recommended for patients with ulcerative gingivitis, gingivitis stomatitis, or asthma [5]. Silver diamine fluoride has also been reported as a material which inhibits demineralization, promoting the remineralization of both enamel and dentin, but staining on the areas where it is applied limits its use [6,7].

On the other hand, nanotechnology and its application in dentistry have allowed for the identification of some characteristics, properties, and benefits in the management of oral diseases such as dental caries [7]; silver nanoparticles (AgNPs) have been extensively used due to their antibacterial properties with sustained activity [8]. Pichaiaukrit et al. (2019) and Ghafar et al. (2020) studied combinations of chitosan with calcium fluoride and sodium fluoride nanoparticles, trying to obtain a material with a sustained release of fluoride for the treatment of incipient lesions of dental caries [9,10]. Hydroxyapatite NPs have been used in toothpaste [11] and other dentifrices to promote remineralization, but their abrasive effect on enamel still causes concerns [12]. In another in vitro study, Surija et al. (2018) found satisfactory results on the effects of chitosan; it inhibits the enamel demineralization process in vitro by increasing the environmental pH and reducing the solubility of the enamel [13].

This study aimed to evaluate the impact of the application of AgNPs and chitosan gel (CG) in combination with a commercial FV on the remineralization of dental enamel through macroscopic evaluation, ICDAS II, laser fluorescence (DIAGNOdent pen), scanning electron microscopy (SEM), and X-ray diffraction (XRD). The following null hypothesis is proposed: There are no differences in the remineralizing effect between the proposed FV + AgNPs + CG treatment and the conventional protocol with FV for initial lesions on smooth surfaces.

2. Materials and Methods

2.1. Participants

The protocol of this study was approved by the Research Ethics Committee of the Faculty of Dentistry of the University of El Salvador with the following document number: CEI-FOUES/2022/003. Participants provided verbal consent, and the recruitment of the patients ranged from 1 April 2022 until 3 March 2023. The 61 healthy teeth used in this study were extracted for orthodontic reasons without any relation to this protocol. The patients were informed, and they donated their pieces for this study with verbal consent.

2.2. Preparation and Characterization of Chitosan Gel with AgNPs

For the preparation of CG, a published protocol was used [14]. In brief, 2 mL of acetic acid was added to 47.8 mL of a dispersion of AgNPs (0.1% by weight concentration). To this mixture, 3.5 wt.% chitosan was added, together with 0.2 mL of glycerol. The mixture was stirred under the same conditions mentioned above until the gel was formed.

2.3. Sample Preparation

The sample consisted of 122 blocks of 61 healthy teeth extracted for orthodontic reasons, preserved in chloramine-T at −17 °C in the Biobank of Dental Organs (BBOD) of the Research Center of the School of Dentistry, University of El Salvador. Previously, soft tissue and calculus residues were removed from the teeth and cleaned with a low-speed micromotor Lynx Classic motor MTI Dental, USA, and pumice stone. A coronal cut of the teeth was made with a diamond disc at low speed, and tooth blocks of vestibular and/or lingual surface were obtained with a proportion of 4 mm long × 3 mm wide, corresponding to the middle third of the oral surface. Previous studies on in vitro remineralization were considered to determine the dimensions of the samples [15–17]. The thickness of the blocks was approximately 2 mm. This was considered to specifically evaluate the middle third of the teeth given the structural characteristics and the thickness of the enamel identified in this location. The surface of the sample blocks was not polished. The blocks were stored in 5 mL beakers with deionized water.

2.4. Macroscopic Evaluation

The blocks were macroscopically evaluated using a stereomicroscope (Leica EZ4E, Leica Microsystems, Balgach, Switzerland), and a vestibular and/or lingual surface photographic record was/were taken. The visual examination was performed according to the criteria of the International Caries Detection and Assessment System ICDAS II and coded as follows: 0—healthy tooth surface; 1—caries lesion in the outer half of the enamel; and 2—caries lesion in the inner half of the enamel. Two previously calibrated researchers performed macroscopic evaluations. Weighted Kappa was calculated to evaluate agreement between examiners. High intra-examiner agreement was obtained for macroscopic evaluation/ICDAS II (Kappa 1.00) and macroscopic evaluation/DIAGNOdent pen (Kappa 0.967). The investigators individually evaluated each of the blocks while the surface was wet. Ninety-six enamel blocks, which met the characteristics of healthy enamel according to published evaluations, were selected [18].

2.5. Laser Fluorescence Measurement

The DIAGNOdent pen 2190 (DIAGNOdent pen, Kavo, Biberach, Germany) was used according to the manufacturer's recommendations; this is a device which measures the depth of the caries lesion and gives a unitless numerical value. A cylindrical sapphire tip was fitted for smooth surfaces and calibrated using a ceramic reference block. For each measurement, the device was zeroed; the sapphire tip was moved over the selected evaluation sites at a 90-degree angle, as indicated by the manufacturer. Evaluations were completed by two examiners. Each examiner made two measurements for each test site, and the mean values were recorded (reported as initial evaluation in Table 1). The manufacturer's suggested cut-off points were used to interpret the measurements and were coded as follows: 0–12 healthy tooth surfaces; 13–24 superficial caries in enamel; >25 lesions in the inner half of enamel [18].

Table 1. DIAGNOdent pen scores were obtained at different exposure times and different remineralization treatments (n = 16/group), expressed as mean ± SD.

Group		Initial [a]		After [b]			Exposition Time								
							24 h		48 h		120 h		168 h		
	n	Mean	SD	Mean	SD	p *	Mean	SD	Mean	SD	Mean	SD	Mean	SD	p **
1	16	6.44	1.79	66.00	29.49		61.75 **	16.03 **	81.5 **	21.76 **	61.25 **	32.36 **	36.50 **	11.21 **	
2	16	6.94	2.41	64.56	32.77		46.75 **	12.04 **	48.25	35.38	80.5 **	37.00 **	34.75 **	11.35 **	
3	16	7.25	2.44	65.06	30.55	<0.05	47.00 **	6.22 **	77.50 **	28.44 **	12.5	4.43	37.00 **	22.14 **	<0.05
4	16	8.31	2.55	76.75	21.92		81.75 **	8.06 **	36.25	11.12	30.5	13.77	33.50 **	1.91 **	
5	16	7.06	2.29	70.50	29.35		14.25	6.95	10.50	3.70	27.75	12.04	12.25	5.68	
6	16	6.00	1.41	61.44	29.83		28.25	12.04	13.50	1.73	7.5	1.73	4.75	1.50	

[a] Evaluation with DIAGNOdent pen; [b] after demineralization; * associations evaluated via Student's t-test $p < 0.05$. ** Bonferroni correction $p < 0.05$. Group 1, FV; fluoride varnish; 2, fluoride varnish + chitosan gel; 3, fluoride varnish + AgNPs; 4, fluoride varnish + AgNPs + chitosan gel; 5, AgNPs + chitosan gel; 6, artificial saliva.

2.6. *Artificial Demineralization of the Sample Surfaces*

The vestibular and/or lingual surfaces of the blocks were demineralized following the application of 35% phosphoric acid 3M, ESPE, St. Paul, MN, USA, for 1 min (Figure 1B,D). The samples were washed with deionized water for acid removal and evaluated again using the DIAGNOdent pen (Table 1) to establish differences between measurements and an adequate demineralization process [19].

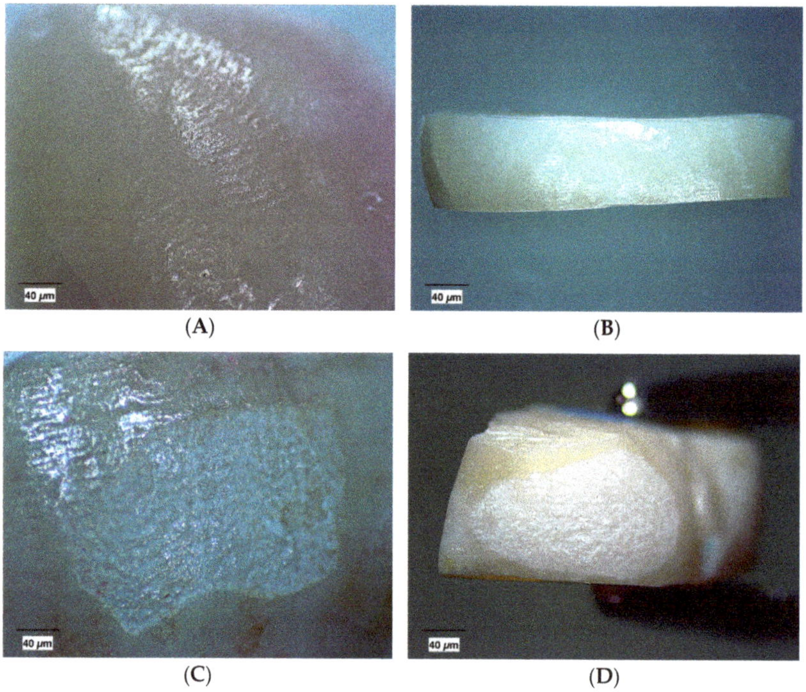

Figure 1. Macroscopic evaluation of vestibular/lingual surfaces by stereomicroscopy: (**A**) smooth and (**B**) continuous surfaces without white spots in healthy enamel; (**C**) smooth and (**D**) continuous surfaces showing porous opaque demineralized enamel.

2.7. *Application of Remineralizing Agents*

The samples were divided into six different groups (n = 16 for each group) according to the remineralizing agent: (1) FV (Clinpro, 3M ESPE, St. Paul, MN, USA); (2) FV + CG; (3) FV + AgNPs; (4) FV + AgNPs + CG; (5) AgNPs + CG; and (6) artificial saliva (Kim

Hidrat, Spain) composed of Xylitol 1%, potassium thiocynate, potassium chloride, sodium chloride, calcium chloride, magnesium chloride, potassium dihydrogen phosphate, sodium saccharin, hydrogenated castor oil, sodium methylparaben, sodium propylparaben, bonoprol, menthol, aroma, citric acid, and water. For the preparation of the different treatment combinations, a ratio of 900 μL:20 μL FV–composite was used. The application volume was 1000 microliters, with an FV ratio of 90% and 2% compound. This ratio was considered based on a previous study on the bacteriological efficacy of chitosan gel and silver nanoparticles [14,20].

A thin layer of remineralizing agent was applied using a microbrush on the demineralized enamel surfaces. The remineralizing agent applications were completed according to the manufacturer's directions. After application, the samples were allowed to stand for 5 min and then immersed in a commercial artificial saliva solution with a pH of 7.2 and kept in an incubator at 37 °C for the exposure time. Four exposure times were considered (24, 48, 120, and 168 h), and treatment was applied every 24 h. At the end of the exposure time, the varnish was carefully removed, and the samples were washed with deionized water. After the application of remineralizing agents, the samples were rinsed with deionized water, removing excess material with a scalpel blade. Group 6 was immersed in the artificial saliva solution without any treatment. The pH value of the artificial saliva solution for the groups was 7.2, and the solution was renewed daily. Evaluations with a DIAGNOdent pen 2190 were performed at the end of each exposure time to record the surface changes.

2.8. X-Ray Diffraction

For the analysis of the crystal structure of tooth enamel, a Rigaku miniFlex 300/600 System X-ray diffractometer (Rigaku Corporation, Tokyo, Japan) using CuKα radiation was used. The operating conditions were as follows: 600 W X-ray tube; wavelength of $\lambda = 0.154056$ nm at 40 kV. Diffraction spectra were acquired in the range of 10–80 degrees. Thirty-two tooth blocks were evaluated which, were pulverized with mortar, obtaining 30 mg; this content was dehydrated using ethanol and placed in the sample holder for the respective analysis in the X-ray diffractor.

2.9. Scanning Electron Microscopy and Energy-Dispersive X-Ray Analyses

Scanning electron microscopy (SEM) and energy dispersive X-ray analysis (EDX) were performed at the Centre for High-Throughput Phenogenomics (University of British Columbia) on an FEI Helios Nanolab 650 SEM/FIB equipped with an EDAX Octane Pro EDX detector (FEI Company, Hillsboro, OR, USA). SEM was used to characterize morphological changes in enamel surfaces after the remineralization procedures. Surface morphology images of the enamel samples were acquired using a 1 kV electron beam with accelerating voltage. Compositional information was obtained via EDX analysis to determine the presence of calcium and phosphorous and their relative concentrations in the sample surfaces. The EDX analyses were performed using a 10 kV electron beam with accelerating voltage.

2.10. Statistical Analysis

The data obtained were subjected to statistical analysis using SPSS V25 software. Kolmogorov–Smirnov was used to identify the normality of the data. Data analysis was performed using the Student's T and Bonferroni correction test. The Bonferroni correction test was used to analyze the comparison of means between laser fluorescence assessments. The significance level was set at $p < 0.05$, with a 95% confidence interval.

3. Results

3.1. Macroscopic Evaluation

A total of 122 blocks of vestibular/lingual surfaces were examined in this study, and 96 of them were included in this study after the microscope examination. Selected images of the blocks were evaluated for healthy enamel characteristics (smooth and continuous surface without white spots) via stereomicroscopy and ICDAS II system assessments (Figure 1A,B). In contrast, images of enamel after the demineralization process are shown in Figure 1C,D. The demineralization process was confirmed using fluorescence.

3.2. Laser Fluorescence

After different times of each treatment, the assessments obtained by DIAGNOdent pen (demineralized/post-treatment) were compared, and we observed that the laser fluorescence values decreased significantly after 168 h of applying the remineralization process in all treatment groups. Table 1 indicates the average values obtained with laser fluorescence before and after demineralization and after the application of remineralizing agents. Significant differences ($p < 0.05$) were obtained between the initial evaluations and after artificial demineralization. Likewise, the comparison of means between the exposure groups was established, being significant for the FV groups (24, 48, 120, and 168 h), FV+ CG (24, 120, and 168 h), FV + AgNPs (24, 48, and 168 h), and FV + AgNPs + CG (24 and 168 h). The lowest value was achieved using the FV + AgNPs + CG treatment, with a reduction from 72.5 to 4.75. The means values obtained with laser fluorescence before and after demineralization were recorded (Table 1). The individual values obtained for the group at 48 and 120 h are scattered around the mean, which represents the variation in the data obtained.

3.3. X-Ray Diffraction

For X-ray diffraction (XRD) evaluation, the results were analyzed in SmartLab Studio II and OriginPro software, Version 2023b (OriginLab Corporation, Northampton, MA, USA). Refinement was performed via the Rietveld method. Lattice parameters indicating the presence of hydroxyapatite and fluorapatite were observed in the samples exposed to FV + CG + AgNPs at 24, 48, and 120 h. The phases identified in the diffractogram (Figure 2) indicate the presence of apatite in healthy enamel and calcium hydroxide phosphate in demineralized enamel. Table 2 indicates the crystalline phases and crystallographic parameters identified in the samples through refinement via the Rietvelt method. The phases of apatite and modified hydroxyapatite are mainly identified, highlighting the identification of fluorapatite in the FV + AgNPs + CG group at 24 h of exposure. The crystallographic axes (a, b, and c) are also indicated, which indicate that the structures evaluated are anisotropic in the "c" direction, which means that the properties and arrangement of the atoms varies in this direction with a hexagonal configuration.

3.4. Scanning Electron Microscopy and EDX Analysis

Surface analysis using scanning electron microscopy showed a characteristic smooth surface in healthy enamel (Figure 3) but not in the artificial saliva group (Figure 4), where eroded prismatic enamel and interprismatic enamel protrusions are identified as a result of the demineralization process. The concentration of the crystalline structure of the enamel shows variations in all groups with respect to normal enamel, with an increase concerning the eroded enamel of the control group. In the groups to which the remineralizing agent was applied at 120 h and 168 h (Figures 4 and 5), morphological changes were observed on the surface where the presence of small crystals was identified in all the groups; however, they do not have a homogeneous distribution.

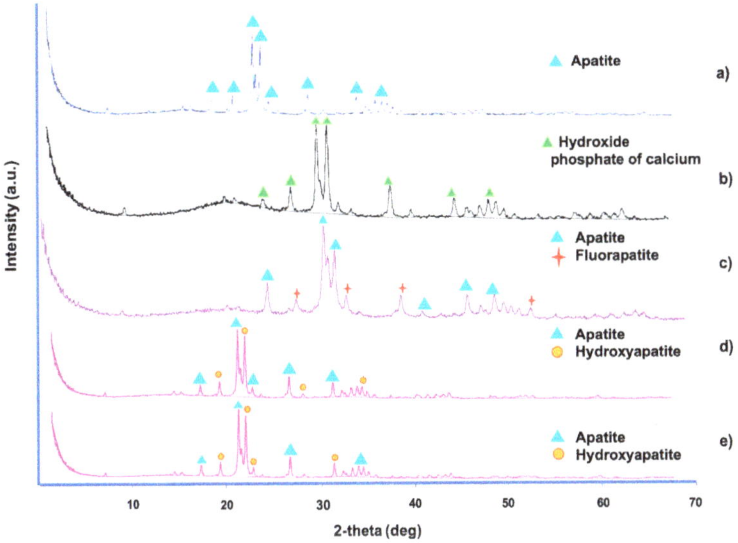

Figure 2. X-ray diffraction spectra: (**a**) healthy enamel; (**b**) enamel demineralized for 1 min with 35% phosphoric acid; (**c**) FV + CG + AgNPs-24 h; (**d**) FV + CG + AgNPs-48 h; (**e**) FV + CG + AgNPs-120 h.

Table 2. Identified crystalline phase.

Exposition Time	Treatment	Phase	a/Å *	b/Å *	c/Å *
	Healthy enamel	Apatite	9.43	9.43	6.87
	Demineralized enamel	Calcium phosphate hydroxide	9.45	9.45	6.89
24 h	Fluoride varnish	Carbonate Hidroxiapatite	9.44	9.44	6.88
	Fluoride varnish + chitosan gel	Apatite	9.43	9.43	6.88
	Fluoride varnish + NpAg + chitosan gel	Apatite	9.45	9.45	6.89
		Fluorapatite	9.45	9.45	6.88
	NpAg + chitosan gel	Apatite	9.43	9.43	6.87
48 h	Fluoride varnish	Apatite	9.45	9.45	6.88
	Fluoride varnish + chitosan gel	Apatite	9.45	9.45	6.88
	Fluoride varnish + NpAg + chitosan gel	Hidroxiapatite	9.43	9.45	6.87
	NpAg + chitosan gel	Apatite	9.43	9.43	6.87
120 h	Fluoride varnish	Apatite	9.42	9.42	6.86
	Fluoride varnish + chitosan gel	Apatite	9.42	9.42	6.87
	Fluoride varnish + NpAg + chitosan gel	Apatite	9.42	9.42	6.86
		Hidroxiapatite	9.42	9.42	6.85
	NpAg + chitosan gel	Apatite	9.47	9.47	6.89
		Potasium, calcium	3.88	3.88	16.63
168 h	Fluoride varnish	Carbonate Hidroxiapatite	9.44	9.44	6.87
	Fluoride varnish + chitosan gel	Apatite	9.43	9.43	6.86
	Fluoride varnish + NpAg + chitosan gel	Apatite	9.42	9.42	6.86
	NpAg + chitosan gel	Apatite	9.45	9.45	6.88

* lattice parameters.

The EDX elemental analysis (Figure S1 in Supplementary File) confirmed that the surface consisted of calcium, phosphorus, carbon, and oxygen. Within the comparison of the groups exposed at 120 h and 168 h, the results of the present study show that after remineralization, the carbon and oxygen content increased in all exposed groups, except in the control group. F content was identified in the FV group (0.92 weight %) and the combination of FV + CG + AgNPs group (0.91) at 120 h and in the groups exposed to 168 h FV + CG + AgNPs and CG gel + AgNPs at 0.19 and 3.5, respectively, with higher F content in the samples combined with CG + AgNPs for the 120 h and 168 h groups. The detection limit of the technique could be the reason why there were groups where FV was used but F was not detected. The Ca/P ratio was higher in the samples treated with FV + CG + AgNPs, CG, and AgNPs in both exposure groups. The values closer to normal enamel indicate that remineralizing treatment achieves a great percentage recovery of enamel mineralization.

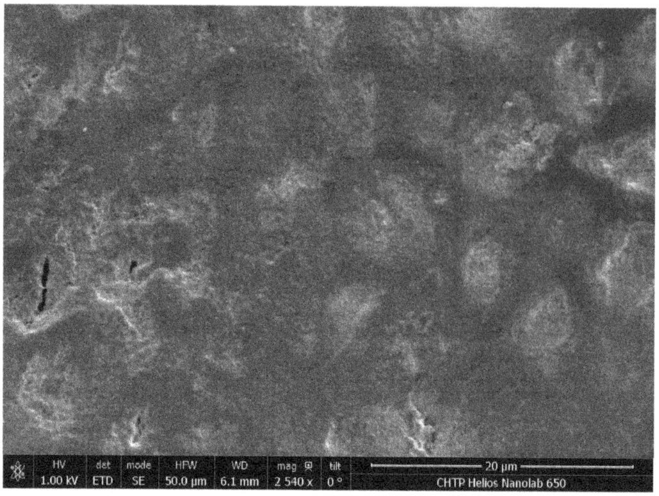

Figure 3. Scanning electron microscopy images of healthy enamel.

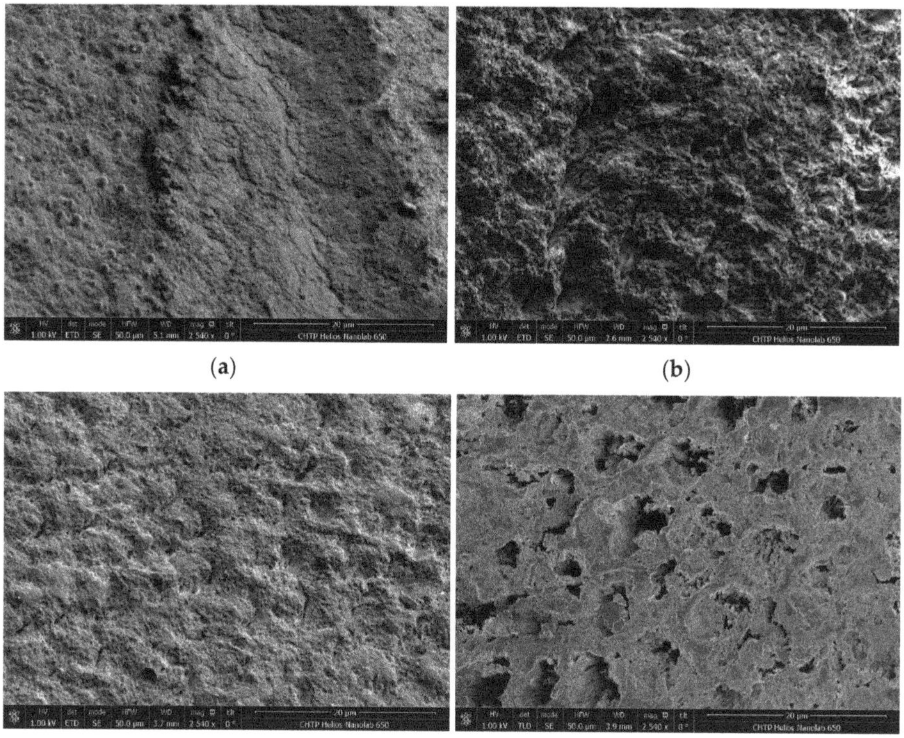

(**a**) (**b**)

(**c**) (**d**)

Figure 4. Scanning electron microscopy images of the enamel surface. Exposure to (**a**) FV, (**b**) FV + CG + AgNPs, (**c**) CG + AgNPs, and (**d**) artificial saliva after 120 h of exposure.

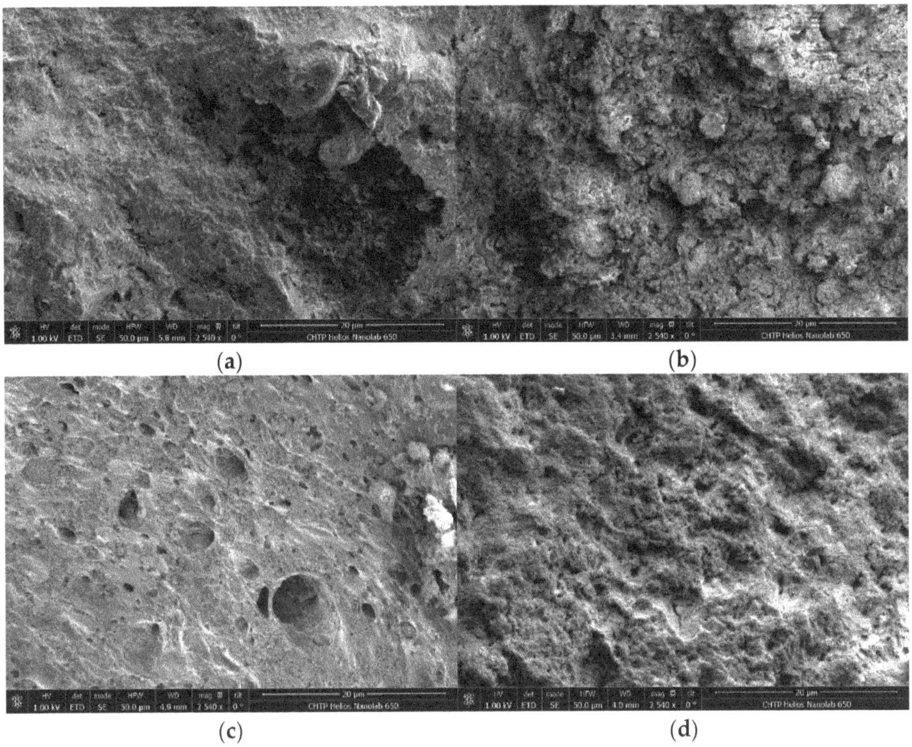

Figure 5. Scanning electron microscopy images of enamel surface. Exposure to (a) FV, (b) FV + CG + AgNPs, (c) CG + AgNPs, and (d) artificial saliva after 168 h of exposure.

4. Discussion

Remineralization is a natural process by which lost minerals are redeposited in the enamel. However, this process is often not sufficient to counteract the continuous loss of minerals, as conditions may vary according to oral conditions. In this regard, biomaterials such as AgNPs and CG play an important role in the remineralization process due to their characteristics and properties [12–14,18,19]. AgNPs have a broad spectrum of antibacterial properties which can inactivate enzymes and prevent DNA replication in bacteria, thereby reducing the risk of caries [20,21]. Moreover, several studies indicate that chitosan products enhance remineralization in artificial lesions in vitro [22–24]. Changes in the surface of demineralized enamel indicate a negative charge due to the leaching of Ca++, and the positive charge left from the chitosan favors adhesion to the demineralized surface [25].

Diagnostic evaluation using the DIAGOdent pen laser is a complementary method for the diagnosis of lesions on the smooth surface of the enamel. Likewise, it is capable of detecting in vitro changes in the surface of dental enamel after exposure to a demineralizing agent. DIAGNOdent pen titrations obtained pre- and post-artificial demineralization indicate significant differences. This is congruent with the XRD analysis and SEM analysis showing important variations among the samples evaluated.

X-ray diffraction results indicate changes related to significant mineral loss and surface changes in the artificially demineralized samples, indicating the presence of calcium hydroxide phosphate. This denotes variations in the composition of the enamel surface in an in vitro context. This allows us to establish that laser fluorescence is capable of detecting in vitro changes in the surface of dental enamel after exposure to a demineralizing agent.

The identified differences between initial and post-demineralization values were measured with the DIAGNOdent pen on the smooth surface of permanent teeth [25]. However, slight variations were measured with the DIAGNOdent pen on the enamel surface after exposure to different acid solutions [26].

The mechanism of remineralization under in vitro conditions has been studied elsewhere, but there is still no agreement among researchers. According to Enax et al. (2024), "...It has been conclusively shown by such in vitro crystallization studies under well-defined conditions that additives can influence nucleation and crystal growth of calcium phosphate even at low concentrations..." [27]. In our research, we found that the presence of silver nanoparticles promotes a different sort of compound in the crystallization products (see DRX results) due to a heterogeneous precipitation process, and also, the great affinity of silver to fluoride is well known, and the presence of silver fluoride in low concentrations—not detectable via XRD—can influence the remineralization process. Research in this area is also needed.

For this present study, the comparison between combinations of remineralizing treatments, considering measurements of demineralized surface and post-treatment with the DIAGNOdent pen, indicate significant differences. Additionally, an increase in means is identified for the FV, FV + AgNPs, FV + GC, and FV + CG + AgNPs groups at 24 and 68h post-exposure, compared to values recorded post-desmineralization. The evaluations stand out for groups containing CG (FV + CG and FV + CG + AgNPs) compared to the conventional FV treatment group; this could be related to CG's adhesion capability on the enamel surface. Other studies identified differences in the means obtained in the groups where chitosan was incorporated, showing significantly lower levels compared to other remineralizing agents [16,28]. Similar conclusions were identified with the same system in groups of enamel blocks exposed to remineralizing treatment [29].

The results related to the exposure time show significant differences in the four exposed groups. It is noteworthy that the XRD results identify crystalline phases of hydroxyapatite in the FV + CG + AgNPs treatment for 120 h groups, congruent with remineralization. The control group, in which only artificial saliva was used, shows remineralizing potential, indicating a decrease in the measurements obtained in this group.

The results of the commercial FV show a predictable behavior concerning its remineralizing effects; however, the means obtained after the application of the treatment groups that included chitosan indicate significant variations with respect to the measurements in demineralized samples.

The samples processed in XRD indicate important variations in the composition and phases identified in the samples submitted to artificial demineralization compared to healthy samples. This allows us to establish that the variations identified by the post-remineralization evaluations allow us to identify phases that coincide with a remineralization process.

Other studies reported similar characteristics in the diffraction spectra obtained in healthy enamel and enamel exposed to F [30–33]. During the evaluation, a loss of crystallinity in demineralized enamel was recorded. Although the samples exposed to F denote characteristic apatite peaks, they indicate that the apatite crystal lattice was recovered to some extent. However, samples exposed to nanocomplexes of phosphorylated chitosan and amorphous calcium phosphate (Pchi-ACP) showed narrower and sharper peaks in the diffractogram. Similar results were reported with the use of mesoporous bioactive glasses loaded with amorphous calcium phosphate to treat white spot lesions [34]. Diffraction analysis indicated that remineralized deposits in all four groups showed a diffraction pattern similar to that of normal enamel and shared a crystal structure similar to hydroxyapatite. Likewise, as a result of the refinement through the Rietveld method, variations in the crystallographic parameters were identified, registering the presence of fluorapatite and hydroxyapatite in the group of blocks exposed to FV + CG + AgNPs at 48 h and 120 h of exposure, which is congruent with remineralization on the enamel surface.

On the other hand, previous studies have used SEM-EDX analysis to study the changes associated with the demineralization–remineralization of dental enamel [3,28,35–37]. In the current study, enamel remineralization included the evaluation of samples on EDX, identifying the weight percentage of different minerals such as calcium and phosphate because they represent the main composition of hydroxyapatite. Ca/P titrations are important indicators of the effects of agents on enamel remineralization [38]. For example, changes in the Ca/P ratio were monitored for samples exposed at 120 h and 168 h. The results indicate that remineralizing agents such as FV, combined with CG and AgNPs, improved enamel remineralization. It is considered that the remineralization process is related to the replacement of minerals, mainly through the redeposition of calcium and phosphate in the enamel [3]. Research on the release of F and its applications in the remineralization process indicates the capacity of F release after an initial application followed by continuous diffusion of F, which allows us to determine that it is one of the treatments with better results for the management of white spot lesions in enamel; however, its effect decreases after seven days of treatment [4,38]. On the other hand, other minerals or compounds besides Ca and P could have a great influence on the remineralization process; for example, chitosan, which, due to its high nitrogen content, represents an excellent vehicle for transporting ions such as Ca and P for biomineralization. Likewise, chitosan adheres to negatively charged surfaces such as those of demineralized enamel [30,39–41]. The results of this study indicate a higher Ca/P ratio in the groups in which commercial FV was combined with CG and AgNPs.

Table 3 shows the results of some studies concerning remineralization using nanoparticles and chitosan in recent years. As we can see, there is active research seeking a coating material with sustained activity without the complications of the use of fluoride, as we stated in the introduction to this paper.

Table 3. Comparison of findings on remineralization using nanoparticles and chitosan in selected articles.

Author(s)	Aim	Bioactive Materials	Main Results	Ref.
Deokar et al.	To evaluate and compare the efficacy of the remineralizing potential of acidulated phosphate fluoride gel, chitosan nanoparticles, and silver diamine fluoride on the microhardness of artificial carious lesions created on extracted teeth.	Chitosan nanoparticles, silver diamine fluoride, and acidulated phosphate fluoride gel	• Chitosan nanoparticles showed the highest increase in remineralization followed by SDF and APF gel	[42]
Hanafy et al.	Chitosan was employed as a novel biomimetic mineralization model to repair damaged enamel to compare its performance with that of bioinspired zinc-doped nanohydroxyapatite.	Chitosan–hydrogel and a zinc-doped nanohydroxyapatite	• Marked reduction in the Ca/P ratio and the lack of the characteristic hydroxyapatite diffraction peaks were detected for demineralized enamel. • The remineralization stage revealed evident recovery of the mineral contents with apparent distinctive XRD patterns of hydroxyapatite in both groups. • SEM analysis showed the absence of etched enamel porosity with the formation of a newly formed rod-like apatite layer.	[43]
Muşat et al.	To induce the biomimetic remineralization of acid-etched human enamel in artificial saliva containing fluoride under agarose and novel chitosan–agarose hydrogels action without a toxic crosslinking agent for chitosan.	Chitosan (CS) and agarose (A) in a biopolymer-based hydrogel	• The values for the CS-A series are closer to the theoretical value of the carbonated hydroxyapatite of natural enamel; those for the series-A indicate the formation of calcium deficient phosphates.	[40]

Table 3. *Cont.*

Author(s)	Aim	Bioactive Materials	Main Results	Ref.
Magalhães et al.	Synthesize, characterize, and determine the effects of a ChNPs suspension on human enamel after cariogenic challenge via pH cycling.	Chitosan nanoparticles	• ChNPs suspension minimized human enamel demineralization after a cariogenic challenge.	[44]
Yamakami et al.	Evaluate the effects of an experimental chitosan/casein gel on enamel demineralization/remineralization in an environment with a high cariogenic challenge.	Chitosan/casein gel	• Chitosan seems to be the major component responsible for reducing the volume loss of demineralized enamel.	[16]
Xu et al.	Review the types, properties, and potential uses of non-metallic nanomaterials systematically for managing dental caries.	Biological organic nanomaterials, synthetic organic nanomaterials, carbon-based nanomaterials, and selenium nanomaterials	• Chitosan nanoparticles have shown a remineralization ability on demineralized enamel.	[45]
Wahied et al.	Evaluate the effect of chitosan nanoparticles on the remineralization of the demineralized enamel surface after being added to nano-hydroxyapatite and nano-calcium phosphate materials.	Nano-hydroxyapatite and nano-calcium phosphate	• Addition of chitosan nanoparticles had a significant effect on the remineralization of the demineralized enamel surface. • The nano NHA + chitosan group had the highest significant value, while the Nβ-TCP group exhibited the lowest significant value.	[46]
Zhang et al.	The purpose of this literature review is to provide an overview of current bioactive materials for caries management.	Bioactive materials for caries management	• Combination of AgNPs and NaF or nano-silver fluoride does not cause black staining and metallic taste, unlike SDF. • AgNPs' properties suggest their potential as an effective anticaries agent without black staining.	[47]

Finally, some questions about the biocompatibility of silver nanoparticles in dentistry have been raised by researchers; in previous studies, we have evaluated the biocompatibility of these gallic acid stabilized-silver nanoparticles, finding cell viability percentages higher than 95% when tested against human dermal fibroblast, and we have also reported that AgNPs concentrations less than 4 μg/mL showed a cytotoxic effect that resulted in a death rate of 13.8% or less. We also found that AgNPs at 500 ppm results in a viability above 75% in human fibroblasts. In an in vivo study, it was found that silver nanoparticles suspensions administrated to Wistar rats did not generate alterations among clinical characteristics, and clinical chemistry and hematology values showed slight alterations; however, at the end of the treatment, most were inside the normal concentration [48].

Considering that the present study was carried out in a laboratory environment, some of the limitations can be attributed to the difficulty of imitating the in vivo characteristics of saliva in the oral cavity. It is necessary to continue the evaluations assessing the microhardness of the enamel after exposure to remineralizing agents. Likewise, based on the results, in vivo evaluations could be considered.

5. Conclusions

The results of the present in vitro study suggest that the combination of FV + CG + AgNPs and CG + AgNPs could induce changes in the surface of the dental enamel related to remineralization, extending its effect up to 168 h compared to commercial FV. Although all the tested materials produce a remineralization effect, it is concluded that they do it at different proportions, producing different remineralization products as observed in X-ray diffraction analysis. Future research could include atomic force microscopy and microhardness analysis to confirm these results. The main contributions of this work are as follows:

- The changes evaluated on the enamel surface after applying FV + CG + AgNPs are related to the remineralization process, extending its effect up to 168 h compared to FV.
- The absence of pigmentation on the enamel surface is also highlighted.

Although these results are remarkable, further investigation into the variations in the microhardness of the enamel surface after exposure to the treatment is needed. Additionally, its clinical evaluation should be considered, considering the variations in salivary pH in vivo.

Supplementary Materials: The following supporting information can be downloaded at: https://www.mdpi.com/article/10.3390/coatings15010040/s1, Figure S1. EDX results for all groups at 120 h and 168 h of treatment.

Author Contributions: Conceptualization, N.N.-M. and G.-A.M.-C.; formal analysis, K.A.A.d.R.; funding acquisition, N.N.-M.; investigation, K.A.A.d.R., W.Y.E.d.G., V.C.M., S.M., N.N.-M. and G.-A.M.-C.; methodology, K.A.A.d.R., W.Y.E.d.G., V.C.M., N.N.-M. and G.-A.M.-C.; writing—original draft preparation, K.A.A.d.R., N.N.-M. and G.-A.M.-C.; writing—review and editing, H.B. and G.-A.M.-C. All authors have read and agreed to the published version of the manuscript.

Funding: This research received no external funding.

Institutional Review Board Statement: This study was conducted in accordance with the Declaration of Helsinki and approved by Research Ethics Committee of the Faculty of Dentistry of the University of El Salvador with the following document number: CEI-FOUES/2022/003.

Informed Consent Statement: Informed consent was obtained from all subjects involved in this study. Participants provided verbal consent, and the recruitment of the patients ranged from 1 April 2022 until 3 March 2023.

Data Availability Statement: Data are contained within the article.

Conflicts of Interest: The authors declare no conflicts of interest.

References

1. John, H.; Franklin, G.-G.; Catherine, F. Biological factors in dental caries: Role of saliva and dental plaque in the dynamic process of demineralization and remineralization (part 1). *J. Clin. Pediatr. Dent.* **2003**, *28*, 47–52.
2. Chen, H.; Gu, L.; Liao, B.; Zhou, X.; Cheng, L.; Ren, B. Advances of Anti-Caries Nanomaterials. *Molecules* **2020**, *25*, 5047. [CrossRef]
3. Kim, M.-J.; Li, M.-J.; Kim, K.-M.; Yang, S.-Y.; Seo, Y.-J.; Chul, C.S.; Ren, B. Enamel Demineralization Resistance and Remineralization by Various Fluoride-Releasing Dental Restorative Materials. *Materials* **2021**, *14*, 4554. [CrossRef]
4. Chhatwani, S.; Hoppe, J.; Naumova, E.A.; Arnold, W.H.; Möhlhenrich, S.C.; Bizhang, M.; Danesh, G. Fluoride ion release characteristics of fluoride-containing varnishes-an in vitro study. *Appl. Sci.* **2021**, *11*, 1452. [CrossRef]
5. Parisay, I.; Boskabady, M.; Bagheri, H.; Babazadeh, S.; Hoseinzadeh, M.; Esmaeilzadeh, F. Investigating the efficacy of a varnish containing gallic acid on remineralization of enamel lesions: An in vitro study. *BMC Oral Health* **2024**, *24*, 175. [CrossRef] [PubMed]
6. Oliveira, B.H.; Rajendra, A.; Veitz-Keenan, A.; Niederman, R. The effect of silver diamine fluoride in preventing caries in the primary dentition: A systematic review and meta-analysis. *Caries Res.* **2019**, *53*, 24–32. [CrossRef]
7. Tejaswi, B.; Gopal Sree, V.; Sivapriya, E.; Archana, D.; PradeepKumar, A.R. Nanoparticles in caries prevention: A review. *J. Glob. Oral Health* **2021**, *4*, 56–66. [CrossRef]
8. Salas-López, E.K.; Pierdant-Pérez, M.; Hernández-Sierra, J.F.; Ruíz, F.; Mandeville, P.; Pozos-Guillén, A.J. Effect of silver nanoparticle-added pit and fissure sealant in the prevention of dental caries in children. *J. Clin. Pediatr. Dent.* **2017**, *41*, 48–52. [CrossRef]
9. Pichaiaukrit, W.; Thamrongananskul, N.; Siralertmukul, K.; Swasdison, S. Fluoride varnish containing chitosan demonstrated sustained fluoride release. *Dent. Mater. J.* **2019**, *38*, 1036–1042. [CrossRef]
10. Ghafar, H.; Khan, M.I.; Sarwar, H.S.; Yaqoob, S.; Hussain, S.Z.; Tariq, I.; Madni, A.U.; Shahnaz, G.; Sohail, M.F. Development and Characterization of Bioadhesive Film Embedded with Lignocaine and Calcium Fluoride Nanoparticles. *AAPS PharmSciTech* **2020**, *21*, 60. [CrossRef]
11. Memarpour, M.; Shafiei, F.; Rafiee, A.; Soltani, M.; Dashti, M.H. Effect of hydroxyapatite nanoparticles on enamel remineralization and estimation of fissure sealant bond strength to remineralized tooth surfaces: An in vitro study. *BMC Oral. Health* **2019**, *19*, 92. [CrossRef] [PubMed]

12. Abedi, M.; Ghasemi, Y.; Nemati, M.M. Nanotechnology in toothpaste: Fundamentals, trends, and safety. *Heliyon* **2024**, *10*, e24949. [CrossRef] [PubMed]
13. Surija, I.; Gunawan, H.A.; Amir, L.R. Effect of chitosan on the enamel demineralization process in vitro: An enamel solubility test. *J. Phys. Conf. Ser.* **2018**, *1073*, 052005. [CrossRef]
14. Sámano-Valencia, C.; Martinez-Castanon, G.A.; Martínez-Gutiérrez, F.; Ruiz, F.; Toro-Vázquez, J.F.; Morales-Rueda, J.A.; Espinosa-Cristóbal, L.F.; Alonso, N.V.Z.; Martínez, N.N. Characterization and biocompatibility of chitosan gels with silver and gold nanoparticles. *J. Nanomater.* **2014**, *2014*, 543419. [CrossRef]
15. Rajan, R.; Krishnan, R.; Bhaskaran, B.; Kumar, S.V. A Polarized Light Microscopic Study to Comparatively evaluate Four Remineralizing Agents on Enamel viz CPP-ACPF, ReminPro, SHY-NM and Colgate Strong Teeth. *Int. J. Clin. Pediatr. Dent.* **2015**, *8*, 42–47. [CrossRef] [PubMed]
16. Yamakami, S.A.; Faraoni, J.J.; Lia, N.S.N.D.; Regula, F.B.; Ohyama, H.; Palma-Dibb, R.G. Effect of an experimental chitosan/casein gel on demineralized enamel under a cariogenic challenge. *Dent. Med. Probl.* **2022**, *59*, 531–538. [CrossRef]
17. Seredin, P.; Goloshchapov, D.; Kashkarov, V.; Emelyanova, A.; Buylov, N.; Barkov, K.; Ippolitov, Y.; Khmelevskaia, T.; Mahdy, I.A.; Mahdy, M.A.; et al. Biomimetic Mineralization of Tooth Enamel Using Nanocrystalline Hydroxyapatite under Various Dental Surface Pretreatment Conditions. *Biomimetics* **2022**, *7*, 111. [CrossRef]
18. Shaalan, O.O. DIAGNOdent versus International Caries Detection and Assessment System in detection of incipient carious lesions: A diagnostic accuracy study. *J. Conserv. Dent.* **2023**, *26*, 199–206. [CrossRef] [PubMed]
19. El Moshy, S.; Abbass, M.M.S.; El-Motayam, A.M. Biomimetic remineralization of acid etched enamel using agarose hydrogel model. *F1000Res* **2018**, *7*, 1476. [CrossRef]
20. Sámano-Valencia, C.; Martinez-Castanon, G.A.; Martínez-Martínez, R.E.; Loyola-Rodríguez, J.P.; Reyes-Macías, J.F.; Ortega-Zarzosa, G.; Niño-Martínez, N. Bactericide efficiency of a combination of chitosan gel with silver nanoparticles. *Mater. Lett.* **2013**, *106*, 413–416. [CrossRef]
21. Noronha, V.T.; Paula, A.J.; Durán, G.; Galembeck, A.; Cogo-Müller, K.; Franz-Montan, M.; Durán, N. Silver nanoparticles in dentistry. *Dent. Mater.* **2017**, *33*, 1110–1126. [CrossRef]
22. Ruan, Q.; Zhang, Y.; Yang, X.; Nutt, S.; Moradian-Oldak, J. An amelogenin-chitosan matrix promotes assembly of an enamel-like layer with a dense interface. *Acta Biomater.* **2013**, *9*, 7289–7297. [CrossRef] [PubMed]
23. Ruan, Q.; Liberman, D.; Bapat, R.; Chandrababu, K.B.; Phark, J.H.; Moradian-Oldak, J. Efficacy of amelogenin-chitosan hydrogel in biomimetic repair of human enamel in pH-cycling systems. *J. Biomed. Eng. Inform.* **2015**, *2*, 119. [CrossRef] [PubMed]
24. Nimbeni, S.B.; Nimbeni, B.S.; Divakar, D.D. Role of chitosan in remineralization of enamel and dentin: A systematic review. *Int. J. Clin. Pediatr. Dent.* **2021**, *14*, 562–568. [CrossRef] [PubMed]
25. NouhzadehMalekshah, S.; Fekrazad, R.; Bargrizan, M.; Kalhori, K.A. Evaluation of laser fluorescence in combination with photosensitizers for detection of demineralized lesions. *Photodiag. Photodyn. Ther.* **2019**, *26*, 300–305. [CrossRef]
26. Kolumban, A.; Moldovan, M.; Ţig, I.A.; Chifor, I.; Cuc, S.; Bud, M.; Badea, M.E. An evaluation of the demineralizing effects of various acidic solutions. *Appl. Sci.* **2021**, *11*, 8270. [CrossRef]
27. Enax, J.; Fandrich, P.; Schulze zur Wiesche, E.; Epple, M. The Remineralization of Enamel from Saliva: A Chemical Perspective. *Dent. J.* **2024**, *12*, 339. [CrossRef] [PubMed]
28. Zhang, J.; Lynch, R.J.M.; Watson, T.F.; Banerjee, A. Remineralisation of enamel white spot lesions pre-treated with chitosan in the presence of salivary pellicle. *J. Dent.* **2018**, *72*, 21–28. [CrossRef]
29. Üstün, N.; Aktören, O. Analysis of efficacy of the self-assembling peptide-based remineralization agent on artificial enamel lesions. *Microsc. Res. Tech.* **2019**, *82*, 1065–1072. [CrossRef]
30. Zhang, X.; Li, Y.; Sun, X.; Kishen, A.; Deng, X.; Yang, X.; Wang, H.; Cong, C.; Wang, Y.; Wu, M. Biomimetic remineralization of demineralized enamel with nano-complexes of phosphorylated chitosan and amorphous calcium phosphate. *J. Mater. Sci. Mater. Med.* **2014**, *25*, 2619–2628. [CrossRef]
31. Bijle, M.N.; Abdalla, M.M.; Ashraf, U.; Siu, K.W.; Tsoi, J.; Yiu, C.K.Y. The acid-resistance potential of arginine-fluoride varnish treated enamel. *J. Mech. Behav. Biomed. Mater.* **2022**, *125*, 104763. [CrossRef]
32. Zhao, I.S.; Yin, I.X.; Mei, M.L.; Lo, E.C.M.; Tang, J.; Li, Q.; So, L.Y.; Chu, C.H. Remineralising dentine caries using sodium fluoride with silver nanoparticles: An in vitro study. *Int. J. Nanomed.* **2020**, *15*, 2829–2839. [CrossRef]
33. Dai, L.L.; Mei, M.L.; Chu, C.H.; Lo, E.C.M. Remineralizing effect of a new strontium-doped bioactive glass and fluoride on demineralized enamel and dentine. *J. Dent.* **2021**, *108*, 103633. [CrossRef]
34. Ren, J.; Rao, J.; Wang, H.; He, W.; Feng, J.; Wei, D.; Zhao, B.; Wang, X.; Bian, W. Synergistic remineralization of enamel white spot lesions using mesoporous bioactive glasses loaded with amorphous calcium phosphate. *Front. Bioeng. Biotechnol.* **2023**, *11*, 1109195. [CrossRef] [PubMed]
35. Bijle, M.N.; Abdalla, M.M.; Ashraf, U.; Ekambaram, M.; Yiu, C.K.Y. Enamel remineralization potential of arginine-fluoride varnish in a multi-species bacterial pH-cycling model. *J. Dent.* **2021**, *104*, 103528. [CrossRef] [PubMed]

36. Zhang, Y.Y.; Wong, H.M.; McGrath, C.P.J.; Li, Q.L. In vitro and in vivo evaluation of electrophoresis-aided casein phosphopeptide-amorphous calcium phosphate remineralisation system on pH-cycling and acid-etching demineralised enamel. *Sci. Rep.* **2018**, *8*, 8904. [CrossRef] [PubMed]
37. Wassel, M.O.; Sherief, D.I. Ion release and enamel remineralizing potential of miswak, propolis and chitosan nano-particles based dental varnishes. *Pediatr. Dent. J.* **2019**, *29*, 1–10. [CrossRef]
38. Vitiello, F.; Tosco, V.; Monterubbianesi, R.; Orilisi, G.; Gatto, M.L.; Sparabombe, S.; Memé, L.; Mengucci, P.; Putignano, A.; Orsini, G. Remineralization Efficacy of Four Remineralizing Agents on Artificial Enamel Lesions: SEM-EDS Investigation. *Materials* **2022**, *15*, 4398. [CrossRef] [PubMed]
39. Zhang, J.; Boyes, V.; Festy, F.; Lynch, R.J.M.; Watson, T.F.; Banerjee, A. In-vitro subsurface remineralisation of artificial enamel white spot lesions pre-treated with chitosan. *Dent. Mater.* **2018**, *34*, 1154–1167. [CrossRef]
40. Muşat, V.; Anghel, E.M.; Zaharia, A.; Atkinson, I.; Mocioiu, O.C.; Buşilă, M.; Alexandru, P. A chitosan–agarose polysaccharide-based hydrogel for biomimetic remineralization of dental enamel. *Biomolecules* **2021**, *11*, 1137. [CrossRef] [PubMed]
41. Chu, S.; Wang, J.; Gao, F. The application of chitosan nanostructures in stomatology. *Molecules* **2021**, *26*, 6315. [CrossRef] [PubMed]
42. Deokar, K.K.; Shashikiran, N.; Maurya, A.; Gaonkar, N.; Gugwad, S.; Hadakar, S.; Taur, S. Comparative Evaluation of Chitosan Nanoparticles, Silver Diamine Fluoride and Acidulated Phosphate Fluoride Gel on Microhardness of Artificial Carious Lesions Created on Extracted Teeth. *J. Clin. Diagn. Res.* **2020**, *14*, 20. [CrossRef]
43. Hanafy, R.A.; El-Fattah, A.A.; Mostafa, D.; Kandil, S. Biomimetic chitosan against bioinspired nanohydroxyapatite for repairing enamel surfaces. *Bioinspired Biomim. Nanobiomater.* **2020**, *9*, 85–94. [CrossRef]
44. Magalhães, T.C.; Teixeira, N.M.; França, R.S.; Denadai, Â.M.L.; Dos Santos, R.L.; Carlo, H.L.; Munchow, E.A.; de Carvalho, F.G. Synthesis of a Chitosan nanoparticle suspension and its protective effects against enamel demineralization after an in vitro Cariogenic challenge. *J. Appl. Oral Sci.* **2021**, *29*, e20210120. [CrossRef]
45. Xu, V.W.; Nizami, M.Z.I.; Yin, I.X.; Lung, C.Y.K.; Yu, O.Y.; Chu, C.H. Caries Management with Non-Metallic Nanomaterials: A Systematic Review. *Int. J. Nanomed.* **2022**, *17*, 5809–5824. [CrossRef] [PubMed]
46. Wahied, D.M.; Ezzeldin, N.; Abdelnabi, A.; Othman, M.S.; El Rahman, M.H.A. Evaluation of Surface Properties of Two Remineralizing Agents after Modification by Chitosan Nano Particles: An In vitro Study. *Contemp. Clin. Dent.* **2023**, *14*, 265–271. [CrossRef] [PubMed]
47. Zhang, O.L.; Niu, J.Y.; Yin, I.X.; Yu, O.Y.; Mei, M.L.; Chu, C.H. Bioactive Materials for Caries Management: A Literature Review. *Dent. J.* **2023**, *11*, 59. [CrossRef]
48. Tristán-López, J.-D.; Niño-Martínez, N.; Kolosovas-Machuca, E.-S.; Patiño-Marín, N.; De Alba-Montero, I.; Bach, H.; Martínez-Castañón, G.-A. Application of Silver Nanoparticles to Improve the Antibacterial Activity of Orthodontic Adhesives: An In Vitro Study. *Int. J. Mol. Sci.* **2023**, *24*, 1401. [CrossRef] [PubMed]

coatings

MDPI

Article

Evaluation of Bone–Implant Interface: Effects of Angiotensin II Receptor Blockade in Hypertensive Rats

Gabriel Mulinari-Santos [1,*], Jaqueline Silva dos Santos [1], Fábio Roberto de Souza Batista [2], Letícia Pitol-Palin [2], Ana Cláudia Ervolino da Silva [2], Paulo Roberto Botacin [1], Cristina Antoniali [1] and Roberta Okamoto [1,*]

[1] Department of Basic Sciences, Araçatuba School of Dentistry, São Paulo State University, Araçatuba 16018-805, SP, Brazil; jaqueline.santos@unesp.br (J.S.d.S.); paulo.botacin@unesp.br (P.R.B.); cristina.antoniali@unesp.br (C.A.)

[2] Department of Diagnosis and Surgery, Araçatuba School of Dentistry, São Paulo State University, Araçatuba 16015-050, SP, Brazil; fabiorsbatista@gmail.com (F.R.d.S.B.); leticia.p.palin@unesp.br (L.P.-P.); ana.ervolino@unesp.br (A.C.E.d.S.)

* Correspondence: gabriel.mulinari@unesp.br (G.M.-S.); roberta.okamoto@unesp.br (R.O.)

Abstract: Hypertension is a global health concern not only correlated with cardiovascular complications, but also with impaired bone metabolism, potentially affecting healing at the bone–implant interface. Losartan, an angiotensin II receptor blocker (ARB) commonly prescribed for hypertension, has shown beneficial effects on bone healing in spontaneously hypertensive rats (SHRs). However, the influence of hypertension and ARBs like losartan on the bone cellular response at the bone–implant interface remains underexplored. Methods: A total of 32 rats were included in this study: 16 SHRs, with 8 receiving losartan (30 mg/kg daily) and 8 receiving no treatment, and 16 normotensive Wistar rats, with 8 receiving losartan and 8 receiving no treatment. After one week of treatment, titanium implants were placed into the tibia of all the animals. The bone–implant interface was assessed 60 days post-implantation using micro-computed tomography (μCT) and an immunohistochemical analysis. Results: (i) The ARB treatment significantly increased the bone volume and bone–implant contact in the SHRs receiving losartan compared to the untreated SHRs. (ii) Consistent with the μCT findings, the immunohistochemistry further confirmed regular bone turnover and increased osteocalcin (OC) mineralization in the treated SHRs. In contrast, no alterations in the bone microarchitecture were noted in the Wistar rats treated with losartan. Conclusions: The results suggest that losartan, an ARB drug, improves bone volume and bone turnover at the bone–implant interface in SHRs.

Keywords: losartan; osseointegration; spontaneously hypertensive rats; implants

1. Introduction

Hypertension is a global health concern, often regarded as a pandemic disease due to its high prevalence worldwide [1]. Alarmingly, hypertension can impact as many as 60% of individuals over the age of 60 [2]. It is the highest cause of medical consults and death before the age of life expectancy [3]. However, a large portion of hypertensive subjects do not adhere to their antihypertensive treatment [4]. If left untreated, hypertension can lead to serious complications, including cardiovascular disorders [5] and cerebrovascular disease [6]. In addition to these, hypertension also disrupts bone metabolism, causing impaired calcium levels [7] and bone density [8,9], osteoporosis, and an increased risk of fractures [10–12]. Notably, these hypertensive effects not only act on long bones but also on the alveolar bone [13]. Preclinical studies have evidenced a delay in alveolar bone

https://doi.org/10.3390/coatings15010073

healing and a reduction in alveolar bone quality in untreated hypertension [13]. Hence, hypertension can influence the bone–implant interface during osseointegration [14].

The renin–angiotensin system is the essential mechanism of blood pressure control [15]. The final step in the hypertensive cascade involves the production of angiotensin II, a peptide hormone that induces vasoconstriction, leading to elevated blood pressure [15]. As a result, many antihypertensive treatments aim to block this cascade, specifically the action of angiotensin II [16]. Losartan, a commonly prescribed antihypertensive, is categorized as an angiotensin II receptor blocker (ARB) [16]. However, recently studies have shown that AT1 receptors are not confined only to the endothelium [17]; they are also present in bone cells, where angiotensin II can stimulate osteoclastogenesis under hypertensive conditions [18]. When AT1 receptors are blocked by losartan, studies have shown reduced osteoclast proliferation, leading to improved bone graft healing [19], enhanced fracture consolidation [20], and increased bone mass [21]. However, the effect of ARB drugs on the bone cellular response at the bone–implant interface remains uncertain.

Recent research supports the proposal that antihypertensive treatment may benefit alveolar bone health in humans [22,23]. Two clinical studies of patients using antihypertensive drugs showed an increase in implant survival in these patients when compared with normotensives [22,23]. Another clinical study evidenced the similar histologic and morphometric characteristics of the alveolar bone of normotensive and hypertensive patients treated with ARB drugs [24]. A recent cohort study observed a reduced early implant failure in patients treated with antihypertensives. Also, our prior studies revealed impaired alveolar bone repair [25] and bone graft consolidation [19] in spontaneous hypertensive rats (SHRs). Interestingly, losartan was shown to improve the healing process in SHRs [19]. Therefore, the aim of this study is to test if the ARB drug losartan influences the bone–implant interface under hypertensive conditions. To test this hypothesis, we use genetically hypertensive SHRs, a well-established model for hypertension [26].

2. Materials and Methods

2.1. Research Design and Ethical Approval

This study was authorized by the Ethics Committee on the Use of Animals of Araçatuba Dental School (CEUA-2016-404). This study was conducted following the ARRIVE guidelines 2.0 [27]. A total of 32 male rats were utilized, 16 adult male Wistar rats (six-month-old Rattus norvegicus; albinus; body weight, 350 to 450 g) and 16 adult male SHRs (six-month-old; body weight, 275 to 350 g). The animals were housed in cages under conditions of stable temperature (22 °C $\pm$ 2 °C), controlled light cycle (12 h light, 12 h dark), balanced feed (Ração Mogiana Alimentos SA, Campinas, Brazil), and controlled amounts of water. The animals were divided into four groups: Wistar, Wistar Losartan, SHR, and SHR Losartan. In total, 16 normotensive Wistar rats were included, with 8 receiving losartan and 8 receiving no treatment. Similarly, 16 SHRs were included, with 8 receiving losartan and 8 receiving no treatment. Randomization was performed by a computer-generated list. All assessments were executed under calibration and blinding check.

2.2. ARB Treatment

Losartan (Biosintetica, São Paulo, Brazil) was utilized daily at 30 mg per kg body weight in consumption water, from seven days prior to implant installation until euthanasia [19]. Systolic blood pressure was checked before placing the implant, confirming it was controlled below 150 mmHG. This check was performed to replicate a controlled pressure, since uncontrolled hypertension before surgery is a contraindication for implant placement. For this measurement, we employed the tail-cuff indirect plethysmography technique using a Physiograph® MK-III-S (Narco Bio-Sistemas, Houston, TX, USA), following the procedure outlined in a previous study [19,25].

2.3. Implant Installation

Animals were anesthetized using a combination of 50 mg/kg ketamine (Coopers Brazil Ltd.a, Cotia, SP, Brazil) and 5 mg/kg xylazine (Coopers Brazil Ltd.a, Cotia, SP, Brazil) administered intramuscularly, along with local anesthesia using 0.3 mL/kg of 2% mepivacaine with adrenaline (1:100,000; Septodont, Saint-Maur-des Fossés, France). The implants were placed in the right tibias. For that, bone exposure was achieved through an incision at the proximal metaphysis, 10 mm below the knee joint. A 1.4 mm diameter spiral bur mounted on an electric motor (BLM 600®; Driller, São Paulo, SP, Brazil) created a hole at 1000 rpm, with irrigation provided by 0.9% sodium chloride solution (Fisiológico®; Laboratórios Biosintética Ltd.a, Ribeirão Preto, SP, Brazil). Titanium implants (Grade 4, 1.5 mm diameter, 3.5 mm length, acid-etched surface; Emfills, Itu, São Paulo, Brazil) were implanted bicortically in the tibias. Wounds were closed using resorbable sutures (Poliglactin 910, Vicryl, Ethicon, Johnson & Johnson, São José dos Campos, Brazil). Post-surgery, animals were administered a single intramuscular injection of pentabiotic (0.1 mL/kg; Fort Dodge Saúde Animal Ltd.a, Campinas, SP, Brazil) and sodium dipyrone (1 mg/kg; Ariston, Indústrias Químicas e Farmacêuticas Ltd.a, São Paulo, Brazil). Subsequently, the animals were euthanized 60 days post-implantation via a fatal dose of thiopental (150 mg/kg body weight; Cristália Ltd., Itapira, SP, Brazil). The implant placement followed established method according to a prior study [28].

2.4. Micro-CT Analysis

The tibias from four rats per group were preserved in 10% buffered formalin for 48 h. Afterward, they were rinsed in running water for 24 h and then stored in 70% ethanol. The tibias were scanned in the longitudinal plane using a SkyScan 1172 micro-CT scanner (Bruker, Aartselaar, Belgium) with settings of 70 kV, 114 mA, and an integration time of 1 × 380 ms. A standard scanning protocol was applied (tube current 165 μA, pixel size 9.92 μm, aluminum–copper filter for beam hardening with a thickness of 0.5 mm, frame averaging of 4, and a rotation step of 0.6 degrees). The region of interest (ROI) was defined as the rectangular area positioned on the first two medullary implant threads, 0.5 mm in height and 0.8 mm in width, totaling 50 slices in both the proximal and distal directions. This resulted in an analysis based on 100 slices, as outlined in previous research [29]. The images were converted to grayscale values between 70 and 100, representing trabecular bone while excluding titanium. Bone volume (BV, mm^3), intersection surface (IS, mm^2), and total porosity (Po(tot), %) were quantified using CTAnalyzer Software (SkyScan CTAn, v1.12.4.0; Leuven, Belgium), with standard settings including a smoothing factor of 1, ring artifact correction set to 8, and a cone beam hardening correction of 24%. The analysis followed the protocols set out in a previous guideline [30].

2.5. Immunohistochemical Analysis

The tibias of the other four rats were further processed for immunohistochemical evaluation. They were first reduced and fixed in 10% buffered formalin (Reagentes Analíticos®, Dinâmica Odonto-Hospitalar Ltd.a, Catanduva, SP, Brazil) for 48 h, followed by immersion in running water for 24 h. The bones were then decalcified in 10% EDTA for five weeks. Once decalcification was complete, the implants were carefully removed using a manual wrench by applying removal torque. The specimens then underwent routine laboratory processing, which included dehydration through a graded series of alcohols and clearing in xylene. The processed tibias were embedded in paraffin separately, and 5 μm semi-serial sections were cut using a microtome.

Endogenous peroxidase activity was inhibited using hydrogen peroxide, and antigen retrieval was performed using phosphate-buffered citrate (pH 6.0). The expression of osteoprotegerin (OPG, catalog number SC 390518), receptor activator of nuclear factor kappa-B ligand (RANKL, catalog number SC 7627), osteocalcin (OC, catalog number SC 18319), and vascular endothelial growth factor (VEGF, catalog number SC 7269) were examined using primary goat antibodies from Santa Cruz Biotechnology (Santa Cruz Biotechnology, Santa Cruz, CA, USA). These markers were chosen to evaluate bone remodeling (OPG and RANKL), mineralization (OC), and vascularization (VEGF). Immunohistochemical staining was carried out using the immunoperoxidase detection method, employing biotinylated anti-goat secondary antibodies (Pierce Biotechnology, Waltham, MA, USA), streptavidin–biotin amplification (Dako, Carpinteria, CA, USA), and diaminobenzidine (Dako, Carpinteria, CA, USA).

For each antibody, protein expression was evaluated semi-quantitatively by assigning scores based on the proportion of positive cells within the first two medullary implant threads. The immunohistochemical reactions were scored qualitatively as follows: "light", indicating 25% of the area showed positive staining for the antigen of interest; "moderate", with 50% of the area positively labeled; and "intense", with 75% of the area showing positive staining. A single evaluator (R.O.) assigned the scores, and the reliability of the scoring was confirmed using the Kappa test ($\kappa > 0.9$) after three consecutive reviews. The most frequent scores across all specimens were considered representative of the overall expression pattern. Images were captured using a photomicroscope (Leica R DMLB, Heerbrugg, Switzerland) connected to a computer, as previously described [19,25].

2.6. Statistical Analysis

All quantitative micro-CT data (BV; IS; Po(tot)) were submitted for tests of normality and homogeneity of variances to confirm the data distribution (Shapiro–Wilk test). Later, two-way ANOVA followed by Tukey's post hoc test was utilized to analyze micro-CT measurements. A significance level of 95% was used ($p < 0.05$). The statistical analyses were carried out using GraphPad Prism 7 software (GraphPad Software; La Jolla, CA, USA).

3. Results

3.1. Micro-CT Results

Losartan significantly increased the mean BV in the hypertensive SHRs group compared to the untreated SHRs (0.083 ± 0.006 mm^3 vs. 0.070 ± 0.006 mm^3; $p = 0.02$; Figure 1A), but had no effect on the normotensive Wistar control group ($p = 0.98$; Figure 1A). Also, losartan increased the area of contact between the bone–implant surface as shown by the IS, comparing the SHRs Losartan to SHRs (0.93 ± 0.01 mm^2 vs. 0.83 ± 0.02 mm^2; $p = 0.03$; Figure 1B). This anabolic effect of losartan on the bone volume was principally due to a reduction in the porosity of the bone trabeculae in the losartan-treated SHRs, compared to the untreated SHRs (54% $\pm$ 2.1 vs. 47 $\pm$ 1%; $p = 0.01$; Figure 1C). Taken together, these μCT structural analyses suggest that losartan promotes anabolic changes in peri-implant bone in a hypertensive rat model (Figure 2).

3.2. Immunohistochemical Results

The positive effect of losartan was observed in the hypertensive animals, as shown in Table 1 and Figure 3. The RANKL expression was intensely positive in the SHRs, but the losartan treatment reduced the staining to a light intensity in the SHRs Losartan group. The OPG staining was light in the SHRs Losartan group, showing a balance with the light RANKL staining. In contrast, the untreated SHRs exhibited light OPG staining and intense RANKL staining, an imbalance that could be correlated with the lower bone volume at the bone–implant interface. In the Wistar Losartan group, both OPG and RANKL showed intense positive staining, while in the Wistar group, the staining was light for OPG and moderate for RANKL. The OC expression showed light staining in the SHRs group, but moderate staining was observed in all other groups, including the SHRs Losartan group. The VEGF expression was particularly high in the Wistar Losartan group, exhibiting moderate staining, while it showed light staining in the Wistar group and moderate staining in both SHRs groups.

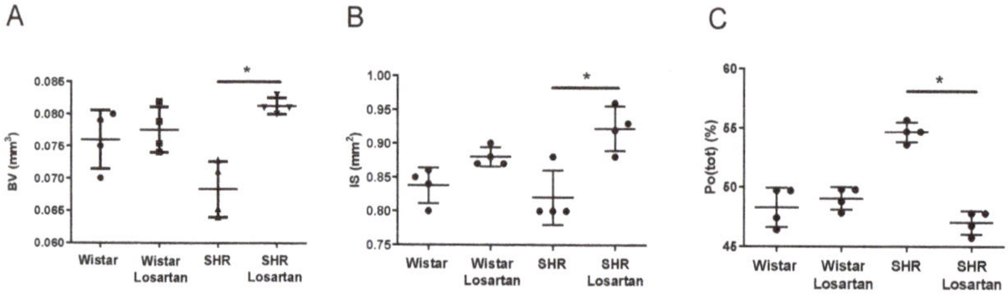

Figure 1. Effect of losartan on bone volume (BV), intersection surface (IS), and trabecular porosity in hypertensive SHRs. (**A**) Scatter plot showing bone volume (BV/TV), (**B**) IS, and (**C**) trabecular bone porosity (Po(tot)). * indicates statistical difference ($p < 0.05$).

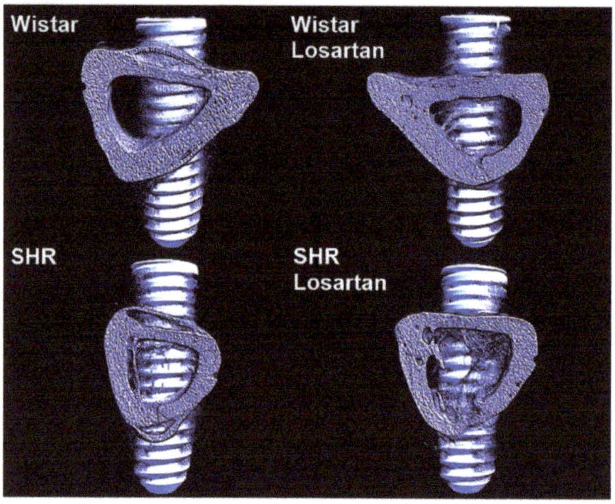

Figure 2. Representative microtomography are provided for all four groups: Wistar, Wistar Losartan, SHRs, and SHRs Losartan. The images demonstrate that losartan leads to an increase in the medullary bone volume in contact with the implant in the SHRs Losartan group. The micro-CT was completed using CTvox software (SkyScan, Version 2.7).

Table 1. Semi-quantitative immunohistochemical scores for OPG, RANKL, OC, and VEGF in peri-implant bone from Wistar, Wistar Losartan, SHRs, SHRs Losartan. Scores are based on the most frequent scores: light, moderate, and intense staining.

	Wistar	Wistar Losartan	SHR	SHR Losartan
OPG	Light	Intense	Light	Light
RANKL	Moderate	Intense	Intense	Light
OC	Moderate	Moderate	Light	Moderate
VEGF	Light	Moderate	Moderate	Moderate

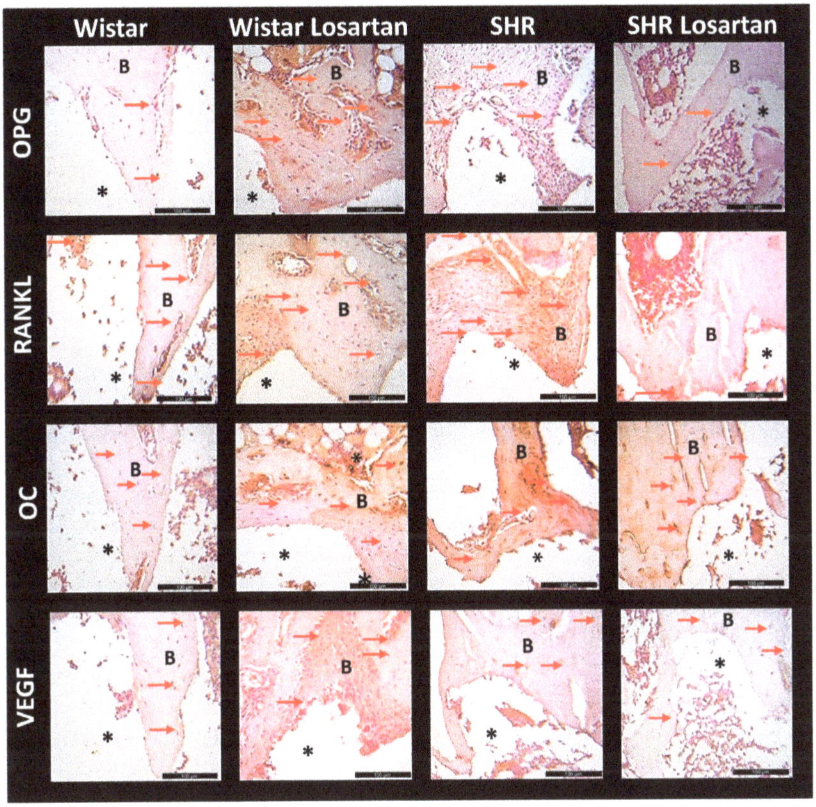

Figure 3. Immunohistochemical staining of bone remodeling markers at the medullary bone of the two first implant threads at the bone–implant interface. Positive staining for OPG, RANKL, OC, and VEGF in peri-implant bone of normotensive (Wistar) and hypertensive (SHRs) rats, with or without losartan treatment. Red arrows indicate positive cell staining. "B" indicates bone, and "*" marks the implant interface where the implant was removed. Images were captured using a photomicroscope (Leica R DMLB, Heerbrugg, Switzerland) Objective 20×.

4. Discussion

Hypertension is a leading cause of medical consultations worldwide, and when left untreated, it leads to severe cardiovascular problems [1] as well as reduced bone metabolism [17]. This disrupted bone metabolism increases the risk of complications in procedures involving bone support [31,32]. Osseointegration depends on an adequate bone support for the primary stylization and a satisfactory bone remodeling in the bone–implant interface [32]. In untreated hypertension, elevated levels of renin and angiotensin II can further compromise bone health [33]. Losartan, an ARB drug, has been shown to reduce osteoclast activity induced by angiotensin II, thereby mitigating bone loss [21]. In consensus with previous studies [21,34–36], we hypothesized that losartan may positively influence the expression of the bone proteins involved in bone remodeling, maturation, mineralization, and vascularization at the bone–implant interface. Therefore, our study demonstrates as the main result that losartan improves the bone volume in contact with the implant at the bone–implant interface in treated SHRs.

The immunohistochemical analyses support this key finding, showing a balanced low OPG and RANKL with moderate OC staining in the SHRs Losartan group. The SHRs group had intense RANKL staining, while the SHRs Losartan group showed light staining. The OC staining was light in the SHRs and moderate in the SHRs Losartan. In addition, OC is synthesized by mature osteoblasts, and recognized as a marker of mineralization [37]. The increased OC can also imply increased bone density [37,38], confirming that the losartan treatment appears to enhance bone formation in SHRs by reducing porosity and increasing bone contact with the implant, as exhibited by the micro-CT data. These results suggest that ARB drugs may modulate bone formation and exhibit an anti-catabolic effect, as previously reported [21,39]. Prior studies have also demonstrated the effect of losartan [18] and the ARB drug telmisartan [40] on reducing periodontal bone loss by lowering osteoclast activity. Similarly, losartan-treated rats have showed reduced orthodontic movement due to lower osteoclastogenesis [41]. No changes in the microarchitecture were observed in the Wistar Losartan group, but bone turnover increased, as shown by the intense OPG and RANKL staining. The VEGF staining shifted to moderate, suggesting accelerated turnover due to elevated VEGF [42].

Clinical evidence supports the positive effects of ARB drugs on compromised bone health. Losartan, for instance, reduces the fracture risk in hemodialysis patients by preventing decreased bone vascularization [11]. Clinical studies have also shown that ARBs improve implant survival [23], even after sinus lift surgery [22]. Our micro-CT data further indicate greater bone contact around the implants in the SHRs treated with losartan, which aligns with improved implant survival in hypertensive patients [14,43]. These studies suggest that antihypertensive agents may support optimal bone formation, stable bone turnover, and enhanced implant stability [14,43]. Furthermore, a recent study found similar alveolar bone morphology in hypertensive subjects treated with ARBs and normotensives [24]. This same report indicated a moderate-to-intense OC staining for the normotensive and hypertensive-treated groups, thus indicating the same level of maturated alveolar bone in both [24]. Dentists cannot prescribe drugs without a dental indication; however, this result suggests that ARBs may enhance bone remodeling at the bone–implant interface, potentially benefiting dental implant success in hypertensive-treated patients. Additionally, systemic monitoring would be necessary to track potential adverse effects or interactions with other treatments.

There are several limitations of this study. For instance, we did not investigate other potential pathways involved in bone healing, such as the Wnt signaling pathway. Additionally, factors like calcium and hormones levels, which can directly influence bone metabolism, were not investigated in this study. Also, we relied on only two analyses, and the inclusion of molecular and immunofluorescence analyses could have further supported our findings. Our findings from the Wistar Losartan group suggest that the beneficial effects observed in the hypertensive rats were not fully replicated in the normotensive rats. Specifically, as seen in the Wistar Losartan group, losartan increased the staining of OPG and RANKL. This accelerated bone turnover may have contributed to the lack of bone improvement observed in the normotensive rats [44]. However, this indicates that the enhancement is also closely related to blood pressure control in hypertension. This premise is supported by the lack of significant effects in the Wistar Losartan group, reinforcing the idea that blood pressure plays a crucial role in peri-implant bone. It is noteworthy that hypertension causes a negative influence on vascular growth [45,46], although here losartan only improved the VEGF staining in the normotensive but not in the SHRs. Future research should explore whether other ARB drugs have similar effects. Additionally, the effects of ARB treatments should be investigated at multiple time points, rather than just one, in animal and clinical studies.

Finally, the results of this study not only deepen our understanding of the impact of hypertension and its treatment on bone healing, but also highlight the potential benefits of incorporating antihypertensive control into dental treatment plans. This suggests that losartan, through its ability as an ARB dug, plays a beneficial role in mitigating the negative impact of hypertension on bone metabolism, particularly in the context of bone remodeling and the bone–implant interface. These outcomes contribute to the evolving knowledge base regarding the intersection of systemic health, pharmacotherapy, and dental care, emphasizing the need for a multidisciplinary approach to managing patients with systemic conditions. As further research elucidates the long-term effects of antihypertensive treatments on osseointegration and bone regeneration, it may become increasingly clear that optimizing blood pressure control is crucial for enhancing the success of implant-based dental treatments and potentially promote further clinical guidelines for optimizing implantology in hypertensive populations. Lastly, although normotensive rats were used as the baseline, no significant effects on bone volume were observed in this group when treated with losartan. Additionally, the use of losartan in normotensive individuals could potentially lead to side effects, such as hypotension and renal complications [47,48], highlighting the need for caution in its application to normotensive populations.

5. Conclusions

In conclusion, our findings indicate that hypertensive rats treated with losartan, an ARB drug, exhibited improved bone volume and bone turnover at the bone–implant interface.

Author Contributions: Conceptualization, G.M.-S. and R.O.; methodology, P.R.B. and F.R.d.S.B.; software, L.P.-P. and A.C.E.d.S.; validation, C.A. and R.O.; formal analysis, G.M.-S.; investigation, G.M.-S and J.S.d.S.; resources, R.O.; data curation, G.M.-S.; writing—original draft preparation, G.M.-S. and R.O.; writing—review and editing, G.M.-S. and R.O.; visualization, P.R.B.; supervision, R.O.; project administration, R.O.; funding acquisition, R.O. All authors have read and agreed to the published version of the manuscript.

Funding: This research was funded by São Paulo Research Foundation (FAPESP) under grant numbers #2017/23082-3 and #2017/16912-0. This study was financially supported by the Coordination for the Improvement of Higher Education Personnel (CAPES), "National Council for Scientific and Technological Development—CNPQ", through a fellowship for Roberta Okamoto (process: 308408/2023-3).

Institutional Review Board Statement: The animal study protocol was approved by the Ethics Committee on the Use of Animals of Araçatuba Dental School (CEUA-2016-404).

Informed Consent Statement: Not applicable.

Data Availability Statement: The original contributions presented in this study are included in the article.

Acknowledgments: This study was also supported by CNPq—National Council for Scientific and Technological Development—INCT Saúde Oral e Odontologia—Grant n. 406840/2022-9. We would like to thank the Multiuser Laboratory of FOA-UNESP and FINEP (FINEP/CT-INFRA—FINEP Agreement: 01.12.0530.00—PROINFRA 01/2011). The authors are grateful for the dental implants provided by the "Medens Indústria e Comércio".

Conflicts of Interest: The authors declare no conflicts of interest.

References

1. Tibazarwa, K.B.; Damasceno, A.A. Hypertension in Developing Countries. *Can. J. Cardiol.* **2014**, *30*, 527–533. [CrossRef]
2. Oliveros, E.; Patel, H.; Kyung, S.; Fugar, S.; Goldberg, A.; Madan, N.; Williams, K.A. Hypertension in Older Adults: Assessment, Management, and Challenges. *Clin. Cardiol.* **2019**, *43*, 99–107. [CrossRef] [PubMed]

3. Ong, K.L.; Cheung, B.M.; Man, Y.B.; Lau, C.P.; Lam, K.S. Prevalence, Awareness, Treatment, and Control of Hypertension among United States Adults 1999–2004. *Hypertension* **2007**, *49*, 69–75. [CrossRef]
4. Hamrahian, S.M.; Maarouf, O.H.; Fülöp, T. A Critical Review of Medication Adherence in Hypertension: Barriers and Facilitators Clinicians Should Consider. *Patient Prefer. Adherence* **2022**, *16*, 2749–2757. [CrossRef] [PubMed]
5. Kearney, P.M.; Whelton, M.; Reynolds, K.; Muntner, P.; Whelton, P.K.; He, J. Global Burden of Hypertension: Analysis of Worldwide Data. *Lancet* **2005**, *365*, 217–223. [CrossRef] [PubMed]
6. Musini, V.M.; Gueyffier, F.; Puil, L.; Salzwedel, D.M.; Wright, J.M. Pharmacotherapy for Hypertension in Adults Aged 18 to 59 Years. *Cochrane Database Syst. Rev.* **2017**, *8*, CD008276. [CrossRef]
7. Wright, G.L.; Rankin, G.O. Concentrations of Ionic and Total Calcium in Plasma of Four Models of Hypertension. *Am. J. Physiol.* **1982**, *243*, H365–H370. [CrossRef]
8. Manrique, N.; Pereira, C.C.; Garcia, L.M.; Micaroni, S.; Carvalho, A.A.; Perri, S.H.; Okamoto, R.; Sumida, D.H.; Antoniali, C. Alveolar Bone Healing Process in Spontaneously Hypertensive Rats (SHR). A Radiographic Densitometry Study. *J. Appl. Oral Sci. Rev. FOB* **2012**, *20*, 222–227. [CrossRef]
9. Javed, F.; Khan, S.A.; Ayers, E.W.; Aziz, E.F.; Akram, M.S.; Nadkarni, G.N.; Sabharwal, M.S.; Ahmad, Z.; Benjo, A.M.; Herzog, E. Association of Hypertension and Bone Mineral Density in an Elderly African American Female Population. *J. Natl. Med. Assoc.* **2012**, *104*, 172–178. [CrossRef]
10. Cappuccio, F.P.; Kalaitzidis, R.; Duneclift, S.; Eastwood, J.B. Unravelling the Links between Calcium Excretion, Salt Intake, Hypertension, Kidney Stones and Bone Metabolism. *J. Nephrol.* **2000**, *13*, 169–177.
11. Yamamoto, S.; Kido, R.; Onishi, Y.; Fukuma, S.; Akizawa, T.; Fukagawa, M.; Kazama, J.J.; Narita, I.; Fukuhara, S. Use of Renin-Angiotensin System Inhibitors Is Associated with Reduction of Fracture Risk in Hemodialysis Patients. *PLoS ONE* **2015**, *10*, e0122691. [CrossRef]
12. Vestergaard, P.; Rejnmark, L.; Mosekilde, L. Hypertension Is a Risk Factor for Fractures. *Calcif. Tissue Int.* **2009**, *84*, 103–111. [CrossRef] [PubMed]
13. Bastos, M.F.; Brilhante, F.V.; Goncalves, T.E.; Pires, A.G.; Napimoga, M.H.; Marques, M.R.; Duarte, P.M. Hypertension May Affect Tooth-Supporting Alveolar Bone Quality: A Study in Rats. *J. Periodontol.* **2010**, *81*, 1075–1083. [CrossRef] [PubMed]
14. Jones, D.; Khan, R.S.; Thompson, J.D.; Ucer, C.; Wright, S. The Effect of Antihypertensive Agents on Dental Implant Stability, Osseointegration and Survival Outcomes: A Systematic Review. *Surgeries* **2024**, *5*, 297–341. [CrossRef]
15. See, S. Angiotensin II Receptor Blockers for the Treatment of Hypertension. *Expert Opin. Pharmacother.* **2001**, *2*, 1795–1804. [CrossRef] [PubMed]
16. Al-Majed, A.R.; Assiri, E.; Khalil, N.Y.; Abdel-Aziz, H.A. Losartan: Comprehensive Profile. *Profiles Drug Subst. Excip. Relat. Methodol.* **2015**, *40*, 159–194. [CrossRef]
17. Zhou, Y.; Guan, X.; Chen, X.; Yu, M.; Wang, C.; Shi, J.; Liu, T.; Wang, H. Angiotensin II/Angiotensin II Receptor Blockade Affects Osteoporosis via the AT1/AT2-Mediated cAMP-Dependent PKA Pathway. *Cells Tissues Organs* **2017**, *204*, 25–37. [CrossRef] [PubMed]
18. Santos, C.F.; Morandini, A.C.; Dionisio, T.J.; Faria, F.A.; Lima, M.C.; Figueiredo, C.M.; Colombini-Ishikiriama, B.L.; Sipert, C.R.; Maciel, R.P.; Akashi, A.P.; et al. Functional Local Renin-Angiotensin System in Human and Rat Periodontal Tissue. *PLoS ONE* **2015**, *10*, e0134601. [CrossRef]
19. Gealh, W.C.; Pereira, C.C.; Luvizuto, E.R.; Garcia-Junior, I.R.; Antoniali, C.; Okamoto, R. Healing Process of Autogenous Bone Graft in Spontaneously Hypertensive Rats Treated with Losartan: An Immunohistochemical and Histomorphometric Study. *J. Oral Maxillofac. Surg.* **2014**, *72*, 2569–2581. [CrossRef] [PubMed]
20. Rajkumar, D.S.; Faitelson, A.V.; Gudyrev, O.S.; Dubrovin, G.M.; Pokrovski, M.V.; Ivanov, A.V. Comparative Evaluation of Enalapril and Losartan in Pharmacological Correction of Experimental Osteoporosis and Fractures of Its Background. *J. Osteoporos.* **2013**, *2013*, 325693. [CrossRef]
21. Chen, S.; Grover, M.; Sibai, T.; Black, J.; Rianon, N.; Rajagopal, A.; Munivez, E.; Bertin, T.; Dawson, B.; Chen, Y.; et al. Losartan Increases Bone Mass and Accelerates Chondrocyte Hypertrophy in Developing Skeleton. *Mol. Genet. Metab.* **2015**, *115*, 53–60. [CrossRef] [PubMed]
22. Garcia-Denche, J.T.; Wu, X.; Martinez, P.P.; Eimar, H.; Ikbal, D.J.; Hernandez, G.; Lopez-Cabarcos, E.; Fernandez-Tresguerres, I.; Tamimi, F. Membranes over the Lateral Window in Sinus Augmentation Procedures: A Two-Arm and Split-Mouth Randomized Clinical Trials. *J. Clin. Periodontol.* **2013**, *40*, 1043–1051. [CrossRef]
23. Wu, X.; Al-Abedalla, K.; Eimar, H.; Arekunnath Madathil, S.; Abi-Nader, S.; Daniel, N.G.; Nicolau, B.; Tamimi, F. Antihypertensive Medications and the Survival Rate of Osseointegrated Dental Implants: A Cohort Study. *Clin. Implant Dent. Relat. Res.* **2016**, *18*, 1171–1182. [CrossRef] [PubMed]

24. Fabris, A.L.d.S.; Mulinari-Santos, G.; Hassumi, J.S.; Freire, A.R.; Faverani, L.P.; Gruber, R.; Okamoto, R. Morphometric and Histologic Characterization of Alveolar Bone from Hypertensive Patients. *Clin. Implant Dent. Relat. Res.* **2017**, *19*, 1106–1113. [CrossRef]
25. Manrique, N.; Pereira, C.C.; Luvizuto, E.R.; Sanchez Mdel, P.; Okamoto, T.; Okamoto, R.; Sumida, D.H.; Antoniali, C. Hypertension Modifies OPG, RANK, and RANKL Expression during the Dental Socket Bone Healing Process in Spontaneously Hypertensive Rats. *Clin. Oral Investig.* **2015**, *19*, 1319–1327. [CrossRef] [PubMed]
26. Okamoto, K.; Aoki, K. Development of a Strain of Spontaneously Hypertensive Rats. *Jpn. Circ. J.* **1963**, *27*, 282–293. [CrossRef] [PubMed]
27. Percie du Sert, N.; Hurst, V.; Ahluwalia, A.; Alam, S.; Avey, M.T.; Baker, M.; Browne, W.J.; Clark, A.; Cuthill, I.C.; Dirnagl, U.; et al. The ARRIVE Guidelines 2.0: Updated Guidelines for Reporting Animal Research. *PLoS Biol.* **2020**, *18*, e3000410. [CrossRef]
28. Mulinari-Santos, G.; dos Santos, J.S.; Kitagawa, I.L.; de Souza Batista, F.R.; Botacin, P.R.; Antoniali, C.; Lisboa-Filho, P.N.; Okamoto, R. Estrogen Deficiency Impairs Osseointegration in Hypertensive Rats Even Treated with Alendronate Coated on the Implant Surface. *J. Funct. Biomater.* **2023**, *14*, 471. [CrossRef] [PubMed]
29. Faverani, L.P.; Polo, T.O.; Ramalho-Ferreira, G.; Momesso, G.A.; Hassumi, J.S.; Rossi, A.C.; Freire, A.R.; Prado, F.B.; Luvizuto, E.R.; Gruber, R.; et al. Raloxifene but Not Alendronate Can Compensate the Impaired Osseointegration in Osteoporotic Rats. *Clin. Oral Investig.* **2017**, *22*, 255–265. [CrossRef]
30. Dempster, D.W.; Compston, J.E.; Drezner, M.K.; Glorieux, F.H.; Kanis, J.A.; Malluche, H.; Meunier, P.J.; Ott, S.M.; Recker, R.R.; Parfitt, A.M. Standardized Nomenclature, Symbols, and Units for Bone Histomorphometry: A 2012 Update of the Report of the ASBMR Histomorphometry Nomenclature Committee. *J. Bone Miner. Res.* **2013**, *28*, 2–17. [CrossRef]
31. Isidor, F. Influence of Forces on Peri-Implant Bone. *Clin. Oral Implants Res.* **2006**, *17* (Suppl. S2), 8–18. [CrossRef]
32. Schenk, R.K.; Buser, D. Osseointegration: A Reality. *Periodontology 2000* **1998**, *17*, 22–35. [CrossRef] [PubMed]
33. Zhang, R.; Yin, H.; Yang, M.; Lei, X.; Zhen, D.; Zhang, Z. Advanced Progress of the Relationship Between Antihypertensive Drugs and Bone Metabolism. *Hypertension* **2023**, *80*, 2255–2264. [CrossRef] [PubMed]
34. Thatcher, S.E.; Zhang, X.; Howatt, D.A.; Lu, H.; Gurley, S.B.; Daugherty, A.; Cassis, L.A. Angiotensin-Converting Enzyme 2 Deficiency in Whole Body or Bone Marrow-Derived Cells Increases Atherosclerosis in Low-Density Lipoprotein Receptor$^{-/-}$ Mice. *Arterioscler. Thromb. Vasc. Biol.* **2011**, *31*, 758–765. [CrossRef] [PubMed]
35. Strawn, W.B.; Ferrario, C.M. Angiotensin II AT1 Receptor Blockade Normalizes CD11b+ Monocyte Production in Bone Marrow of Hypercholesterolemic Monkeys. *Atherosclerosis* **2008**, *196*, 624–632. [CrossRef]
36. Donmez, B.O.; Ozdemir, S.; Sarikanat, M.; Yaras, N.; Koc, P.; Demir, N.; Karayalcin, B.; Oguz, N. Effect of Angiotensin II Type 1 Receptor Blocker on Osteoporotic Rat Femurs. *Pharmacol. Rep. PR* **2012**, *64*, 878–888. [CrossRef] [PubMed]
37. Kuo, T.-R.; Chen, C.-H. Bone Biomarker for the Clinical Assessment of Osteoporosis: Recent Developments and Future Perspectives. *Biomark. Res.* **2017**, *5*, 18. [CrossRef] [PubMed]
38. Zoch, M.L.; Clemens, T.L.; Riddle, R.C. New Insights into the Biology of Osteocalcin. *Bone* **2016**, *82*, 42–49. [CrossRef] [PubMed]
39. Bayar, A.; Turan, A.; Gulle, K.; Akpolat, M.; Turan, I.; Turhan, E. The Effects of the Angiotensin Converting Enzyme Inhibitor Enalapril and the Angiotensin II Type 1 Receptor Blocker Losartan on Fracture Healing in Rats. *Clin. Investig. Med.* **2015**, *38*, E164–E172. [CrossRef] [PubMed]
40. Brito, V.G.B.; Patrocinio, M.S.; de Sousa, M.C.L.; Barreto, A.E.A.; Frasnelli, S.C.T.; Lara, V.S.; Santos, C.F.; Oliveira, S.H.P. Telmisartan Prevents Alveolar Bone Loss by Decreasing the Expression of Osteoclasts Markers in Hypertensive Rats with Periodontal Disease. *Front. Pharmacol.* **2020**, *11*, 579926. [CrossRef] [PubMed]
41. Moura, A.P.; Montalvany-Antonucci, C.C.; Rodrigues de Albuquerque Taddei, S.; Queiroz-Junior, C.M.; Biguetti, C.C.; Garlet, G.P.; Ferreira, A.J.; Teixeira, M.M.; Silva, T.A.; Andrade, I., Jr. Effects of Angiotensin II Type I Receptor Blocker Losartan on Orthodontic Tooth Movement. *Am. J. Orthod. Dentofac. Orthop.* **2016**, *149*, 358–365. [CrossRef]
42. Street, J.; Bao, M.; deGuzman, L.; Bunting, S.; Peale, F.V.; Ferrara, N.; Steinmetz, H.; Hoeffel, J.; Cleland, J.L.; Daugherty, A.; et al. Vascular Endothelial Growth Factor Stimulates Bone Repair by Promoting Angiogenesis and Bone Turnover. *Proc. Natl. Acad. Sci. USA* **2002**, *99*, 9656–9661. [CrossRef] [PubMed]
43. Masri, D.; Bar-Hai, D.; Masri-Iraqi, H.; Kahn, A.; Chaushu, G.; Chaushu, L. Early Implant Failure in Patients Using Antihypertensive Medications: A Retrospective Cohort Study. *Int. Dent. J.* **2024**, *in press*. [CrossRef] [PubMed]
44. Shetty, S.; Kapoor, N.; Bondu, J.D.; Thomas, N.; Paul, T.V. Bone Turnover Markers: Emerging Tool in the Management of Osteoporosis. *Indian J. Endocrinol. Metab.* **2016**, *20*, 846–852. [CrossRef] [PubMed]
45. Rizzoni, D.; Pasini, E.; Flati, V.; Rodella, L.F.; Paiardi, S.; Assanelli, D.; De Ciuceis, C.; Porteri, E.; Boari, G.E.; Rezzani, R.; et al. Angiotensin Receptor Blockers Improve Insulin Signaling and Prevent Microvascular Rarefaction in the Skeletal Muscle of Spontaneously Hypertensive Rats. *J. Hypertens.* **2008**, *26*, 1595–1601. [CrossRef]
46. Petnehazy, T.; Cooper, D.; Stokes, K.Y.; Russell, J.; Wood, K.C.; Granger, D.N. Angiotensin II Type 1 Receptors and the Intestinal Microvascular Dysfunction Induced by Ischemia and Reperfusion. *Am. J. Physiol. Gastrointest. Liver Physiol.* **2006**, *290*, G1203–G1210. [CrossRef]

47. Racasan, S.; Hahnel, B.; Van Dergiezen, D.M.; Blezer, E.L.; Goldschmeding, R.; Braam, B.; Kriz, W.; Koomans, H.A.; Joles, J.A. Temporary Losartan or Captopril in Young SHR Induces Malignant Hypertension despite Initial Normotension. *Kidney Int.* **2004**, *65*, 575–581. [CrossRef] [PubMed]
48. Collister, J.P.; Hornfeldt, B.J.; Osborn, J.W. Hypotensive Response to Losartan in Normal Rats. Role of Ang II and the Area Postrema. *Hypertension* **1996**, *27*, 598–606. [CrossRef] [PubMed]

coatings

Article

Nanostructured Coatings for Spinal Fixation Screws: A Dual-Function Approach Against Biofilm Formation and Implant Failure

Tiberiu Gabriel Panaitescu [1], Adelina-Gabriela Niculescu [1,2], Valentina Grumezescu [3], Bogdan Costăchescu [4,5,*], Alexandra Cătălina Bircă [1], Paul Cătălin Balaure [1], Ovidiu Cristian Oprea [6], Ionela Cristina Voinea [7], Miruna S. Stan [7], Alina Maria Holban [8], Bogdan Ștefan Vasile [1], Alexandru Mihai Grumezescu [1,2] and Daniel Mihai Teleanu [9]

1 Department of Science and Engineering of Oxide Materials and Nanomaterials, National University of Science and Technology POLITEHNICA Bucharest, 1-7 Polizu Street, 011061 Bucharest, Romania; gabriel.panaitescu17@gmail.com (T.G.P.); adelina.niculescu@upb.ro (A.-G.N.); ada_birca@yahoo.com (A.C.B.); paul.balaure@upb.ro (P.C.B.); bogdan.vasile@upb.ro (B.Ș.V.); agrumezescu@upb.ro (A.M.G.)
2 Research Institute of the University of Bucharest—ICUB, University of Bucharest, 90-92 Panduri, 050663 Bucharest, Romania
3 National Institute for Laser, Plasma and Radiation Physics, 409 Atomistilor Street, 077125 Magurele, Romania; valentina.grumezescu@inflpr.ro
4 Department of Neurosurgery, "Grigore T. Popa" University of Medicine and Pharmacy Iasi, 11 Universitatii Street, 700115 Iasi, Romania
5 Clinical Emergency Hospital "Prof. Dr. Nicolae Oblu", 2 Ateneului Street, 700309 Iasi, Romania
6 Department of Inorganic Chemistry, Physical Chemistry and Electrochemistry, National University of Science and Technology POLITEHNICA Bucharest, 1-7 Polizu Street, 011061 Bucharest, Romania; ovidiu.oprea@upb.ro
7 Department of Biochemistry and Molecular Biology, Faculty of Biology, University of Bucharest, 91-95 Splaiul Independentei, 050095 Bucharest, Romania; ionela-cristina.voinea@bio.unibuc.ro (I.C.V.); miruna.stan@bio.unibuc.ro (M.S.S.)
8 Department of Botany and Microbiology, Faculty of Biology, University of Bucharest, 050095 Bucharest, Romania; alina_m_h@yahoo.com
9 Faculty of Medicine, Carol Davila University of Medicine and Pharmacy, 8 Eroii Sanitari Street, 050474 Bucharest, Romania; daniel.teleanu@umfcd.ro
* Correspondence: bogdan.costachescu@umfiasi.ro

Abstract: Implant-associated infections represent challenging complications following orthopedic surgeries, with spinal fixation procedures being particularly linked with increased risks. Thus, urgent research is required to develop enhanced solutions to avoid bacterial colonization, associated implant failure, and severe issues. Our study is based on the laser coating of surfaces with a composite mixture of PLA/Fe_3O_4@CEF that can fight against infectious agents and preserve their activity for a prolonged time. In the present study, we synthesized Fe_3O_4@Ceftriaxone (CEF) nanoparticles by co-precipitation and blended them into polylactic acid (PLA)-based coatings that were thoroughly evaluated from physicochemical and biological points of view. The novelty of this work is the dual functionality of these coatings, combining localized, sustained antibiotic delivery with enhanced biocompatibility for spinal screw applications. The coatings exhibited substantial anti-biofilm effects, reducing *Staphylococcus aureus* colonization from 1.8×10^8 to 1.6×10^5 CFU/mL and *Pseudomonas aeruginosa* from 1.2×10^{11} to 1.9×10^6 CFU/mL after 24 h. Furthermore, in vitro assays with murine preosteoblasts and human osteoblasts demonstrated excellent biocompatibility, maintaining >95% cell viability and showing no significant cytotoxicity or inflammatory response. These results highlight the potential of PLA/Fe_3O_4@CEF composite coatings in preventing implant-associated infections and promoting osseointegration, offering a multifunctional strategy for improving spinal fixation screw longevity and patient outcomes.

 https://doi.org/10.3390/coatings15050584

Keywords: polylactic acid; magnetite; ceftriaxone; coatings; MAPLE; antimicrobial activity; spinal fixation screws

1. Introduction

Spinal instrumentation has been employed in surgical procedures for more than 6 decades, allowing for the enhancement of clinical outcomes in patients requiring spine support [1]. Spinal fixation screws are especially regarded as critical components for stabilizing the spine following trauma, degenerative conditions, or surgical interventions, providing structural support and facilitating spinal fusion [2]. Pedicle screw fixation is particularly important for procedures requiring internal stabilization, such as stabilization of thoracolumbar fractures and dislocations, correction of spinal deformities, and ensuring rigidity in oncological and fusion surgeries [3,4]. In contrast to other types of fixation devices, pedicle screws can be used in laminectomy or traumatic disruption of posterior spinal elements, as their anchorage is independent of dorsal structures [5].

Despite their advantages, pedicle screws also present certain drawbacks. These implantable devices have been associated with complications such as fixation loss, fatigue failure, misplacement, tissue rejection, dural tears, cerebrospinal fluid leakage, nerve injury, and infection, imposing careful surgical planning and implant optimization to enhance outcomes and avoid future revision interventions [6–8]. An ideal spinal fixation system must exhibit biostability, biocompatibility, and appropriate biomechanical properties to ensure long-term functional success and patient safety [2]. Even though metallic implants are the preferred option for fixing the spine due to their mechanical strength, these devices are limited by poor osseointegration, inadequate antibacterial properties, and an increased risk of bacterial colonization, ultimately resulting in implant failure and low patient satisfaction [9].

Among the challenges encountered with spinal implants, the risk of infection is a major drawback, posing a tremendous diagnostic and therapeutic burden for spinal surgeons [10]. With the widespread use of invasive medical devices, healthcare-associated infections caused by pathogenic biofilms have emerged as a leading cause of morbidity and mortality [11,12]. Biofilm formation on spinal implants significantly contributes to implant-associated infections, as these sessile microbial communities exhibit extreme resistance to antibiotics and host immune responses, making eradication difficult [13]. Biofilms consist of live and dead bacteria embedded in an extracellular matrix, which hinders the diffusion of antimicrobial agents, resulting in treatment failure and persistent infections. Moreover, bacteria from biofilms can detach and disseminate, seeding new infection sites and exacerbating the clinical burden [14]. In severe cases, the only viable option to control infection is the removal of the affected implant, a measure that not only increases healthcare costs but also exposes patients to additional surgical risks and prolonged recovery periods [15,16].

As infection prevention is considered more valuable than reacting to adverse clinical outcomes, several pre-, peri-, and post-operative procedures have been employed to avoid implant-related infections [10,16–18]. The prevention of implant-associated infections in spinal fixation surgery often relies on a combination of systemic antibiotic prophylaxis, strict aseptic surgical protocols, and the development of antimicrobial implant materials [13,18]. Due to the significant reported infection rates in spinal fusion surgeries (ranging from 1% to 12%), prophylactic antibiotic administration is generally implied [10]. Despite the acknowledged benefits of pre-operative systemic antibiotics, this approach is not without limitations. High doses are often needed to achieve effective therapeutic concentrations, resulting in the rise of a series of issues, including potential tissue toxicity, poor patient compliance, and the emergence of antibiotic-resistant pathogens [19].

Antimicrobial-functionalized implant materials have gained increasing attention as a means of local infection control. Coating spinal fixation screws with antimicrobial agents, nanoparticles, or drug-eluting systems enables sustained localized drug release, effectively reducing the need for high-dose systemic antibiotics and lowering the risk of resistance development [13]. For instance, advanced-generation cephalosporins, such as ceftriaxone, have demonstrated strong efficacy against nosocomial and drug-resistant infections by disrupting bacterial cell wall synthesis. Additionally, the pre-operative bathing of implants in antibiotic solutions has been reported as an effective local prophylactic measure to prevent deep infections in spinal instrumentation surgeries [20]. Thus, functionalizing implant surfaces with such agents offers a promising strategy for prolonged antibacterial activity at the implantation site, improving infection control while minimizing systemic side effects.

Nanostructured coatings have great potential in spinal fixation, as they enhance osseointegration, promote cellular adhesion and proliferation, and reduce infection risks [21]. However, despite the growing interest in antimicrobial coatings for orthopedic implants, there remains a lack of multifunctional systems that combine sustained antibiotic release, biocompatibility, and stable adhesion to clinically relevant substrates.

In this context, this study aims to expand the plethora of infection-preventive modifications for spinal fixation screws, presenting the development of nanostructured anti-infective coatings based on magnetite nanoparticles (Fe_3O_4 NPs), ceftriaxone (i.e., a third-generation cephalosporin antibiotic), and polylactic acid (PLA). Specifically, co-precipitation of iron salts and the antibiotic agent was performed to obtain ceftriaxone-functionalized Fe_3O_4 nanoparticles (Fe_3O_4@CEF), while the matrix-assisted pulsed laser evaporation (MAPLE) technique was used to obtain nanostructured coatings (i.e., Fe_3O_4@CEF and PLA/Fe_3O_4@CEF) (Figure 1). MAPLE has been scarcely applied for coating spinal fixation screws, particularly with composite formulations incorporating both biodegradable polymers and antibiotic-functionalized magnetic nanoparticles.

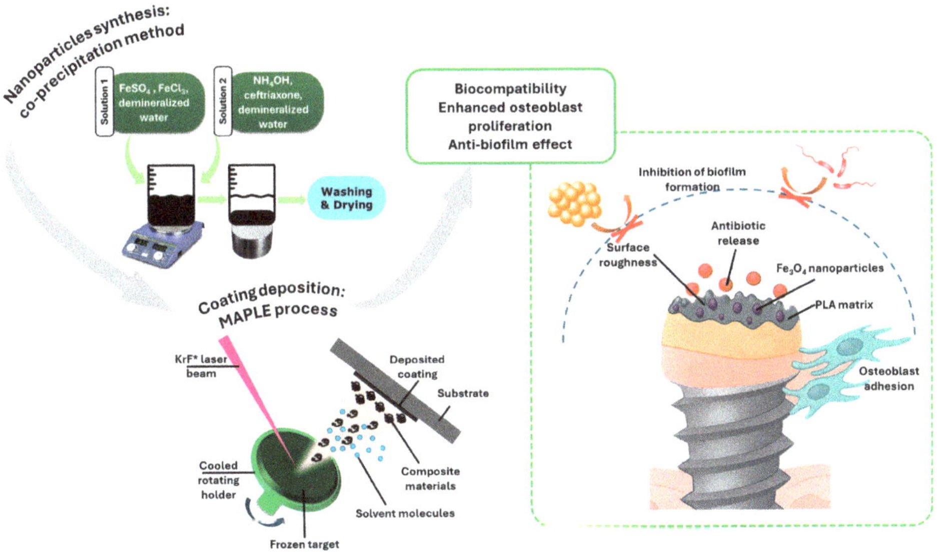

Figure 1. Schematic representation of study workflow and application of developed nanostructured coatings.

Therefore, this study addresses a critical gap by developing and characterizing PLA/Fe_3O_4@CEF nanostructured coatings fabricated by MAPLE, aiming to inhibit bacterial biofilm formation, support osteoblast viability simultaneously, offering a dual-function antimicrobial and biocompatible interface that could enhance clinical outcomes and reduce implant-associated complications.

2. Materials and Methods

2.1. Materials

The experimental work was conducted using analytical-grade reagents purchased from Sigma–Aldrich/Merck (Darmstadt, Germany). All substances were used as received, without further purification.

2.2. Nanoparticle Synthesis

The Fe_3O_4@CEF nanoparticles were synthesized at room temperature by the co-precipitation method. First, Solution 1 (S1) was prepared using a 1:2 molar ratio of $FeSO_4$ (Fe^{2+}) to $FeCl_3$ (Fe^{3+}) by dissolving the appropriate amounts of each salt in 300 mL of demineralized water. The second solution (S2) was prepared with 0.2 g ceftriaxone and 9 mL of NH_4OH in 300 mL of demineralized water. The Fe^{3+}/Fe^{2+} solution was dropped into the basic solution of CEF under magnetic stirring. After precipitation, magnetite–ceftriaxone (Fe_3O_4@CEF) was separated with a strong NdFeB permanent magnet. The nanoparticles were repeatedly washed with deionized water and left to dry at room temperature.

2.3. Coatings Deposition

Homogeneous solutions with 2% consisting of Fe_3O_4@CEF and PLA/Fe_3O_4@CEF in dimethyl sulfoxide (DMSO) were prepared. Before each laser deposition, 3 mL of the freshly prepared solution were frozen in liquid nitrogen (77 K) for 30 min. The frozen targets were maintained at liquid nitrogen temperature in the deposition chamber using a cooling system. The MAPLE technique was performed using a UV KrF* excimer laser source generating pulses of τ_{FWHM} = 25 ns at λ = 248 nm. The laser fluence was set at 200, 300, and 400 mJ/cm^2, with a repetition rate of 15 Hz. The number of applied laser pulses in all cases was 50,000. The targets were rotated with a frequency of 0.4 Hz. The substrates (i.e., spinal screw slices, Si (100)) were placed parallel to the target at a separation distance of 4 cm.

2.4. Characterization Methods

2.4.1. Physicochemical Characterization

The purity and crystallinity of the magnetite powder were investigated through X-ray Diffraction (XRD) analysis performed with an Empyrean diffractometer from PANalytical (Almelo, The Netherlands). Scans were collected at ambient temperature, using $Cu_{K\alpha}$ radiation (λ = 1.056 Å) in the Bragg angle interval between 10° and 80°, with a current of 15 mA and a voltage of 30 kV.

The morphology and dimensions of the obtained materials were determined using a scanning electron microscope (SEM). In this respect, an FEI SEM (Hillsboro, OR, USA) was used. Samples were sectioned using a diamond blade, mounted on sample holders, and introduced into the analysis chamber. Imaging was performed in secondary electron mode at an accelerating voltage of 30 keV.

The Transmission Electron Microscopy (TEM) analysis was performed on the powdery sample (nanoparticles). A small quantity of sample was dispersed in pure ethanol and subjected to ultrasonication for 15 min. A drop of the resulting suspension was placed on a carbon-coated copper grid and air-dried at room temperature. The sample was examined using a Tecnai^{TM} G2 F30 S-TWIN high-resolution TEM (FEI, Hillsboro, OR, USA) equipped with a SAED accessory. The microscope was operated in transmission mode at 300 kV, with a point resolution of 2 Å and a line resolution of 1 Å. SAED analysis was further conducted in bright field mode to obtain information on the crystallinity of the nanomaterials.

FT-IR spectroscopy was used to investigate the integrity of functional groups within the synthesized material. A Nicolet 6700 FT-IR spectrometer (Thermo Fisher Scientific, Waltham, MA, USA) equipped with a ZnSe crystal was utilized for the analysis. A small amount of particle suspension was placed on the crystal, and measurements were performed at ambient temperature. Spectra were acquired in the 4000–1000 cm^{-1} range, with 32 scans per sample at a resolution of 4 cm^{-1}. Data acquisition and processing were conducted using Omnic software (version 8.2, Thermo Fisher Scientific).

Thermal stability and degradation behavior were assessed using an STA 449C Jupiter device (NETZSCH-Gerätebau GmbH, Selb, Germany). A small amount of powdered sample was placed in an alumina crucible and analyzed under an air atmosphere over a temperature range of 30–500 °C. The heating rate was set to 10 °C/min.

Information regarding the investigation of the chemical alteration and material transfer of the obtained coatings was shown using an IRM analysis with a Nicolet iN10 MX FT-IR microscope (Thermo Fischer Scientific Company, Waltham, MA, USA). Spectral acquisition was performed in reflection mode with 32 scans per sample at a 4 cm^{-1} resolution, in the 4000–600 cm^{-1} range. The obtained data were processed using Omnic Picta 8.2 software.

2.4.2. Biological Characterization

The biocompatibility of the tested materials was evaluated using murine MC 3T3-E1 preosteoblasts (catalog number CRL-2593^{TM}) and hFOB 1.19 human osteoblasts (catalog number CRL-3602^{TM}), purchased from the American Type Culture Collection (ATCC, Rockville, MD, USA) and cultured in high glucose Dulbecco's Modified Eagle's Medium (DMEM) supplemented with 10% fetal bovine serum (FBS) at 37 °C in a humidified 5% CO_2 incubator, and in DMEM–Ham's F12 (1:1) with 2.5 mM L-glutamine (without phenol red), 10% FBS, and 0.3 mg/mL G418 at 34 °C, 5% CO_2 humidified atmosphere, respectively. The osteoblasts were cultured in direct contact with the thin coatings deposited on spinal screw slices. All samples were sterilized under ultraviolet light for 1 h prior to cell seeding, which was performed at a density of 100,000 cells/cm^2. After 24 h of incubation, the biocompatibility tests were performed as previously described in detail, an Olympus IX71 inverted microscope (Olympus, Tokyo, Japan) being used to monitor cell adhesion and proliferation throughout the experiment.

Briefly, the culture medium collected at 24 h was used to quantify the lactate dehydrogenase (LDH) release as an indicator of cell membrane damage using a commercial kit from Roche (Basel, Switzerland) and to measure the nitric oxide (NO) production with the help of Griess reagent (0.1% naphthylethylenediamine dihydrochloride, 1% sulfanilamide in 5% H_3PO_4) at 550 nm. For the MTT assay, cells cultured on slides were incubated with 1 mg/mL MTT solution for 2 h at 37 °C. The resulting formazan crystals were solubilized in isopropanol, and absorbance was measured at 595 nm using a FlexStation microplate reader (Molecular Device, Sunnyvale, CA, USA).

The staining of live and dead osteoblasts was assessed by LIVE/DEAD Viability/Cytotoxicity Kit (Thermo Fischer Scientific, USA). After washing with PBS, cells were incubated for 30 min at 37 °C with calcein-AM (2 μM) and ethidium homodimer-1 (4 μM) solutions, and the fluorescence images were acquired.

F-actin cytoskeletal organization was assessed by staining with Alexa Fluor 488-conjugated phalloidin (Invitrogen, Thermo Fisher Scientific, Waltham, MA, USA) for 30 min at room temperature after the cells were previously fixed in 70% ethanol for 15 min and permeabilized with 0.1% Triton X-100. Following the nuclei counterstaining with DAPI, representative images were captured on Olympus IX71 microscope.

The results of experiments performed in triplicate were subjected to statistical analysis performed by one-way ANOVA with Bonferroni post-hoc test (GraphPad Prism 5), with $p < 0.05$ being considered significant.

To assess the effect of nanostructured surfaces on biofilm formation, the samples were first sterilized under UV radiation for 20 min on each side. The samples were placed in 6-well plates, and 2 mL of liquid medium supplemented with 50 µL of microbial suspension (0.5 McFarland standard) were added. Plates were incubated at 37 °C for 24 h. After incubation, samples were washed with artificial seawater (AFS), and the culture medium was replaced to allow further biofilm development. Incubation was continued for 24 or 48 h. At the end of each incubation period, the samples were washed with AFS and transferred to sterile tubes containing 1 mL of AFS. The tubes were vortexed for 30 s to detach the biofilm cells, and serial dilutions were plated on solid culture media for colony-forming unit (CFU/mL) quantification.

The bacterial strains used (*S. aureus* and *Ps. aeruginosa*) were obtained from the Microbiology Laboratory, Faculty of Biology, University of Bucharest.

Biofilm formation on nanostructured surfaces was evaluated using 1 cm × 1 cm sterilized samples, which were placed individually in 6-well plates containing 2 mL of liquid medium and 50 µL of microbial suspension (0.5 McFarland standard). Plates were incubated at 37 °C for 24 h, followed by washing with AFS and medium replacement to allow further biofilm development. Samples were incubated for an additional 24 h. At the end of each incubation period, biofilm-covered samples were washed, transferred to sterile tubes containing 1 mL of AFS, and vortexed for 30 s to detach the adhered cells. Serial dilutions of the resulting suspension were plated on solid media for CFU/mL quantification.

3. Results

The indexed patterns correspond to the obtained iron oxide-based particles, which was proven using XRD analysis. As depicted (Figure 2), the analyzed sample exhibits high crystallinity, with strong peaks at 30.33°, 35.51°, 43.16°, 53.61°, 57.17°, and 62.82°, correlated with (220), (311), (400), (422), (511), and (440) diffraction planes, respectively. This diffraction pattern corresponds to the crystallographic system of the cubic inverse spinel structure Fe_3O_4, agreeing with JCPDS card No. 19-0629 and prior literature findings [22–24], thus confirming the crystalline structure.

The Fe_3O_4 nanoparticles were further subjected to HR-TEM analysis.

The HR-TEM images (Figure 3) confirmed the synthesized nanoparticles' spherical shape, reduced dimensions, and cluster formation tendency. As can be seen in Figure 3a,b, the interplanar distance is at around 2.52 Å and 1.48 Å, which are values that correspond to the (311) and (440) crystallographic orientation of cubic magnetite Fe_3O_4 [25]. This finding aligns with the data obtained from XRD analysis, where the peak at the (311) diffraction plane was the most intense. The average diameter of the nanoparticles was estimated at around 4.83 ± 0.19 nm with a quasi-spherical morphology (Figure 3c). The SAED pattern depicted in Figure 3d provides information regarding sample crystallinity. The observed concentrical rings correspond to (111), (220), (311), (400), (422), and (511) diffraction planes, highlighting the presence of a highly crystalline magnetite [26]. The most intense diffraction ring is noticed for the (311) plane, in agreement with XRD, reconfirming that this is the dominant crystallographic orientation in the analyzed nanoparticles.

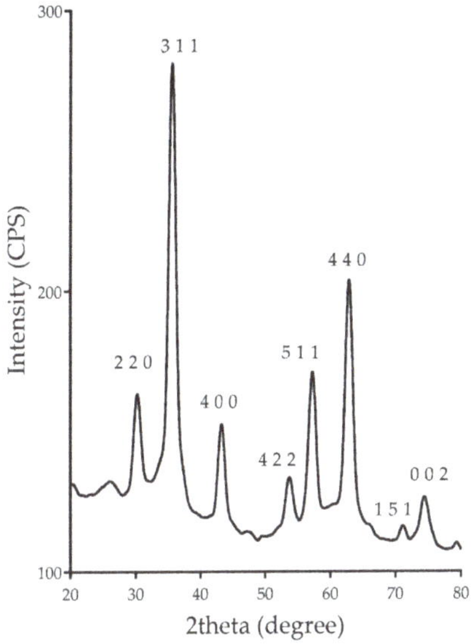

Figure 2. X-ray diffractogram of Fe_3O_4@CEF nanoparticles.

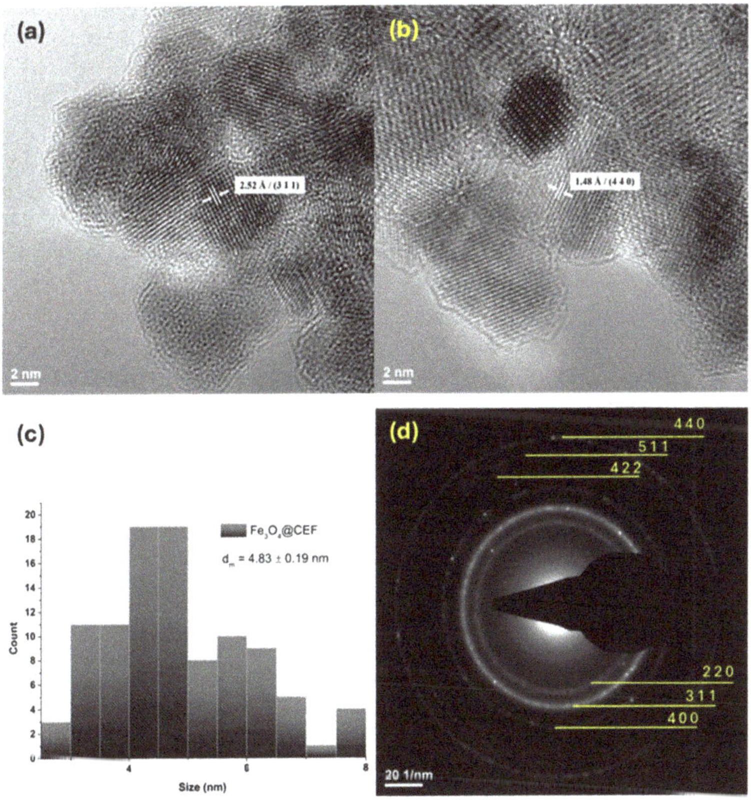

Figure 3. (**a**,**b**) HR-TEM, (**c**) histogram and (**d**) SAED pattern of Fe_3O_4@CEF nanoparticles.

The FTIR spectrum of Fe_3O_4@CEF (Figure 4) revealed a characteristic band near 547 cm^{-1}, corresponding to Fe–O stretching vibrations, confirming the presence of the magnetite core [27–29]. A peak at 1655 cm^{-1} is assigned to the C=O stretching vibration (amide I) from the β-lactam ring of ceftriaxone. The band at 1624 cm^{-1} likely arises from N–H bending or C=C stretching in the aromatic system. Additional absorptions at 1571 cm^{-1} and 1461 cm^{-1} are attributed to C=C/C=N stretching of aromatic and heterocyclic rings and CH_2 deformation, respectively. Peaks at 1110 cm^{-1} and 1032 cm^{-1} are associated with C–N stretching and in-plane C–H bending vibrations, indicating the retention of ceftriaxone's functional groups after surface binding to Fe_3O_4 nanoparticles [30–33].

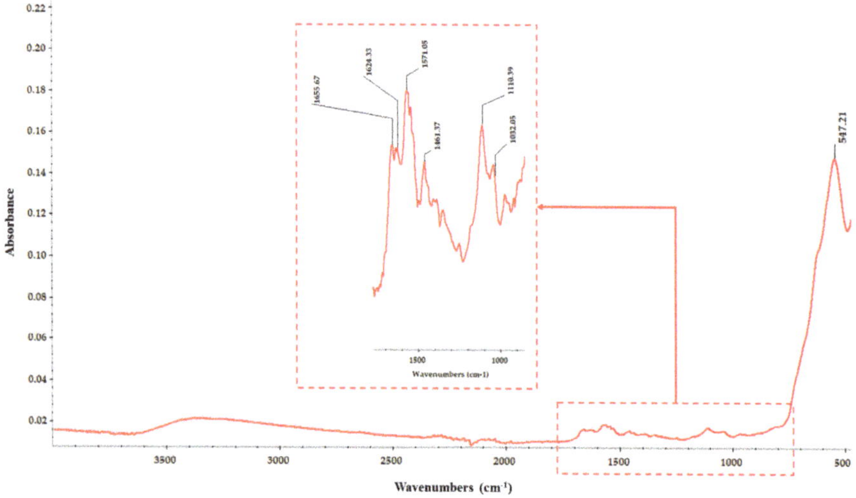

Figure 4. FT-IR spectrum of Fe_3O_4@CEF nanoparticles.

The thermal behavior of the synthesized nanoparticles was assessed through TG-DSC, allowing a comparison of pristine (Fe_3O_4) and Fe_3O_4@CEF nanoparticles. For the pristine material, in the temperature range between room temperature and 120 °C, there occurs the loss of water molecules and –OH groups from the surface of magnetite. In this interval, there is a mass loss of 1.74%, which is accompanied by a slight endothermic effect. With increasing temperature, two other successive mass losses (120–200 °C and 200–300 °C) were observed, accounting for 0.54% and 0.98%, respectively. Both mass losses were associated with exothermic processes and can be attributed to the combustion (oxidation) of organic impurities present on the surface of Fe_3O_4. The strong exothermic peak at 565.8 °C corresponds to the crystalline phase transition from magnetite to maghemite. The mass loss continues at a slower rate up to 900 °C, registering an additional 0.73% loss. The residual mass represents 95.45% of the initial sample amount.

On the other hand, Fe_3O_4@CEF nanoparticles exhibit a mass loss when reaching from room temperature to 155 °C that can be attributed to the removal of adsorbed water molecules or surface hydroxyl (–OH) groups. The first significant mass loss (i.e., 2.84%) occurs between ~200–400 °C, with two exothermic peaks at 306.3 °C and 373.1 °C, corresponding to the oxidation of organic residues or other thermally labile surface contaminants. Further, the sample faces a similar phase transformation as pristine nanoparticles, with the graph displaying a strong exothermic peak at 583.8 °C being attributed to the change from magnetite (Fe_3O_4) to maghemite (γ-Fe_2O_3). During this process, another mass loss is observed, 1.59%, possibly due to the release of oxygen from the crystal lattice. The total residual mass after thermal degradation is 93.71%, indicating a thermally stable inorganic fraction, likely composed of oxidized iron phases. By comparing the results obtained for the two types of iron oxide-based nanomaterials, it was estimated that the ceftriaxone amount loaded on Fe_3O_4@CEF nanoparticles is ~1.82%.

Based on these results, Fe_3O_4@CEF nanoparticles were further utilized to fabricate coatings using the MAPLE technique at different laser fluences. Infrared microscopy (IRM) was used to investigate the stoichiometric laser transfer of the materials.

The optimum laser fluence was identified using comparative IRM analysis (Figure 5) between Fe_3O_4@CEF drop-cast and coatings deposited by MAPLE at 200 mJ/cm^2, 300 mJ/cm^2, and 400 mJ/cm^2 laser fluences. Absorbance intensities of IR spectra maps are proportional to color changes, starting with blue (the lowest intensity) and gradually increasing from green and yellow to red (the highest intensity).

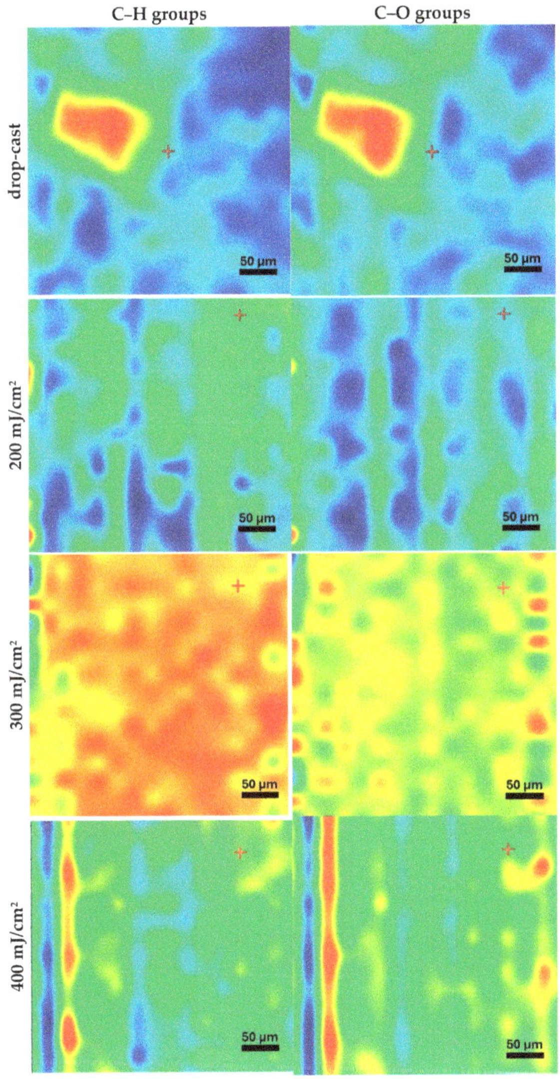

Figure 5. IR maps of Fe_3O_4@CEF drop-cast and coatings obtained at different laser fluences (C–H and C–O bonds distribution characteristic of ceftriaxone).

The laser fluence of 300 mJ/cm^2 was identified as optimal for the synthesis of Fe_3O_4@CEF coatings, with the yellow and red colors being representative, in this case, for the distribution of C–H and C–O groups. This corresponds to a minimum chemical alteration and adequate material transfer. The IR maps show absorption band intensities ranging from yellow–orange (medium) to orange–red (medium–high).

In the case of PLA/Fe_3O_4@CEF coatings (fabricated at the same laser fluences), one can observe that the lowest functional group degradation was recorded at 300 mJ/cm^2 laser fluence. At this fluence, the color distribution is better emphasized than in the case of the other two laser fluences. This is indicative of a superior stoichiometric deposition of the coatings (Figure 6).

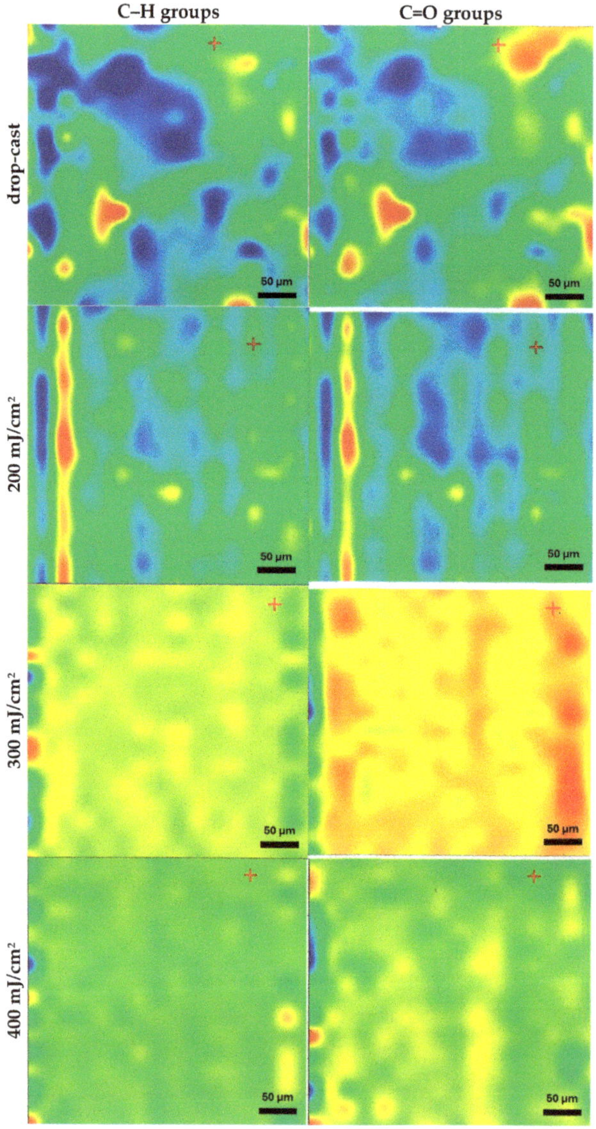

Figure 6. IR maps of PLA/Fe_3O_4@CEF drop-cast and coatings obtained at different laser fluences (C–H and C=O bonds distribution characteristics of ceftriaxone and PLA, respectively).

Along with the IR maps, complementary information was gathered from the IR spectra of Fe_3O_4@CEF and PLA/Fe_3O_4@CEF drop-cast and coatings deposited at 200, 300, and 400 mJ/cm^2 (Figure 7). The lowest degree of functional group degradation was registered at 300 mJ/cm^2. The other two experimental variants led to a significant modification in the absorption bands characteristic of carboxyl groups.

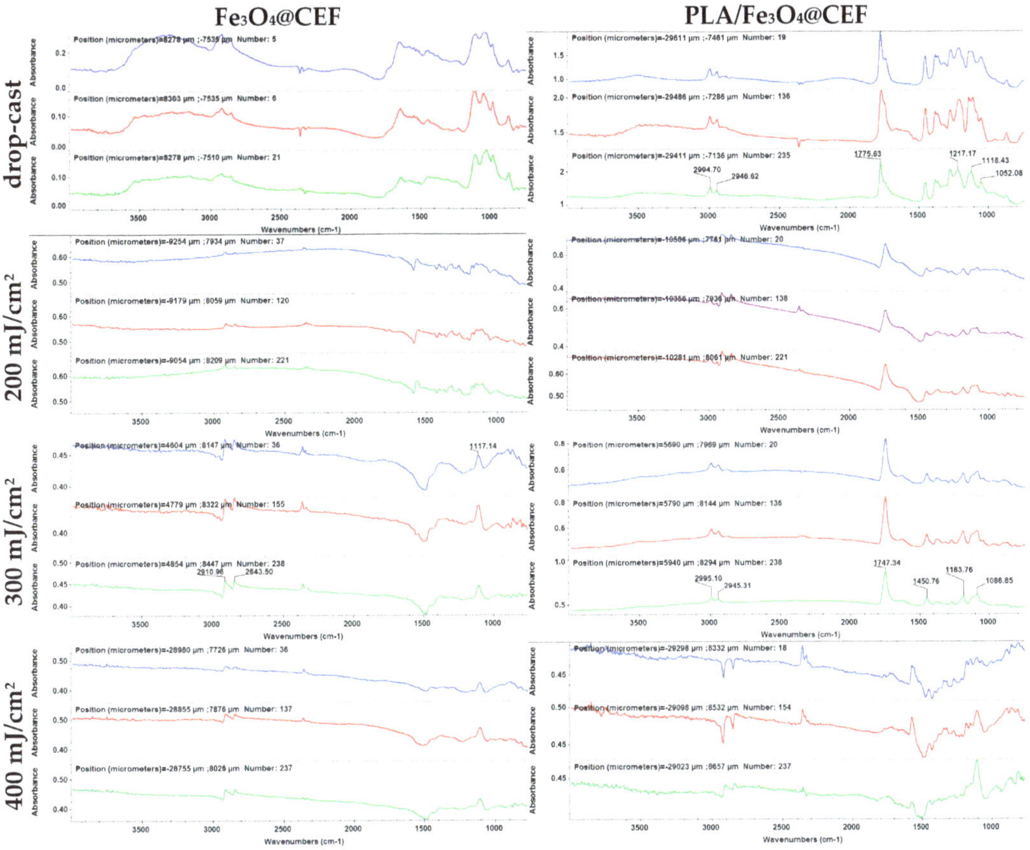

Figure 7. IR spectra of Fe_3O_4@CEF and PLA/Fe_3O_4@CEF drop-cast and coatings obtained at different laser fluences.

Further, the optimum fluence of 300 mJ/cm^2 was used to assess the biological characteristics for both types of composites (i.e., Fe_3O_4@CEF and PLA/Fe_3O_4@CEF).

The cross-sectional SEM images (Figure 8) showed uniform coverage with a smooth morphology for PLA/Fe_3O_4@CEF coatings obtained at 300 mJ/cm^2, with a thickness between 200–300 nm.

The synthesized nanostructured materials were further characterized from a biological point of view. Specifically, the biocompatibility of murine preosteoblasts (Figure 9) and normal human osteoblasts (Figure 10) was evaluated at 24 h after seeding on Fe_3O_4@CEF and PLA/Fe_3O_4@CEF coatings. In the case of MC3T3-E1 murine preosteoblasts, the MTT assay revealed that both types of surfaces led to cell-viability levels close to 100%, indicating that the samples do not exert cytotoxic effects. Moreover, NO production (Figure 9a) remained at baseline levels, suggesting the absence of an inflammatory response in preosteoblasts. These findings are crucial for ensuring that the implants support early-stage bone formation without eliciting excessive immune activation. Additionally, phase-contrast microscopy analysis (Figure 9b) suggested that cells adhered well to the coated surfaces, maintaining a healthy morphology, further reinforcing the notion that the materials provide a suitable for preosteoblast attachment and proliferation. Importantly, the demonstrated biocompatibility of these coatings with murine preosteoblasts suggests their potential application medicine.

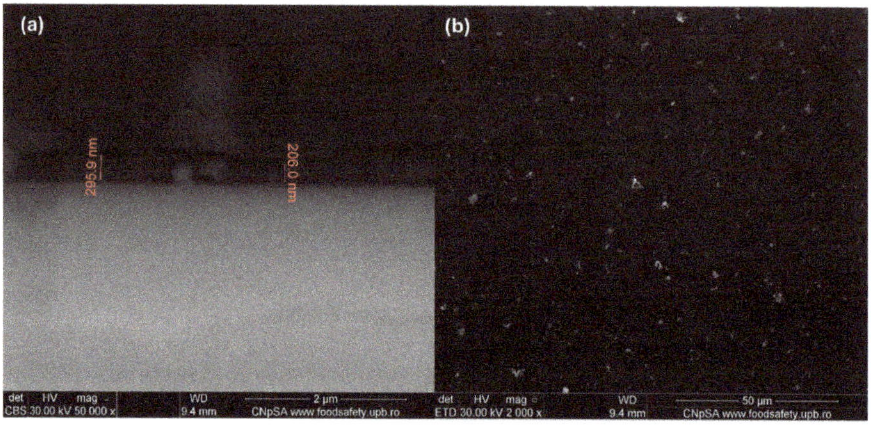

Figure 8. Cross-sectional (**a**) and top-view (**b**) SEM micrographs of PLA/Fe_3O_4@CEF coatings obtained at 300 mJ/cm^2 at 50,000× and 2000× magnifications, respectively.

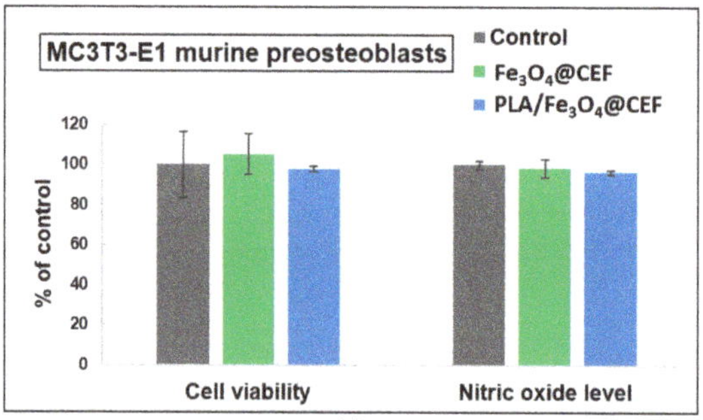

(a)

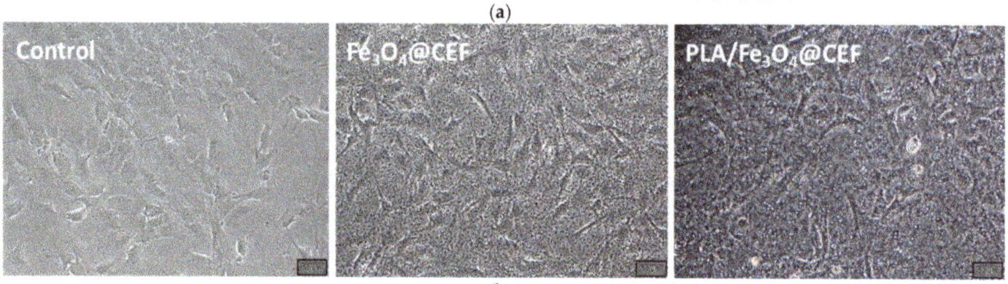

(b)

Figure 9. Biological response of MC3T3-E1 murine preosteoblasts after 24 h of culture on Fe_3O_4@CEF and PLA/Fe_3O_4@CEF surfaces, shown by (**a**) cell viability and nitric oxide production, and (**b**) phase-contrast images. Data of three independent experiments (mean ± standard deviation) are presented as percentage of control (cells grown on uncoated spinal screw slices). No statistically significant differences ($p > 0.05$) were observed between groups, as determined by one-way ANOVA with Bonferroni post hoc test.

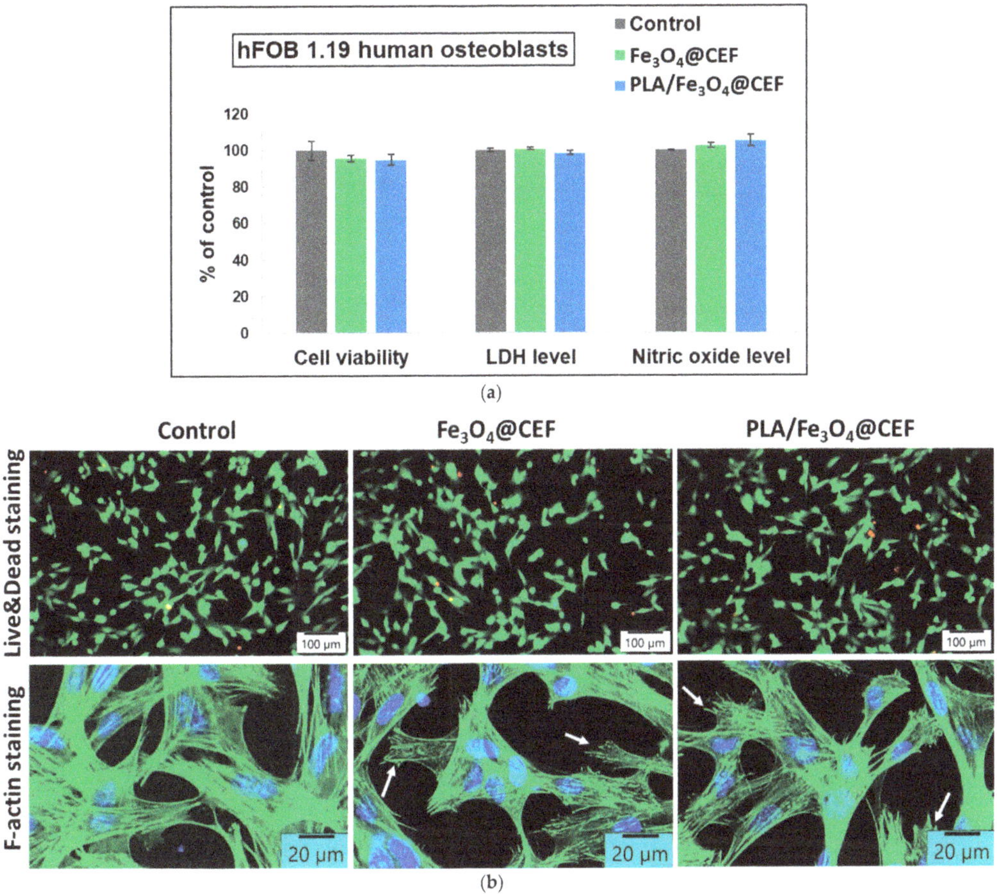

Figure 10. Biocompatibility assessment Fe_3O_4@CEF and PLA/Fe_3O_4@CEF coatings after 24 h of incubation with hFOB 1.19 human osteoblasts, measured by (**a**) viability, LDH and NO levels, and (**b**) staining of live and dead cells (labeled in green with calcein AM and in red by propidium iodide, respectively) and cytoskeleton actin filaments (labeled in green by phalloidin-FITC, and nuclei in blue with DAPI). Note the well-distributed focal adhesion points indicated by white arrows. Data of three independent experiments (mean ± standard deviation) are presented as percentage of control (cells grown on uncoated spinal screw slices). No statistically significant differences ($p > 0.05$) were observed between groups, as determined by one-way ANOVA with Bonferroni post hoc test.

For human osteoblasts, the MTT assay (Figure 10a) confirmed good cell viability across all tested coatings, similar to the findings in preosteoblasts. In addition, LDH measurements supported the non-toxic nature of the materials, as there was no notable increase in membrane damage or cell lysis, and the NO assay demonstrated no significant increase in inflammation. Furthermore, the live/dead staining corroborated these results, showing predominantly viable cells with minimal cell death. F-actin staining highlighted well-organized cytoskeletal structures, with cells maintaining their typical elongated morphology and spreading efficiently over the coated surfaces. The presence of an intact actin cytoskeleton and well-distributed focal adhesion points (indicated by white arrows in Figure 9b), enhanced by the nanostructured surface compared to the control, suggests strong cell–substrate interactions, which are essential for the osseointegration process.

The nanostructured coatings were further evaluated for anti-biofilm properties against two bacterial strains: *Staphylococcus aureus* and *Pseudomonas aeruginosa* (Figure 11), with results being presented in terms of colony-forming units per milliliter (CFU/mL), which indicate the number of viable bacteria forming biofilms on the surfaces after 24 and 48 h of incubation.

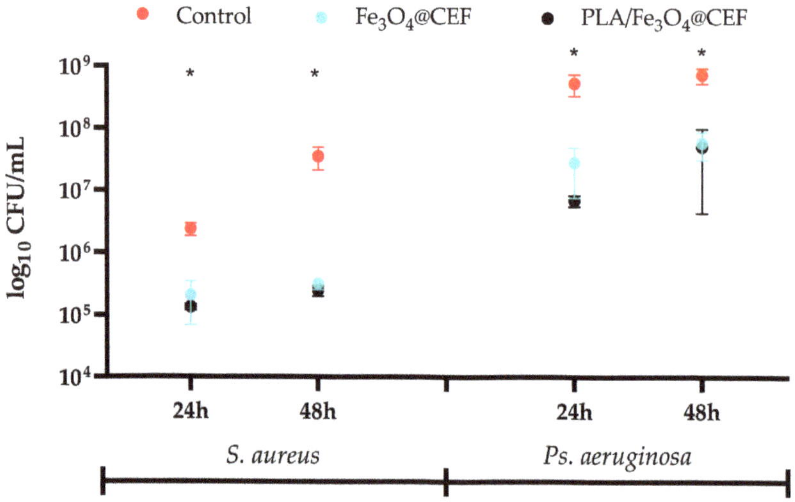

Figure 11. Evaluation of biofilm development after 24 and 48 h of incubation in the presence and absence of Fe_3O_4-based coatings for *S. aureus* and *Ps. aeruginosa* strain. * $p \leq 0.5$, one-way Anova when comparing control vs. Fe_3O_4@CEF samples.

As shown in Figure 11, both Fe_3O_4@CEF and PLA/Fe_3O_4@CEF coatings exhibit pronounced anti-biofilm effects, particularly against *S. aureus*. The control group (uncoated surface) displayed a high bacterial load, with mean values around 10^6 CFU/mL at 24 h and increasing to approximately 10^8 CFU/mL at 48 h, highlighting the robust biofilm-forming capacity of *S. aureus* on untreated materials. In contrast, both coated surfaces demonstrated a significant reduction in bacterial adhesion, with CFU counts reduced to approximately 10^5 CFU/mL at both 24 and 48 h. The difference was statistically significant ($p < 0.05$), confirming the coatings' efficacy in preventing early-stage biofilm formation.

For *Ps. aeruginosa*, although the initial bacterial load on control samples remained high (~10^8–10^9 CFU/mL), both nanostructured coatings reduced bacterial counts to approximately 10^7 CFU/mL at 24 h. After 48 h, Fe_3O_4@CEF maintained this reduction, while PLA/Fe_3O_4@CEF showed slightly greater variability but still a significant decrease compared to the control. These results demonstrate that the coatings not only inhibit initial bacterial adhesion but also prevent sustained biofilm growth over time.

Therefore, the nanostructured coatings developed in this study demonstrated effective anti-adhesive behavior by significantly reducing biofilm formation by *S. aureus* and *Ps. aeruginosa*. The coatings inhibited bacterial adhesion to the surface, thereby limiting the development of biofilms associated with medical devices. These anti-biofilm effects were sustained over both 24 and 48 h incubation periods, indicating stability over time. Inhibiting biofilm formation by *S. aureus* and *Ps. aeruginosa* is relevant for reducing the risk of device-associated infections, as these species are frequently implicated in hospital-acquired infections.

4. Discussion

The use of nanostructured antimicrobial coatings on fixation devices has become particularly relevant in the field of spinal surgery, as these implants are highly susceptible to bacterial colonization and infection-related complications [6,34,35]. Recent studies have explored a broad range of materials for designing infection-resistant coatings, including ceramics [34,36–38], natural and synthetic polymers [19,39–42], metal and metal oxide particles [11,41,43], carbon-based materials [19], antibiotic drugs [19,36,42], natural antimicrobial agents [44], and various combinations among them. From the plethora of possibilities, this study chose to develop nanocomposite coatings with magnetite (iron oxide—Fe_3O_4), ceftriaxone (CEF), and polylactic acid (PLA) toward creating synergistic outcomes.

The Fe_3O_4 nanoparticles have gained significant attention for biomedical applications due to their intrinsic magnetic properties, ability to enhance osteogenesis, and potential for magnetically targeted drug delivery [11,45]. Additionally, Fe_3O_4 nanoparticles influence functional bone tissue regeneration through osteo-mimetic mechanisms, making them promising candidates for enhancing osseointegration in orthopedic implants [46,47]. Fe_3O_4 nanoparticles have been shown to stimulate osteoblast proliferation and differentiation, potentially through the activation of signaling pathways such as BMP-2/Smad and MAPK, as well as by promoting intracellular calcium deposition and mineralization. These effects may be enhanced under magnetic stimulation, enabling magneto-mechanical stimulation of bone tissue, hence supporting bone regeneration and osseointegration, which is especially beneficial in load-bearing spinal applications [48–51].

Moreover, the properties of Fe_3O_4 nanoparticles can be further tuned to ensure multifunctional roles by surface functionalization. Modifying Fe_3O_4 nanoparticles with antibiotic drugs represents a particularly advantageous solution for countering resistant microbial strains [52].

Ceftriaxone (CEF), a broad-spectrum β-lactam antibiotic, has been widely used in spinal surgery infection prevention, with reports suggesting that local administration via implant coatings could serve as an effective alternative to systemic therapy [20,53]. This third-generation cephalosporin antibiotic is known for its activity against both Gram-positive and Gram-negative bacteria [54]. It exerts its bactericidal effect by inhibiting bacterial cell wall synthesis, making it highly effective in treating various infectious diseases [33]. Thus, based on its potent broad-spectrum activity, ceftriaxone was chosen to modify the surface of Fe_3O_4 nanoparticles and endow them with synergistic anti-infective capabilities.

In implantable medical devices, particularly in spinal fixation systems, PLA offers distinct advantages that recommend it as an interface between the screw surface and surrounding bone tissue. This biodegradable polymer serves as a supportive scaffold that gradually degrades into lactic acid, a natural metabolite, further allowing space for new bone tissue formation. Thus, it enables the formation of bioactive composites with osteoconductive and osteoinductive properties to improve spinal fusion rates and enhance tissue regeneration [55].

Henceforth, the incorporation of ceftriaxone-modified Fe_3O_4 nanoparticles into PLA coatings can thus improve bioactivity, antimicrobial efficacy, and mechanical stability, both optimizing the performance of spinal fixation screws and justifying the material choice for this work.

This study reports on the successful synthesis and characterization of Fe_3O_4@CEF nanoparticles, and their subsequent incorporation into PLA-based nanostructured coatings. The aim was to highlight their potential for biomedical applications, particularly in endowing spinal fixation screws with antimicrobial and biofilm-resistant properties. The XRD analysis confirmed the crystallinity, purity and structural integrity of the synthesized Fe_3O_4 nanoparticles, which exhibited a cubic inverse spinel structure. SEM and TEM analyses revealed that the obtained materials displayed a quasi-spherical morphology with nanometric diameters and a slight agglomeration tendency [56–58]. HR-TEM further allowed for a refined size estimation to ~19.5 nm, which corroborated well with the results obtained by XRD and SAED analyses. The presence of characteristic functional groups of both Fe_3O_4 and CEF in the FT-IR spectra confirmed the successful functionalization of the nanoparticles with the antibiotic. The TG-DSC analysis indicated a CEF loading of ~1.82% within the Fe_3O_4@CEF nanoparticles.

The MAPLE deposition technique was used in recent studies to develop biocompatible coatings with enhanced properties [11,59,60]. The advantages of MAPLE as a solvent-free deposition technique include its ability to preserve the antibiotic's bioactivity, ensure homogeneous thin-film formation, and achieve strong adhesion to the substrate, which are essential characteristics for the stability and efficacy of implant coatings [60–62].

By analyzing the transfer efficiency and integrity of the materials, the coatings obtained at 300 mJ/cm^2 laser fluence were selected as the best choice for the current study. Thus, these nanostructured coatings exhibited a smooth surface morphology and uniform distribution.

The biocompatibility assessments demonstrated that both Fe_3O_4@CEF nanoparticles and PLA/Fe_3O_4@CEF composite were non-toxic to both murine and human osteoblasts. This result supports their potential use in biomedical applications (i.e., osseointegration in spinal fixation screws). One should mention that these findings align well with prior reports on biocompatibility, osteoinductive potential, and biodegradability of magnetite nanoparticles [46,63]. PLA matrix is recognized for its biocompatibility, biodegradability, and mechanical stability, which recomend it for spinal fixation implants [64].

The anti-biofilm properties of the coatings were also evaluated in the presence of both *S. aureus* and *Ps. aeruginosa*, two major pathogens responsible for implant-associated infections [65]. The obtained results demonstrated that Fe_3O_4@CEF and PLA/Fe_3O_4@CEF coatings effectively inhibited biofilm formation, with the PLA-based coating demonstrating superior efficacy, particularly against the Gram-negative strain.

The observed anti-biofilm properties of PLA/Fe_3O_4@CEF coatings can be attributed to a multifactorial mechanism based on the synergistic activity of component materials. Firstly, the sustained presence of CEF on the surface enables continuous disruption of bacterial cell wall synthesis, impairing microbial adhesion and proliferation. Secondly, Fe_3O_4 nanoparticles contribute to generating a hostile surface topography that reduces bacterial anchoring. The nanostructured coatings interfere with initial bacterial adhesion by inducing mechanical stress or disrupting the extracellular polymeric substance matrix formation. The combination of chemical (antibiotic) and physical (nanostructured surface roughness) cues is likely to prevent biofilm initiation and maturation, offering great promise for preventing implant-related infections.

Considering the broad context of spinal surgery, a multifaceted approach combining pre-operative infection control protocols, systemic prophylaxis, and antimicrobial coatings is the most effective strategy for reducing implant-related infections. The emergence of multifunctional implant coatings incorporating multiple antibacterial agents, such as combinations of antibiotics, antimicrobial peptides, and bioactive nanoparticles, represents a promising avenue for enhancing implant safety and reducing the risk of antibiotic resistance [66–68].

The PLA/Fe_3O_4@CEF nanostructured coatings developed in this study offer a distinctive combination of antimicrobial efficacy, biocompatibility, and structural integrity. Our herein-presented nanostructured materials maintained over 95% cell viability and successfully supported osteoblast adhesion and proliferation.

With the herein reported approach, CEF incorporation into a PLA-based matrix via MAPLE ensured sustained and localized antibiotic release, minimizing systemic exposure and subsequent side effects.

The findings of this study highlight the potential of PLA/Fe_3O_4@CEF nanocomposite coatings for infection prevention and osseointegration enhancement in spinal implants. The novel nanostructured coatings tailored specifically for spinal implant surfaces have a two-fold potential in creating an antimicrobial biocompatible surface that supports osteoblast proliferation, aiming to reduce implant-associated complications and improve overall clinical outcomes. This multifaceted approach advances our system as a competitive alternative to existing antibacterial coatings for orthopedic applications.

Even though the developed materials were effective within the study window, supplementary evaluations are needed to assess long-term stability under physiological conditions. In particular, further research is required to optimize coating durability, long-term stability, and in vivo performance. Future work should focus on in vivo evaluations, assessing these coatings' osseointegration potential, antimicrobial efficacy, and mechanical performance under dynamic physiological conditions. Moreover, aspects related to drug-release kinetics will also be addressed in future studies, as they are critical for understanding the duration and sustainability of antimicrobial activity at the implant site.

Despite these limitations, the proposed coating strategy enables localized drug delivery, potential magnetic responsiveness, and strong coating adhesion, representing a promising step forward in orthopedic implant surface engineering.

5. Conclusions

This study has presented the synthesis, physicochemical characterization, and biological evaluation of innovative antimicrobial nanostructured materials that can be used to improve spinal fixation devices to reduce infection risk and enable long-term functional success. The experimental results demonstrated the successful functionalization of Fe_3O_4@CEF nanoparticles by XRD, SEM, TEM, and SAED, confirming their crystalline structure, morphology, and composition, while FT-IR and TG-DSC analyses verified the functionalization of Fe_3O_4@CEF and estimated the drug loading at ~1.82%. Further, the nanoparticles were incorporated into PLA-based MAPLE-deposited coatings, with 300 mJ/cm^2 identified as the optimal laser fluence. In terms of biological behavior, both Fe_3O_4@CEF and PLA/Fe_3O_4@CEF coatings demonstrated excellent biocompatibility, with the latter considerably enhancing osteoblast proliferation. The coatings exhibited significant anti-biofilm effects against *S. aureus* and *Ps. aeruginosa*, with PLA/Fe_3O_4@CEF showing superior inhibition properties.

Based on the presented findings, it can be concluded that the mixing of Fe_3O_4 nanoparticles, CEF, and PLA offers a promising strategy for antimicrobial coatings on spinal fixation screws. Building on the results of this work, research can be moved further to in vivo studies to evaluate long-term stability, mechanical performance, and antimicrobial efficacy in physiological conditions, eventually bringing improved solutions to clinical settings. Moreover, we believe this strategy offers a modular platform that could be extended to other implant types or combined with additional therapeutic agents in the future.

Author Contributions: Conceptualization, B.C.; Data curation, B.C., P.C.B., I.C.V., A.M.G. and D.M.T.; Formal analysis, T.G.P., A.-G.N., V.G., B.C., A.C.B., P.C.B., O.C.O., I.C.V., M.S.S., A.M.H., B.Ș.V. and A.M.G.; Investigation, T.G.P., A.-G.N., V.G., A.C.B., O.C.O., I.C.V., M.S.S., A.M.H., B.Ș.V. and A.M.G.; Methodology, V.G. and P.C.B.; Validation, B.C., M.S.S. and D.M.T.; Writing—original draft, T.G.P., A.-G.N., V.G., B.C., A.C.B., P.C.B., O.C.O., I.C.V., M.S.S., A.M.H., B.Ș.V., A.M.G. and D.M.T.; Writing—review and editing, A.-G.N., V.G., B.C., P.C.B., A.M.G. and D.M.T. All authors have read and agreed to the published version of the manuscript.

Funding: This research received no external funding.

Institutional Review Board Statement: Not applicable.

Informed Consent Statement: Not applicable.

Data Availability Statement: The original contributions presented in this study are included in the article. Further inquiries can be directed to the corresponding author.

Conflicts of Interest: The authors declare no conflicts of interest.

References

1. Yoshihara, H. Rods in spinal surgery: A review of the literature. *Spine J.* **2013**, *13*, 1350–1358. [CrossRef]
2. Costăchescu, B.; Niculescu, A.-G.; Grumezescu, A.M.; Teleanu, D.M. Screw Osteointegration—Increasing Biomechanical Resistance to Pull-Out Effect. *Materials* **2023**, *16*, 5582. [CrossRef]
3. Christodoulou, E.; Chinthakunta, S.; Reddy, D.; Khalil, S.; Apostolou, T.; Drees, P.; Kafchitsas, K. Axial pullout strength comparison of different screw designs: Fenestrated screw, dual outer diameter screw and standard pedicle screw. *Scoliosis* **2015**, *10*, 15. [CrossRef] [PubMed]
4. Mu, S.; Wang, J.; Gong, S. Mechanical analysis of posterior pedicle screw system placement and internal fixation in the treatment of lumbar fractures. *Comput. Math. Methods Med.* **2022**, *2022*, 497754. [CrossRef] [PubMed]
5. Nemade, A.; Shikalgar, A.; Sancheti, S.; Wadkar, S.P. Biomechanical analysis of spinal pedicle screws under static compression and tensile bending. *Mater. Today Proc.* **2021**, *47*, 4778–4785. [CrossRef]
6. Warburton, A.; Girdler, S.J.; Mikhail, C.M.; Ahn, A.; Cho, S.K. Biomaterials in spinal implants: A review. *Neurospine* **2019**, *17*, 101. [CrossRef]
7. Ramezani, M.; Ripin, Z.M. An Overview of Enhancing the Performance of Medical Implants with Nanocomposites. *J. Compos. Sci.* **2023**, *7*, 199. [CrossRef]
8. Oladele, I.O.; Onuh, L.N.; Agbeboh, N.I.; Alewi, D.D.; Lephuthing, S.S. The relationship and functional links between human age, growth, and biomedical implants: A review on the application of bulk and nanomaterials. *Nano Sel.* **2023**, *4*, 419–441. [CrossRef]
9. Prodana, M.; Stoian, A.B.; Burnei, C.; Ionita, D. Innovative Coatings of Metallic Alloys Used as Bioactive Surfaces in Implantology: A Review. *Coatings* **2021**, *11*, 649. [CrossRef]
10. Quaile, A. Infections associated with spinal implants. *Int. Orthop.* **2012**, *36*, 451–456. [CrossRef]
11. Pirușcă, I.A.; Balaure, P.C.; Grumezescu, V.; Irimiciuc, S.-A.; Oprea, O.-C.; Bîrcă, A.C.; Vasile, B.; Holban, A.M.; Voinea, I.C.; Stan, M.S.; et al. New Fe_3O_4-Based Coatings with Enhanced Anti-Biofilm Activity for Medical Devices. *Antibiotics* **2024**, *13*, 631. [CrossRef] [PubMed]
12. Alberts, A.; Bratu, A.G.; Niculescu, A.-G.; Grumezescu, A.M. New Perspectives of Hydrogels in Chronic Wound Management. *Molecules* **2025**, *30*, 686. [CrossRef]
13. Li, X.; Huang, T.; Heath, D.E.; O'Brien-Simpson, N.M.; O'Connor, A.J. Antimicrobial nanoparticle coatings for medical implants: Design challenges and prospects. *Biointerphases* **2020**, *15*, 060801. [CrossRef]
14. Grămadă, A.M.; Niculescu, A.-G.; Bîrcă, A.C.; Holban, A.M.; Ciceu, A.; Balta, C.; Herman, H.; Hermenean, A.; Stoica, A.-E.; Ardelean, S.; et al. In Vitro and In Vivo Evaluation of rPET/Cu-Alg Nanofibers for Anti-Infective Therapy. *Polymers* **2025**, *17*, 68. [CrossRef] [PubMed]
15. Koley, S.; Mukherjee, M. Genetic Basis of Biofilm Formation and Spread of Nosocomial Infections. In *Analytical Methodologies for Biofilm Research*; Nag, M., Lahiri, D., Eds.; Springer: New York, NY, USA, 2021; pp. 269–298.
16. Ahmadabadi, H.Y.; Yu, K.; Kizhakkedathu, J.N. Surface modification approaches for prevention of implant associated infections. *Colloids Surf. B Biointerfaces* **2020**, *193*, 111116. [CrossRef]
17. Honda, M.; Matsumoto, M.; Aizawa, M. Potential Application of Protamine for Antimicrobial Biomaterials in Bone Tissue Engineering. *Int. J. Mol. Sci.* **2020**, *21*, 4368. [CrossRef] [PubMed]
18. Kandi, V.; Vadakedath, S. Implant-associated infections: A review of the safety of cardiac implants. *Cureus* **2020**, *12*, e12267. [CrossRef]
19. Huang, Y.; Guo, X.; Wu, Y.; Chen, X.; Feng, L.; Xie, N.; Shen, G. Nanotechnology's frontier in combatting infectious and inflammatory diseases: Prevention and treatment. *Sig. Transduct. Target. Ther.* **2024**, *9*, 34. [CrossRef]
20. Eren, B.; Karagöz Güzey, F.; Kitiş, S.; Özkan, N.; Korkut, C. The effectiveness of pedicle screw immersion in vancomycin and ceftriaxone solution for the prevention of postoperative spinal infection: A prospective comparative study. *Acta Orthop. Et Traumatol. Turc.* **2018**, *52*, 289–293. [CrossRef]
21. Liang, W.; Zhou, C.; Bai, J.; Zhang, H.; Long, H.; Jiang, B.; Dai, H.; Wang, J.; Zhang, H.; Zhao, J. Current developments and future perspectives of nanotechnology in orthopedic implants: An updated review. *Front. Bioeng. Biotechnol.* **2024**, *12*, 1342340. [CrossRef]

22. Niculescu, A.-G.; Munteanu, O.M.; Bîrcă, A.C.; Moroșan, A.; Purcăreanu, B.; Vasile, B.Ș.; Istrati, D.; Mihaiescu, D.E.; Hadibarata, T.; Grumezescu, A.M. New 3D Vortex Microfluidic System Tested for Magnetic Core-Shell Fe_3O_4-SA Nanoparticle Synthesis. *Nanomaterials* **2024**, *14*, 902. [CrossRef]
23. Arévalo, P.; Isasi, J.; Caballero, A.C.; Marco, J.F.; Martín-Hernández, F. Magnetic and structural studies of Fe_3O_4 nanoparticles synthesized via coprecipitation and dispersed in different surfactants. *Ceram. Int.* **2017**, *43*, 10333–10340. [CrossRef]
24. Khan, M.A.M.; Khan, W.; Ahamed, M.; Alhazaa, A.N. Investigation on the structure and physical properties of Fe_3O_4/RGO nanocomposites and their photocatalytic application. *Mater. Sci. Semicond. Process.* **2019**, *99*, 44–53. [CrossRef]
25. Han, R.; Li, W.; Pan, W.; Zhu, M.; Zhou, D.; Li, F.-S. 1D Magnetic Materials of Fe_3O_4 and Fe with High Performance of Microwave Absorption Fabricated by Electrospinning Method. *Sci. Rep.* **2014**, *4*, 7493. [CrossRef]
26. Li, X.; Xiang, Z.; Donghua, Q.; Zhao, Z.; Jiaxiang, C.; Shao, Z. Dissimilatory Iron [Fe(III)] Reduction by a Novel Fermentative, Piezophilic Bacterium Anoxybacter fermentans DY22613T Isolated from East Pacific Rise Hydrothermal Sulfides. *Geomicrobiol. J.* **2019**, *36*, 291–302. [CrossRef]
27. Mercan, D.-A.; Niculescu, A.-G.; Bîrcă, A.C.; Cristea, D.-E.; Moroșan, A.; Tudorache, D.-I.; Purcăreanu, B.; Vasile, B.Ș.; Radu, D.; Grigoroscuta, M.A.; et al. Vortex-Mixing Microfluidic Fabrication of Micafungin-Loaded Magnetite–Salicylic Acid–Silica Nanocomposite with Sustained-Release Capacity. *Materials* **2024**, *17*, 5816. [CrossRef]
28. Cornell, R.M.; Schwertmann, U. *The Iron Oxides: Structure, Properties, Reactions, Occurrences, and Uses*; Wiley-vch Weinheim: Weinheim, Germany, 2003; Volume 664.
29. Fang, J.-M.; Li, S.-H.; Gong, W.-Q.; Sun, Z.-Y.; Yang, H.-G. FTIR study of adsorption of PCP on hematite surface. *Guang Pu Xue Yu Guang Pu Fen Xi* **2009**, *29*, 318–321. [PubMed]
30. Anum, K.; Naureen, H.; Riaz, T.; Sarwar, A.; Ahsan, A.; Butt, E.; Awais, S. Development and characterization of ceftriaxone in-situ gel-forming biodegradable parenteral depot system. *Trop. J. Pharm. Res.* **2021**, *20*, 673–680. [CrossRef]
31. Kawish, M.; Elhissi, A.; Jabri, T.; Muhammad Iqbal, K.; Zahid, H.; Shah, M.R. Enhancement in Oral Absorption of Ceftriaxone by Highly Functionalized Magnetic Iron Oxide Nanoparticles. *Pharmaceutics* **2020**, *12*, 492. [CrossRef]
32. Gunasekaran, S.; Charles, J. Spectral measurements and qualitative analysis of ceftriaxone and cefotaxime. *Asian J. Chem.* **2008**, *20*, 1343.
33. Manimekalai, P.; Manavalan, R. Selection of excipients for the formulation of Ceftriaxone sodium loaded chitosan nanoparticle through drug-excipient compatibility testing. *Int. J. PharmTech Res.* **2015**, *8*, 5–10.
34. Morimoto, T.; Hirata, H.; Eto, S.; Hashimoto, A.; Kii, S.; Kobayashi, T.; Tsukamoto, M.; Yoshihara, T.; Toda, Y.; Mawatari, M. Development of Silver-Containing Hydroxyapatite-Coated Antimicrobial Implants for Orthopaedic and Spinal Surgery. *Medicina* **2022**, *58*, 519. [CrossRef] [PubMed]
35. Sui, J.; Hou, Y.; Chen, M.; Zheng, Z.; Meng, X.; Liu, L.; Huo, S.; Liu, S.; Zhang, H. Nanomaterials for Anti-Infection in Orthopedic Implants: A Review. *Coatings* **2024**, *14*, 254. [CrossRef]
36. AbouAitah, K.; Bil, M.; Pietrzykowska, E.; Szałaj, U.; Fudala, D.; Woźniak, B.; Nasiłowska, J.; Swiderska-Sroda, A.; Lojkowski, M.; Sokołowska, B.; et al. Drug-Releasing Antibacterial Coating Made from Nano-Hydroxyapatite Using the Sonocoating Method. *Nanomaterials* **2021**, *11*, 1690. [CrossRef]
37. O' Sullivan, C.; O' Neill, L.; O' Leary, N.D.; O' Gara, J.P.; Crean, A.M.; Ryan, K.B. Osteointegration, antimicrobial and antibiofilm activity of orthopaedic titanium surfaces coated with silver and strontium-doped hydroxyapatite using a novel blasting process. *Drug Deliv. Transl. Res.* **2021**, *11*, 702–716. [CrossRef]
38. Maskowicz, D.; Maroszek, K.; Jendrzejewski, R.; Sawczak, M. Hydroxyapatite Nanocoatings Deposited by Means of Resonant Matrix-Assisted Pulsed Laser Evaporation. *Materials* **2024**, *17*, 5778. [CrossRef] [PubMed]
39. Dhingra, S.; Sharma, S.; Saha, S. Infection Resistant Surface Coatings by Polymer Brushes: Strategies to Construct and Applications. *ACS Appl. Bio Mater.* **2022**, *5*, 1364–1390. [CrossRef]
40. Perasoli, F.; Silva, L.; Figueiredo, B.; Bastos, J.; Carneiro, S.; Sampaio, J.; Araújo, V.; Beato, F.R.; Araújo, F.; Barboza, A.P. Mechanisms related to inhibition of fungal biofilm formation on medical device coated with poly (methylmethacrylate-co-dimethylacrylamide). *Biointerface Res. Appl. Chem.* **2023**, *13*, 332.
41. Akshaykranth, A.; Jayarambabu, N.; Rao, T.V.; Rakeshkumar, R. Structural, Thermal, Optical, Mechanical, and Antibacterial Properties of PLA/Nanoclay/TiO_2 Nanocomposite Films. *Lett. Appl. NanoBioSci.* **2023**, *12*, 70.
42. Jaworska, J.; Jelonek, K.; Jaworska-Kik, M.; Musiał-Kulik, M.; Marcinkowski, A.; Szewczenko, J.; Kajzer, W.; Pastusiak, M.; Kasperczyk, J. Development of antibacterial, ciprofloxacin-eluting biodegradable coatings on Ti6Al7Nb implants to prevent peri-implant infections. *J. Biomed. Mater. Res. Part A* **2020**, *108*, 1006–1015. [CrossRef]
43. Zhu, L.; Tong, X.; Ye, Z.; Lin, Z.; Zhou, T.; Huang, S.; Li, Y.; Lin, J.; Wen, C.; Ma, J. Zinc phosphate, zinc oxide, and their dual-phase coatings on pure Zn foam with good corrosion resistance, cytocompatibility, and antibacterial ability for potential biodegradable bone-implant applications. *Chem. Eng. J.* **2022**, *450*, 137946. [CrossRef]
44. Mihai, D.-C.; Dițu, L.-M.; Gheorghe, I.; Banu, O.; Georgescu, M.; Mihăescu, G. Investigation of the Antibiofilm Activity of Some Spices and Medicinal Plants Essential Oils. *Biointerface Res. Appl. Chem.* **2023**, *13*, 301.

45. Friedrich, R.P.; Cicha, I.; Alexiou, C. Iron Oxide Nanoparticles in Regenerative Medicine and Tissue Engineering. *Nanomaterials* **2021**, *11*, 2337. [CrossRef] [PubMed]
46. Priyadarshini, B.; Stango, A.X.; Balasubramanian, M.; Vijayalakshmi, U. In situ fabrication of cerium-incorporated hydroxyapatite/magnetite nanocomposite coatings with bone regeneration and osteosarcoma potential. *Nanoscale Adv.* **2023**, *5*, 5054–5076. [CrossRef]
47. Oprea, M.; Pandele, A.M.; Nechifor, A.C.; Nicoara, A.I.; Antoniac, I.V.; Semenescu, A.; Voicu, S.I.; Enachescu, C.I.; Fratila, A.M. Improved Biomineralization Using Cellulose Acetate/Magnetic Nanoparticles Composite Membranes. *Polymers* **2025**, *17*, 209. [CrossRef]
48. Yang, J.; Wu, J.; Guo, Z.; Zhang, G.; Zhang, H. Iron Oxide Nanoparticles Combined with Static Magnetic Fields in Bone Remodeling. *Cells* **2022**, *11*, 3298. [CrossRef] [PubMed]
49. Bin, S.; Wang, A.; Guo, W.; Yu, L.; Feng, P. Micro Magnetic Field Produced by Fe3O4 Nanoparticles in Bone Scaffold for Enhancing Cellular Activity. *Polymers* **2020**, *12*, 2045. [CrossRef]
50. Wang, Q.; Chen, B.; Cao, M.; Sun, J.; Wu, H.; Zhao, P.; Xing, J.; Yang, Y.; Zhang, X.; Ji, M.; et al. Response of MAPK pathway to iron oxide nanoparticles in vitro treatment promotes osteogenic differentiation of hBMSCs. *Biomaterials* **2016**, *86*, 11–20. [CrossRef]
51. Huang, Z.; He, Y.; Chang, X.; Liu, J.; Yu, L.; Wu, Y.; Li, Y.; Tian, J.; Kang, L.; Wu, D.; et al. A Magnetic Iron Oxide/Polydopamine Coating Can Improve Osteogenesis of 3D-Printed Porous Titanium Scaffolds with a Static Magnetic Field by Upregulating the TGFβ-Smads Pathway. *Adv. Healthc. Mater.* **2020**, *9*, e2000318. [CrossRef]
52. Zúñiga-Miranda, J.; Guerra, J.; Mueller, A.; Mayorga-Ramos, A.; Carrera-Pacheco, S.E.; Barba-Ostria, C.; Heredia-Moya, J.; Guamán, L.P. Iron Oxide Nanoparticles: Green Synthesis and Their Antimicrobial Activity. *Nanomaterials* **2023**, *13*, 2919. [CrossRef]
53. Sheikh, S.; Swapnil, K.; Tripathi, C.D.; Meshram, G.G.; Karim, B.A. Intra-operative drug level monitoring of pre-operative antibiotic for surgical prophylaxis in the patients of elective spinal surgery. *J. Basic Clin. Physiol. Pharmacol.* **2023**, *34*, 797–804. [CrossRef]
54. Yue, Q.; Shen, T.; Wang, C.; Gao, C.; Liu, J. Study on the interaction of bovine serum albumin with ceftriaxone and the inhibition effect of zinc (II). *Int. J. Spectrosc.* **2012**, *2012*, 284173. [CrossRef]
55. Chang, S.Y.; Kang, D.-H.; Cho, S.K. Innovative Developments in Lumbar Interbody Cage Materials and Design: A Comprehensive Narrative Review. *Asian Spine J* **2024**, *18*, 444–457. [CrossRef] [PubMed]
56. Niculescu, A.-G.; Chircov, C.; Grumezescu, A.M. Magnetite nanoparticles: Synthesis methods—A comparative review. *Methods* **2022**, *199*, 16–27. [CrossRef] [PubMed]
57. Blachowicz, T.; Grzybowski, J.; Ehrmann, A. Influence of agglomerations on magnetic properties of polymer matrices filled with magnetic nanoparticles. *Mater. Today Proc.* **2022**, *67*, 792–796. [CrossRef]
58. Koo, K.N.; Ismail, A.F.; Othman, M.H.D.; Bidin, N.; Rahman, M.A. Preparation and characterization of superparamagnetic magnetite (Fe3O4) nanoparticles: A short review. *Malays. J. Fundam. Appl. Sci.* **2019**, *15*, 23–31. [CrossRef]
59. Axente, E.; Elena Sima, L.; Sima, F. Biomimetic Coatings Obtained by Combinatorial Laser Technologies. *Coatings* **2020**, *10*, 463. [CrossRef]
60. Florea, D.A.; Grumezescu, V.; Bîrcă, A.C.; Vasile, B.Ș.; Iosif, A.; Chircov, C.; Stan, M.S.; Grumezescu, A.M.; Andronescu, E.; Chifiriuc, M.C. Bioactive Hydroxyapatite-Magnesium Phosphate Coatings Deposited by MAPLE for Preventing Infection and Promoting Orthopedic Implants Osteointegration. *Materials* **2022**, *15*, 7337. [CrossRef]
61. Russo, T.; Peluso, V.; Gloria, A.; Gargiulo, V.; Alfe, M.; Ausanio, G. An integrated design strategy coupling additive manufacturing and matrix-assisted pulsed laser evaporation (MAPLE) towards the development of a new concept 3D scaffold with improved properties for tissue regeneration. *Nanoscale Adv.* **2024**, *6*, 3064–3072. [CrossRef]
62. Alfe, M.; Minopoli, G.; Tartaglia, M.; Gargiulo, V.; Ausanio, G. Biocompatible Hybrid Graphenic Thin Coatings on Flexible Substrates through Matrix-Assisted Pulsed Laser Evaporation (MAPLE). *ACS Appl. Mater. Interfaces* **2024**, *16*, 38956–38967. [CrossRef]
63. Nehra, P.; Chauhan, R.P.; Garg, N.; Verma, K. Antibacterial and antifungal activity of chitosan coated iron oxide nanoparticles. *Br. J. Biomed. Sci.* **2018**, *75*, 13–18. [CrossRef] [PubMed]
64. Şen Karaman, D.; Karakaplan, M.B.; Erdoğan, N. Bacteriostatic polylactic acid coatings enriched with zinc oxide and silica nanoparticles for titanium pedicle screws. *JOM* **2021**, *73*, 4410–4418. [CrossRef]
65. Grămadă, A.M.; Stoica, A.-E.; Niculescu, A.-G.; Bîrcă, A.C.; Vasile, B.Ș.; Holban, A.M.; Mihaiescu, T.; Șerban, A.I.; Ciceu, A.; Balta, C.; et al. Zinc Oxide-Loaded Recycled PET Nanofibers for Applications in Healthcare and Biomedical Devices. *Polymers* **2025**, *17*, 45. [CrossRef]
66. AlQurashi, D.M.; AlQurashi, T.F.; Alam, R.I.; Shaikh, S.; Tarkistani, M.A.M. Advanced Nanoparticles in Combating Antibiotic Resistance: Current Innovations and Future Directions. *J. Nanotheranostics* **2025**, *6*, 9. [CrossRef]

67. Ana, I.D.; Vrana, N.E.; Morita, A.; Satria, G.A.P.; Hathroubi, S. Antibacterial surface functionalization of biomedical scaffolds: A transformation towards more adaptive, resilient regenerative therapy. *Results Surf. Interfaces* **2025**, *19*, 100481. [CrossRef]
68. Akshaya, S.; Rowlo, P.K.; Dukle, A.; Nathanael, A.J. Antibacterial Coatings for Titanium Implants: Recent Trends and Future Perspectives. *Antibiotics* **2022**, *11*, 1719. [CrossRef]

coatings

MDPI

Article

Preparation of Superhydrophobic Hydroxyapatite Coating on AZ31 Mg Alloy by Combining Micro-Arc Oxidation and Liquid-Phase Deposition

Yanqing Hu [1], Xin Liang [1], Yujie Yuan [2], Feiyu Jian [2] and Hui Tang [2,*]

1 Chengdu Aircraft Industry Group Co., Ltd., Chengdu 610092, China
2 School of Materials and Energy, University of Electronic Science and Technology of China, Chengdu 611731, China; yuanyujie2022@163.com (Y.Y.); 2022030901029@stu.uestc.edu.cn (F.J.)
* Correspondence: tanghui@uestc.edu.cn

Abstract: Magnesium as a biodegradable metal implant has garnered attention. Nevertheless, its rapid degradation rate and insufficient osseointegration restrict its clinical applications. In order to enhance the corrosion resistance and bioactivity of magnesium alloys, superhydrophobic hydroxyapatite (HA) layers were synthesized on micro-arc oxidized (MAO)-treated AZ31B magnesium alloy through liquid-phase deposition. This study examined the surface morphology, phase composition, bonding strength, wettability, electrochemical properties, and in vitro mineralization of the synthesized coatings. The study results demonstrated that the improved corrosion resistance of composite coatings in Hank's solution is due to the formation of a protective HA layer. The inclusion of the MAO coating significantly enhances the bonding strength between the hydroxyapatite (HA) layer and the bare magnesium alloy. The concentration of NaH_2PO_4 affects both the microstructure and wettability. The composite coating exhibited excellent osseointegration capabilities, with new HA layers observed after immersing the samples in simulated body fluid (SBF) solution for three days. These findings suggest that the combination of MAO and solution treatment presents a promising method for enhancing biocompatibility and reducing magnesium degradation, thus making it a viable option for biodegradable implant applications.

Keywords: magnesium alloy; micro-arc oxidation; liquid-phase deposition; hydroxyapatite; corrosion behavior

1. Introduction

Magnesium alloys are increasingly recognized as highly promising materials in the biomedical field owing to their unique advantages [1,2]. Primarily, magnesium is naturally present in the human body, which plays an important role in numerous biological functions. Their biocompatibility renders magnesium alloys suitable for medical implants [3,4]. A distinguishing feature of magnesium alloys is their biodegradability, which enables implants to gradually dissolve and can be absorbed or excreted by the body, thus eliminating the need for secondary surgery to remove the implant [5,6]. Magnesium alloys possess mechanical properties that more closely resemble natural bone compared to other metals such as titanium or stainless steel. This reduces the risk of stress shielding, a phenomenon where the implant assumes excessive load, resulting in bone weakening [7]. Furthermore, the degradation process releases magnesium ions, which can stimulate bone growth and healing, rendering them particularly beneficial for orthopedic applications [8]. Magnesium alloys can be tailored to adjust their degradation rates and mechanical properties by modifying their composition and processing methods, thereby offering flexibility for various biomedical applications [9].

Although magnesium alloy biomedical materials offer numerous advantages, they also present specific challenges and limitations that must be addressed. A primary concern is the rapid and often unpredictable degradation rate of magnesium alloys in physiological environments, which could impair implant integrity before the tissue has adequately healed. This rapid degradation can also result in the accumulation of hydrogen gas pockets around the implant site, potentially causing inflammation and impeding the healing process [10]. Moreover, magnesium and its alloys are generally regarded as non-bioactive, indicating that they do not inherently promote specific biological responses such as bone growth or tissue integration [11].

To address these challenges, different surface modification methods, such as chemical conversion coatings [12], electrodeposition [13], and micro-arc oxidation (MAO) [14], among others, are being investigated to improve the bioactivity and corrosion resistance of magnesium alloys. Among these techniques, MAO has attracted significant attention. MAO coatings create a stable ceramic layer on the surface of the magnesium alloy, which can significantly elevate the corrosion resistance of magnesium alloy and prolong its functional lifespan in the body [15]. However, MAO coatings typically exhibit a porous surface structure, which can serve as initiation sites for corrosion and compromise the protective nature of the coating. These defects may permit bodily fluids to reach the underlying magnesium alloy, potentially resulting in localized degradation. Therefore, it is crucial to seal the micro-pores and micro-cracks in the MAO coating to further enhance its anticorrosion performance. This can be achieved by integrating the MAO coating with additional complementary processes [16]. The combination of a hydroxyapatite (HA) layer and an MAO coating has emerged as an effective way of improving the corrosion resistance and bioactivity of MAO coatings on magnesium alloys [17,18]. HA layers are famed for their unique properties, such as promoting bone cell attachment and proliferation, thereby facilitating bone ingrowth and integration with the implant. The MAO coating is essential for creating a strong and durable bond with the substrate alloy within the composite structure. Not only does it form a robust connection, but it also cultivates an environment that fosters the formation of the HA layer within the structure. This specially designed environment facilitates the in situ generation of HA layers, ensuring that they integrate seamlessly. The HA layer is extremely efficient in sealing tiny flaws that may be present within the MAO coating.

Li et al. developed a distinct, double-layered MAO/HA coating. They achieved this by sealing the micro-pores in the MAO coating using a solution treatment, thereby significantly enhancing its corrosion resistance [19]. Tang et al. implemented a straightforward and eco-friendly steam sealing technique that significantly improved the corrosion resistance of the MAO coating on AZ31 alloy. This study also revealed the connection between wettability and osteogenesis [20]. Wang et al. produced needle-shaped fluoridated hydroxyapatite on the surface of the MAO layer. During degradation testing in simulated bodily fluids, the composite coating was found to have a strong ability to induce hydroxyapatite deposition on its surface [21]. Li et al. designed Ca-P nanostructured coatings on the surface of MAO coatings using chemical deposition. The Ca-P nanostructured coatings grew and completely covered the MAO coating. The morphology and thickness of Ca-P nanostructured coatings can be regulated by varying the deposition time [22]. Zhang et al. used a liquid-phase deposition technique to construct a Ca-P sealing coating on the outer surface of the MAO coating [23]. Farshid et al. devised bilayer HA/MAO coatings on WE43 alloy and revealed that these coatings exhibited higher adhesion strength and BMP-2 viability compared to random coatings. Moreover, the wettability of bilayer HA/MAO coatings can enhance osteo-immunomodulatory activity [24]. Recent studies involving MAO/HA bilayer coatings have clearly demonstrated that tuning the surface architecture and chemical compositions can modulate wettability, osteoclastogenesis, and corrosion resistance [25]. Unfortunately, the impact of a changing chemical environment on the morphology, wettability, and performance of the composite coating is seldom reported.

In this study, MAO/HA composite coatings were successfully fabricated on biodegradable AZ31 Mg alloy substrates using micro-arc oxidation and liquid-phase deposition (Scheme 1). Subsequently, the wettability and corrosion resistance of the coatings were evaluated to elucidate the relationship between corrosion performance and coating structure.

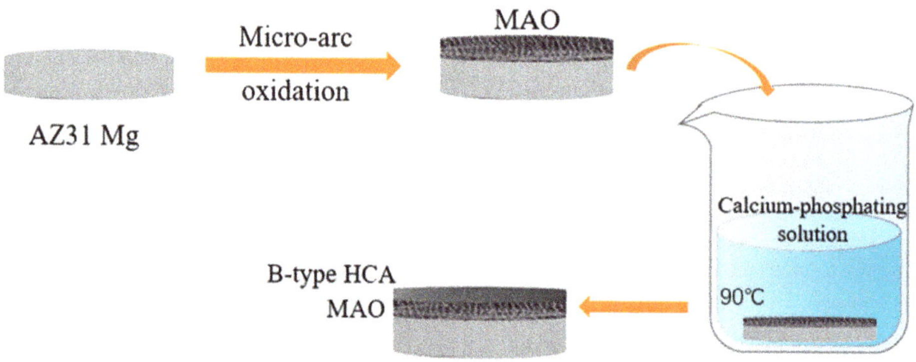

Scheme 1. Schematic representation of the preparation process of MAO-LPD coating.

2. Experiments

The surface modification process for a AZ31 magnesium alloy involves the following two steps: (i) high-potential anodization of the alloy to generate an MAO ceramic coating and (ii) solution treatment in $C_{10}H_{12}N_2O_8Na_2Ca$ and NaH_2PO_4 to form a hydroxyapatite (HA) coating.

2.1. Preparation of MAO Coating

AZ31 magnesium alloy disks were machined to dimensions of $\phi16 \times 6$ mm. The disks were ground using abrasive papers, ultrasonically cleaned with $(CH_3)_2CO$ and H_2O, and subsequently dried. A fresh electrolyte was made by mixing reagent-grade chemicals, specifically 6 g/L NaOH and 12 g/L $Na_3PO_4 \cdot 12H_2O$, in deionized water for the micro-arc oxidation (MAO) process. The applied voltage was set to 400 V, with a frequency of 1000 Hz, a duty cycle of 10%, and an oxidation duration of 10 min. The temperature of the electrolyte was kept at 30 °C using a cooling system. Samples prepared via micro-arc oxidation were designated as MAO coatings.

2.2. Liquid-Phase Deposition

In the subsequent phase, hydroxyapatite (HA) coatings were applied to the MAO-coated samples via liquid-phase deposition. The precursor solution was prepared by dissolving Na_2HPO_4 and EDTACa in H_2O. The pH was adjusted to 9 by adding 0.5 mol/L NaOH, and the Ca/P ratio was maintained at 1.67 by adjusting the concentrations of NaH_2PO_4 and EDTACa. The magnesium alloy coated with MAO was immersed in the solution at 90 °C for eight hours. The concentrations of Na_2HPO_4 were controlled at 0.05, 0.1, and 0.15 mmol, respectively, and the resulting samples were labeled as MAO-LPD-1,

MAO-LPD-2, and MAO-LPD-3. Following the reaction, the samples were rinsed with H_2O and dried.

2.3. Morphologies and Characterizations

The microscopic morphologies of the coated samples were examined via scanning electron microscopy (SEM, FEI, QUANTA-200F). Elemental concentrations were determined using an energy-dispersive X-ray spectrometer (EDS, Oxford Model 7537). The phase composition of the coated samples was analyzed using thin-film X-ray diffraction (TFXRD, Philip X'Pert) with Cu Kα radiation at a glancing angle of 2°. Additionally, Fourier-transform infrared spectroscopy (FT-IR, Bruker Vector 22, Berlin, Germany) was employed to analyze the phase and structure of the specimens. The coating thickness was measured with an eddy current-based thickness gauge (CTG-10, Time Company, China). The surface roughness of the specimens (Ra) was optically measured using a TR200 roughness gauge (Time Company, China). The static contact angles of the various surfaces were measured at 25 °C and 50% relative humidity using the sessile drop method. This was conducted using a 1 μL water droplet within a telescopic goniometer (OCA20, Germany). The average contact angle values were calculated from measurements taken at five different positions, and images were captured using a digital camera.

2.4. Bonding Strength Tests

The interfacial bond strength between the composite coating and the AZ31 magnesium alloy substrate was evaluated using an adhesion–tension test, in accordance with ASTM C633 [26]. Sample preparation involved removing the coating from one side and roughening the exposed substrate. Both sides of the prepared samples were then affixed to cylindrical steel jigs (10 mm diameter) using J-39 adhesive. For each sample type, the reported bonding strength represents the average of five independent measurements.

2.5. Electrochemical Tests

Corrosion measurements were performed using a PARSTAT 2273 potentiostat in a standard three-electrode configuration. The configuration comprised a saturated Ag/AgCl reference electrode, a platinum counter electrode, and a working electrode with an active area of 1 cm^2. Polarization tests were carried out in Hank's solution maintained at 37 °C. Prior to testing, the samples were immersed in Hank's solution for 1 h to stabilize the open circuit potential (OCP). A potential range from −0.5 V to +0.5 V (vs. OCP) was applied at a scan rate of 1 mV/s. The Tafel extrapolation method was utilized to estimate the corrosion current density (Icorr) and corrosion potential. The polarization resistance (Rp) was calculated using the Stern–Geary equation. Electrochemical impedance spectroscopy (EIS) measurements were conducted over a frequency range of 100 kHz to 10 mHz, with a perturbation amplitude of 10 mV. To ensure reproducibility, all electrochemical tests were conducted in triplicate.

2.6. In Vitro Mineralization Tests

The specimens were vertically immersed in simulated body fluid (SBF) contained within a glass beaker to assess their in vitro biomineralization behavior. A volume-to-sample area ratio of 20 mL/cm^2 was maintained. The samples were soaked for 3 days. After the immersion test, the samples were removed from the SBF, rinsed with deionized water, and dried with warm air. The samples' composition and surface microstructure were subsequently analyzed using X-ray diffraction (XRD) and scanning electron microscopy (SEM).

3. Results and Discussions

3.1. Surface Morphologies and Elemental Ratio of the Coatings

Figure 1 illustrates the surface morphologies of the MAO, MAO-LPD-1, MAO-LPD-2, and MAO-LPD-3 coatings. The surface morphology of the MAO-coated Mg (Figure 1a) exhibits a characteristic structure with uniformly distributed agglomerated solid particles and micro-pores. These holes are formed when molten oxide and gas bubbles are expelled from the micro-arc discharge channels during the MAO process [27]. These pores potentially compromise the long-term corrosion resistance of the MAO coating by allowing corrosive fluids to penetrate. Fortunately, the structure provides additional sites for HA nucleation, enhancing interface binding during solution treatment. In the MAO-LPD-1 sample, the porous structure is similar to that of the MAO sample. However, the micropores are smaller, and the solid particles have increased in size (Figure 1b). With increased NaH_2PO_4 concentration, the synthesized coating consists of dandelion flower-like clusters formed by numerous nanoscale needle-like crystals. The layer is uniform and nearly free of pores or cracks, as shown in Figure 1c. This homogeneity prevents contact between the MAO coating and corrosive fluids. At a NaH_2PO_4 concentration of 1.5 mmol, the composite coating surface is uniformly and completely covered with numerous bar-like structures. The high-magnification SEM image inset in Figure 1d reveals a surface morphology featuring numerous nano-plates with thicknesses ranging from approximately 1–3 μm and lengths between 10 and 20 μm. It is evident that the concentration of NaH_2PO_4 significantly affects the morphology.

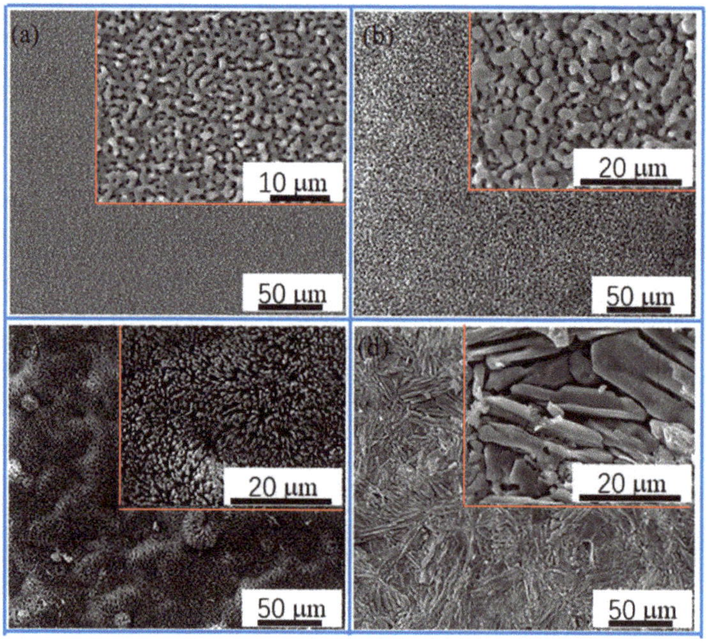

Figure 1. Surface morphologies of MAO and MAO-LPD coatings: (**a**) MAO, (**b**) MAO-LPD-1, (**c**) MAO-LPD-2, and (**d**) MAO-LPD-3.

EDS analysis shows that the composite coatings contain the elements Mg, O, P, and Ca (Figure 2). Contrasted with the MAO coating, the Mg content in the composite coatings is significantly lower due to the formation of solution treatment deposit layers. It can be deduced that Mg originates from the MAO coating or Mg substrate and deposits into the outer layer. The Ca elements originate from the solution. From the EDS results, the relative concentrations of Mg and O decrease with the increase of NaH_2PO_4 concentration in the solution. There is an increase in the relative concentration of P. The relative concentration of Ca increases initially, followed by a decrease. This trend is related to the deposition layer formation process. The Ca/P atomic ratio is calculated and shown in Figure S1. The Ca/P atomic ratio is 1.91, 1.15, and 0.77 for the MAO-LPD-1, MAO-LPD-2, and MAO-LPD-3 coatings, respectively. Notably, the NaH_2PO_4 concentration in the solution significantly affects the Ca/P atomic ratio. This variation is due to the differing deposition rates of calcium and phosphorus ions. Additionally, the high ionization tendency of magnesium (Mg) and the release of Mg^{2+} ions may result in the replacement of calcium (Ca^{2+}) ions. This study shows that Mg^{2+} ions are released from the PEO layer when MgO dissolves.

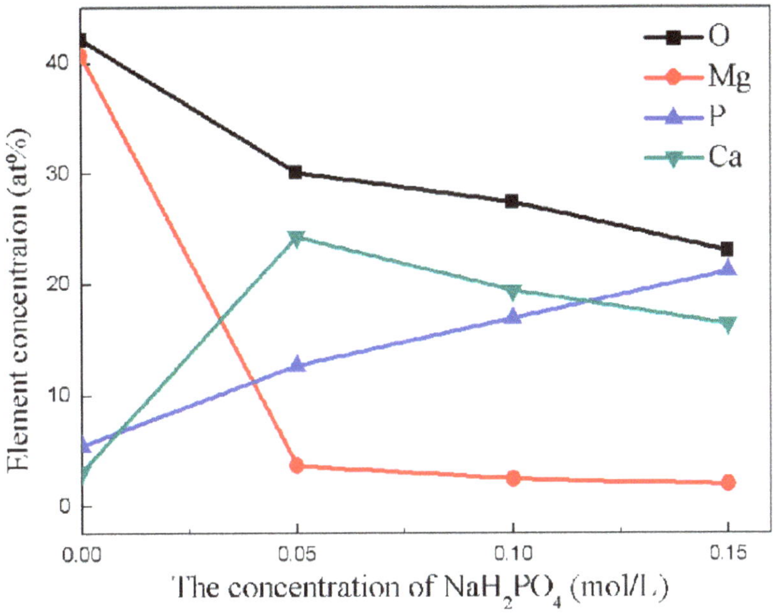

Figure 2. The relative contents of elements in the composite coatings.

3.2. Cross-Sectional Morphology and EDS Analysis

The thickness of the composite coatings formed at different NaH_2PO_4 concentrations via MAO and solution treatment is shown in Figure S2. The thickness of the MAO coating was roughly 13.5 ± 0.7 μm. Measurements of the composite coatings revealed thicknesses of approximately 15.2 μm, 20.8 μm, and 22.4 μm for the MAO-LPD-1, MAO-LPD-2, and MAO-LPD-3, respectively. This increase is attributed to the formation of the HA layer. The cross-sectional images and corresponding EDS mappings of the MAO and MAO-LPD-2 samples are presented in Figure 3. According to its cross-section morphology (Figure 3a), the MAO coating comprises a double layer: a loose, porous outer layer and a dense inner layer. The micropores in the outer layer are closely associated with residual discharge channels and entrapped gas pores. Figure 3b displays a cross-sectional view of the MAO-LPD-2 coatings. The pores in the outer layer of the MAO coating are filled by HA synthesized through solution treatment despite the complex interior shapes of these pores. The LPD layer exhibits a high degree of density and uniformity, exhibiting a close proximity to the MAO ceramic layer. Since the treatment solution contacts the entire surface of the MAO coating, including the pores, the HA layer demonstrates an effective sealing capability. Given the excellent sealing properties and bioactivity of the HA layer, the MAO-LPD composite coating is expected to exhibit improved corrosion resistance and enhanced bioactivity. EDS area scanning (indicated in Figure 3b) for the MAO coating and MAO-LPD-2 coating is shown in Figures 3c and 3d, respectively. Contrasted with the MAO coating, the Mg content decreased markedly, while the Ca and C contents increased significantly. The Ca element originates from the solution, while the C element may derive from CO_2 dissolved in the water. This suggests the formation of carbonate-substituted hydroxyapatite, aligning with previous reports [26].

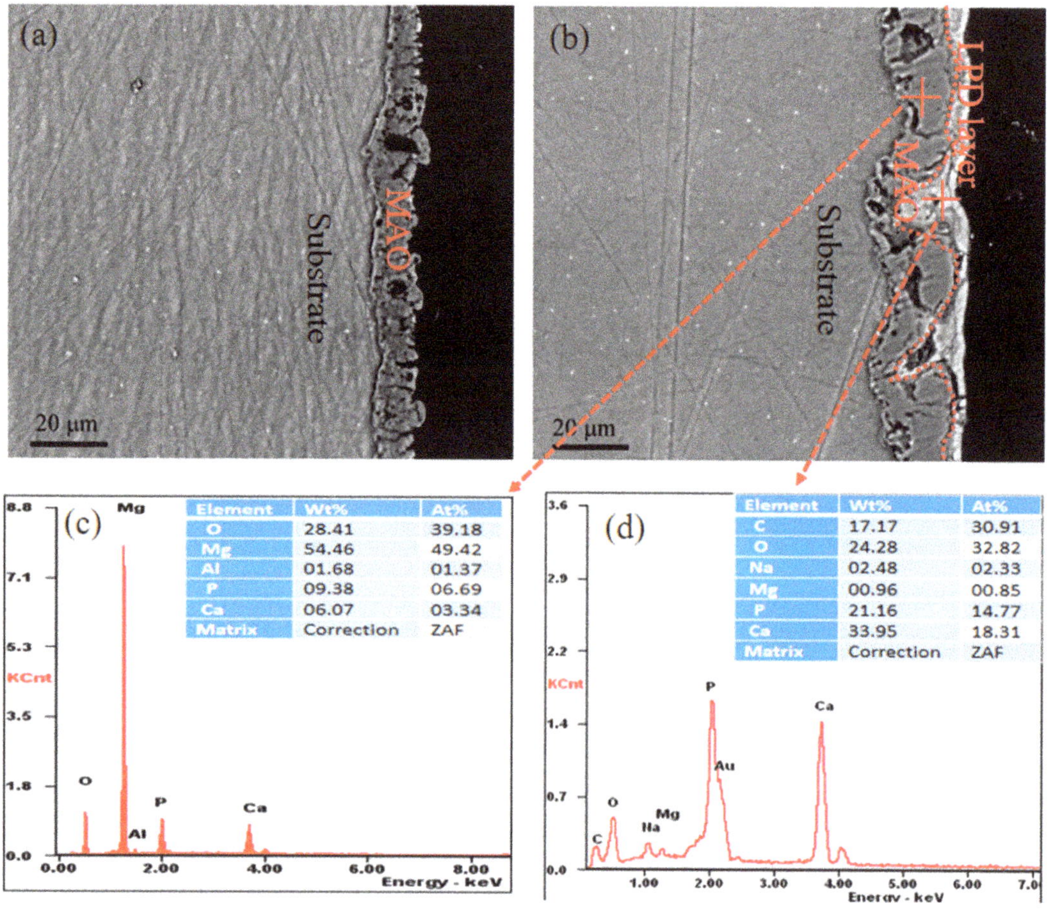

Figure 3. Cross-section morphologies and EDS analysis: (**a**) MAO coating, (**b**) MAO-LPD-2, (**c**) EDS analysis for MAO section, and (**d**) EDS analysis for LPD layer.

Figure 4 presents the EDS analysis of the MAO-LPD-2 coating cross-section. Near the coating/substrate interface, the concentration of P gradually increases from the substrate towards the coating, whereas the concentration of Mg exhibits an opposite trend. A reduction in Mg concentration is observed at the interface between the MAO coating and the LPD layer with decreasing depth. However, the Ca content increases sharply in this interfacial region. The concentration of P remains relatively constant throughout the coatings. The variations in Ca concentration within the MAO coating indicate that the solution penetrated into the coating, depositing in the pores and micro-cracks, thereby sealing them.

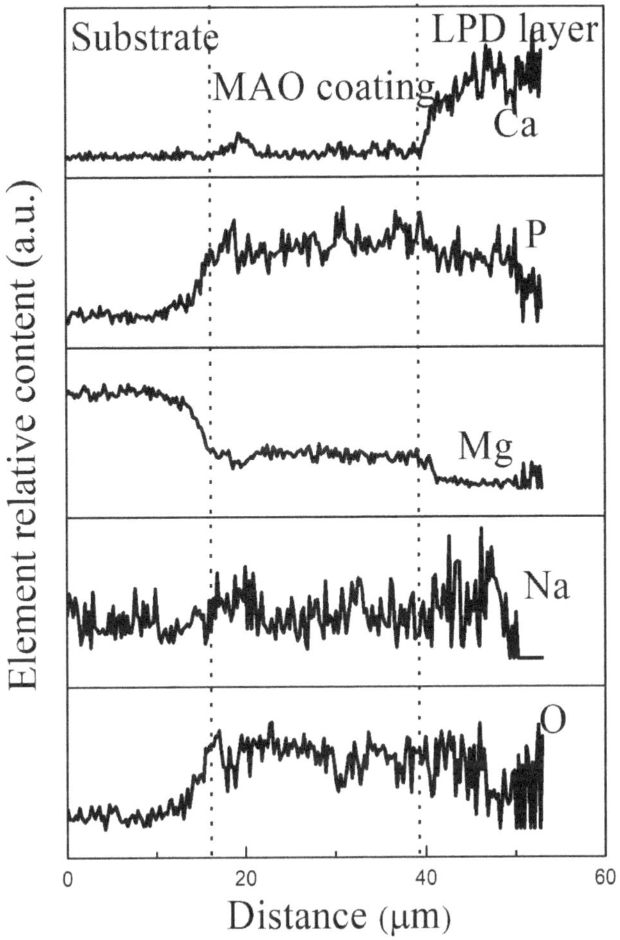

Figure 4. Element line analysis of MAO-LPD-2 coating.

Figure 5 presents the XRD patterns of MAO, MAO-LPD-1, MAO-LPD-2, and MAO-LPD-3. Figure 5a shows the XRD patterns of MAO-treated Mg. Besides the diffraction peaks from Mg, only peaks corresponding to MgO are detected on the MAO substrate. No P-containing phase is identified in the XRD measurements, likely due to its low content and nanocrystalline or amorphous structure. After the solution treatment, HA diffraction peaks (JCPDS No. 54-0022) appear in all coatings, indicating that an HA layer was successfully synthesized on the MAO surface [28]. As the concentration of NaH_2PO_4 in the solution increases, the XRD patterns of the MAO-LPD samples remain similar to that of the HA sample, suggesting that the NaH_2PO_4 concentration does not significantly impact the original HA coatings. Concurrently, the corresponding intensities of Mg and MgO decrease significantly with increasing NaH_2PO_4 concentration, indicating a probable increase in the thickness of the composite coatings.

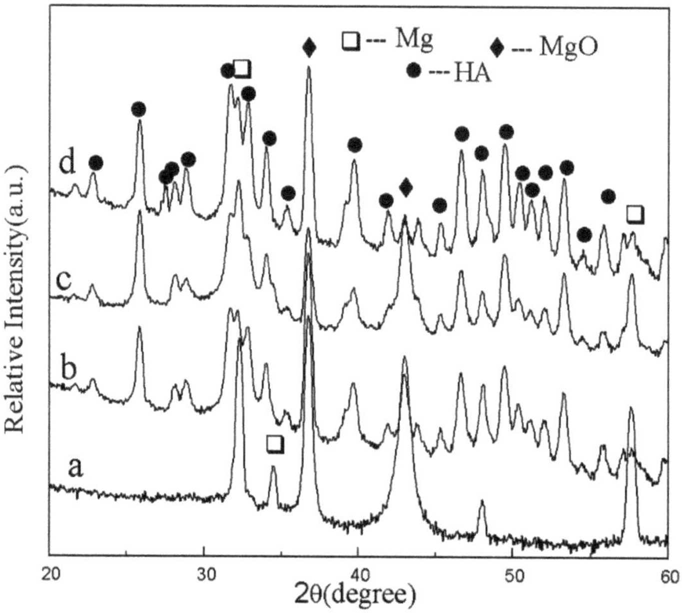

Figure 5. XRD patterns of the coatings: (a) MAO, (b) MAO-LPD-1, (c) MAO-LPD-2, and (d) MAO-LPD-3.

Figure 6 presents the FT-IR spectra of the composite coatings treated with a combination of MAO and solution treatment. The spectra reveal absorption peaks at 1021 cm^{-1} and 598 cm^{-1} corresponding to the vibration modes of the P-O bond in the phosphate group, indicating the formation of an apatite structure [29]. The presence of H_2O in the coating is indicated by two peaks: a broad absorption band at 3437 cm^{-1} and a bending mode at 1646 cm^{-1} [30]. Additionally, absorption peaks at 860 cm^{-1}, 1448.0 cm^{-1}, and 1414.5 cm^{-1} are associated with the symmetric bending and asymmetric stretching vibrations of CO_3^{2-} [31]. The presence of the CO_3^{2-} absorption peak is likely due to the reaction of CO_2 in the air with the solution, suggesting the presence of B-type HCA, wherein some PO_4^{3-} groups are substituted by CO_3^{2-} groups [32].

Figure S3 shows the surface roughness (average Ra values) of the MAO and composite coatings. It is abundantly clear that the surface roughness of the MAO coating significantly increases after the solution treatment. The composite coatings have undergone an increase in surface roughness which is attributed to solution deposition layers on the surface of the MAO coating; furthermore, this observation aligns well with the changes in surface morphologies depicted in Figure 1. As the concentration of NaH_2PO_4 in the solution increases, the Ra value of the composite coatings initially rises and then declines, reaching a maximum when the concentration of NaH_2PO_4 is 0.1 mmol/L.

The wetting behavior of the samples was examined using water contact angle (CA) measurements. Figure 7 presents photographs of 1 μL water droplets on the coated samples. It is evident that the surface of the MAO-coated magnesium alloy is hydrophilic, exhibiting a CA of approximately 87°. This high hydrophilicity, combined with surface defects, makes the MAO film more susceptible to corrosive environments. Upon solution treatment with varying concentrations of NaH_2PO_4, the contact angles for MAO-LPD-1, MAO-LPD-2, and MAO-LPD-3 were found to be 118°, 161°, and 152°, respectively. Both MAO-LPD-2 and MAO-LPD-3 coatings exhibited superhydrophobic characteristics. The consistent changes in contact angles, along with SEM results, indicate that the highly textured morphology is crucial for water repellency. The dandelion flower-like structure, composed of needle-like nanorods, creates a hydrophobic surface, as shown in Figure S4. Furthermore, the superhydrophobic structure can block contact with corrosive agents, thereby providing improved corrosion resistance.

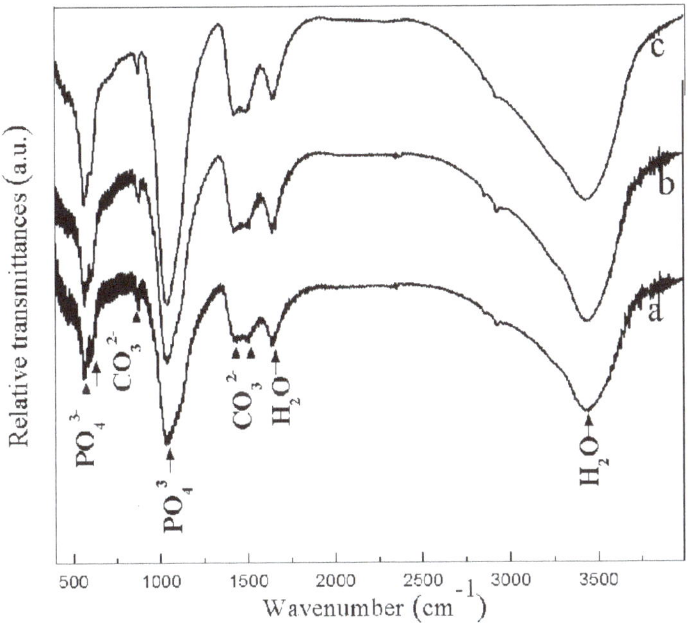

Figure 6. FT-IR spectra of MAO-LPD coatings (a) MAO-LPD-1, (b) MAO-LPD-2, and (c) MAO-LPD-3.

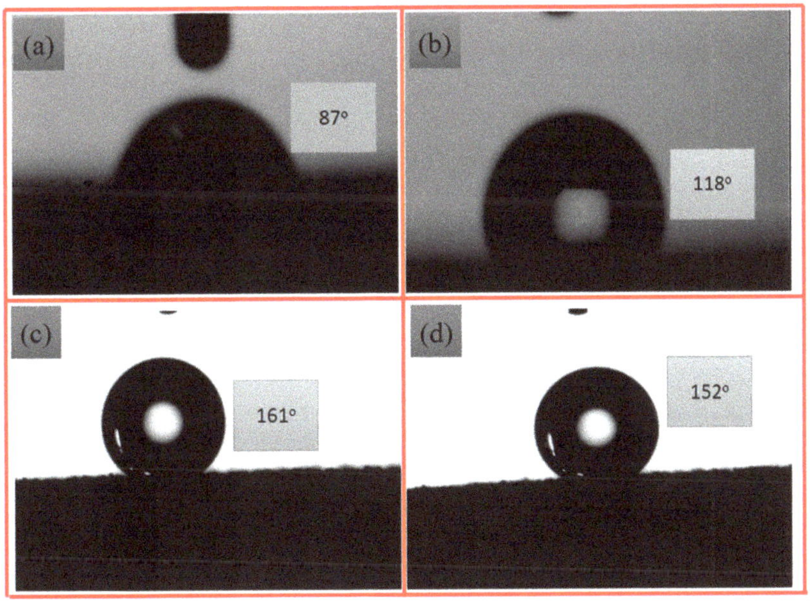

Figure 7. Contact angles on the surface of MAO coating and MAO-LPD coatings: (**a**) MAO, (**b**) MAO-LPD-1, (**c**) MAO-LPD-2, and (**d**) MAO-LPD-3.

The bonding strength of implant materials is crucial for their clinical applications. This strength is assessed by the durability of the coatings on the modified implants [33]. Figure 8a highlights the importance of the bonding strength between the coating and the substrate. The control sample, with an LPD coating, exhibits an average bonding strength of 15 MPa. For the MAO-LPD composite coatings, the MAO interlayer significantly enhances the bonding strength between the LPD coating and the AZ31 magnesium alloy substrate. Specifically, the bonding strengths for MAO-LPD-1, MAO-LPD-2, and MAO-LPD-3 are 172%, 110%, and 75% higher, respectively, compared to the LPD coating alone. This increased bonding strength is attributed to the porous outer layer of the MAO coating, which interlocks with the LPD layer, enhancing mechanical interlocking.

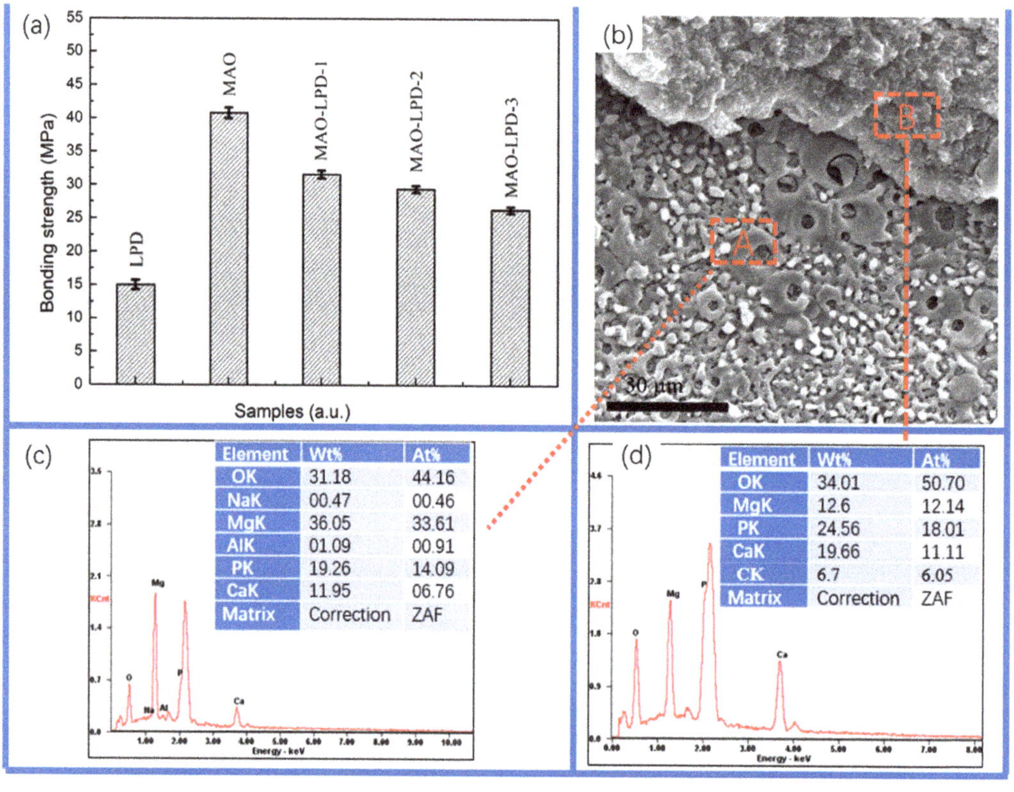

Figure 8. Bonding strength, typical fractured surface, and EDS analysis of the fractures, (**a**) the bonding strength of the coatings, (**b**) SEM photographs of the fractured surface after the bond strength test, (c) EDS analysis on area A, and (**d**) EDS analysis on area B.

Figure 8b illustrates the typical morphology of the MAO-LPD-2 coating, revealing two distinct regions, labeled A and B. These regions represent different planes of crack propagation within the composite coatings and adhesive joints. The majority of the fracture surface corresponds to plane A where the original morphology of the MAO layer has been retained and most of the LPD layer has peeled off, indicating that the tensile fracture occurs at the interface between the MAO and LPD layers. Region B exhibits fewer micropores due to the sealing effect of the LPD layer and the adhesive.

Energy-dispersive X-ray (EDX) elemental maps of the fracture surface at the boundary between Regions A and B are presented in Figure 8c,d. Region A exhibits high concentrations of magnesium and low concentrations of calcium, while Region B exhibits the opposite, reflecting the higher calcium content in the LPD layer compared to the MAO layer. The substantial carbon content in Region B indicates the presence of epoxy adhesive, characteristic of cohesive failure within the coating manifesting as spot flaking on the coating surface.

Metal implants are susceptible to corrosion and the release of metal ions upon contact with tissues and body fluids, which can degrade their performance. The parameters of corrosion resistance in metal implants are critical for determining their efficiency and performance. The corrosion resistance of the AZ31 alloy, MAO, MAO-LPD-1, MAO-LPD-2, and MAO-LPD-3 was evaluated using potentiodynamic polarization tests in Hank's solution, with the corresponding polarization curves presented in Figure 9. From these curves, the corrosion potentials (Ecorr) and corrosion current densities (Icorr) of the samples were derived and summarized in Table 1.

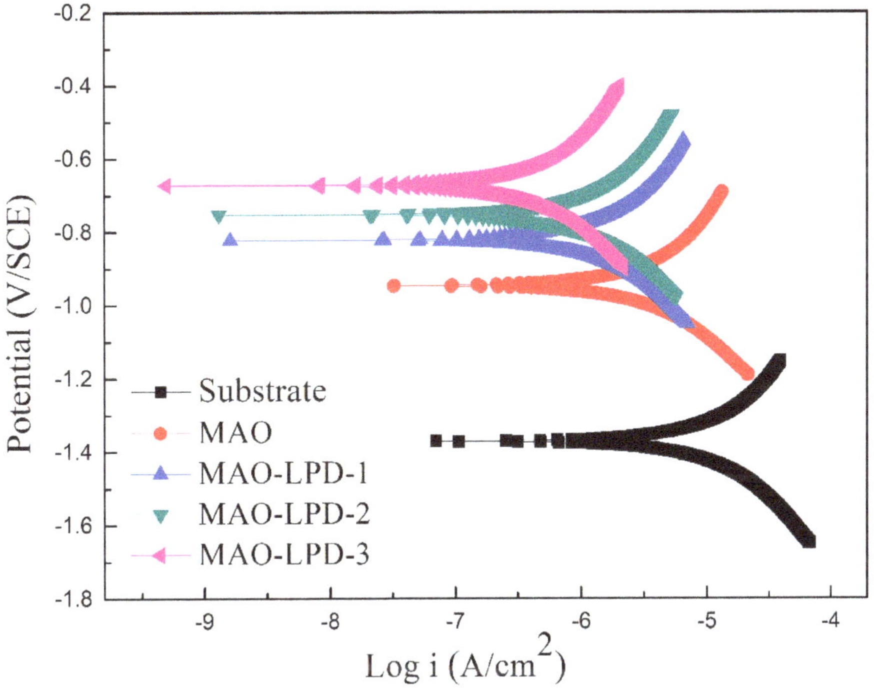

Figure 9. Polarization curves of the coatings.

Table 1. The results of potentiodynamic polarization test in Hank's solution.

Samples	E_{corr} (mV)	I_{corr} (A/cm^2)
AZ31 alloy	−137	86.7
MAO coating	−95	5.23
MAO-LPD-1	−83	0.933
MAO-LPD-2	−77	0.548
MAO-LPD-3	−65	0.281

The MAO coating exhibited a more positive corrosion potential than the uncoated magnesium alloy, with solution treatment further enhancing the coating's corrosion potential. The MAO treatment significantly reduced the Icorr of the magnesium alloy from 86.7 μA/cm^2 to 5.23 μA/cm^2, demonstrating a marked improvement in corrosion resistance. This enhancement is attributed to the MAO coating's ability to effectively impede the penetration of corrosive media, thereby improving the corrosion resistance of the magnesium alloy.

After the PEO samples were treated with the solution, there was a clear improvement in the Ecorr values for all of the samples, MAO-LPD-1 (−83 mV), MAO-LPD-2 (−77 mV), and MAO-LPD-3 (−65 mV), compared to the MAO coating (−95 mV). This shift toward more positive Ecorr values indicates an enhancement in the electrochemical stability of the samples. Furthermore, the Icorr of the samples following solution treatment was significantly reduced, approximately one to two orders of magnitude lower than that of the MAO coating, illustrating the additional improvement in corrosion resistance conferred by the solution treatment.

Moreover, as the NaH_2PO_4 concentration increases, the corrosion current density progressively decreases. The MAO-LPD-3 coating exhibited the lowest Icorr value and the highest Ecorr value, suggesting that it provides the most efficient protection for the Mg alloy. These results highlight the superior corrosion resistance of the MAO-LPD-2 and MAO-LPD-3 coatings, which can be attributed to the coatings' hydrophobic properties and the occlusion of most micro-pores in the MAO layer, thereby improving the overall corrosion protection.

To thoroughly investigate the electrochemical properties of the samples, electrochemical impedance spectroscopy (EIS) was performed using Hank's solution. Figure 10 presents the Nyquist plots for bare AZ magnesium, the MAO-coated sample, MAO-LPD-1, MAO-LPD-2, and MAO-LPD-3 coatings. The experimental data are represented by the symbols, while the fitted data are shown by the solid lines. The Nyquist plots clearly show that the radius of the capacitive loop for the MAO-treated samples is significantly larger than that of the pure magnesium alloy. Furthermore, the radius of the capacitive loop for the solution-treated samples is considerably larger than that of the MAO-coated samples. As the concentration of NaH_2PO_4 in the solution increased, the capacitive loop of the MAO-LPD composite coatings continued to expand. This observation suggests that higher NaH_2PO_4 concentrations in the solution enhanced the sealing effect of the MAO coating and also confirms the MAO-LPD composite coatings' ability to effectively prevent the infiltration of corrosive media, thereby significantly improving the corrosion resistance of magnesium alloys. In general, a larger radius of the capacitive loop corresponds to higher resistance, indicating a greater resistance to electron exchange. For metal surfaces, having a lower electron exchange capacity means less corrosion.

Figure 11 shows the equivalent circuit models that were used to fit the experimental data to different types of electrode/electrolyte interfaces. Capacity-depressed semicircles in the high-frequency range are exhibited by the Nyquist curves of the bare Mg alloy, and the Nyquist plots were simulated by the equivalent circuit shown in Figure 11a. The Nyquist plot of MAO-coated Mg is characterized by a high-frequency capacitive loop, followed by a low-frequency capacitive loop. The outer coating layer's properties are represented by the high-frequency capacitive loop, while the inner coating layer's properties are represented by the low-frequency capacitive loop [34]. With regard to the MAO-treated specimen, two possibilities emerge with respect to the number of time constants in the EIS results, i.e., two or three time constants, which may be ascribed to disparities in electrical parameters and electrolyte composition [35]. Discerning between the external and internal layers is frequently arduous due to the relatively feeble barrier properties of the external layer. Occasionally, the outer and inner layers come together, or the time constant linked to the outer layer cannot be measured [36]. In this study, two well-defined time constants emerge at medium and low frequencies, which are regarded as representations of the responses of the PEO layer and the double electric layer, respectively. Thus, a two-time-constant model, as shown in Figure 11b, was employed to fit the EIS results. Following liquid-phase deposition, a new time constant appears at high frequencies. This is associated with the deposition layer. Consequently, three time constants are distributed across the high-, medium-, and low-frequency ranges. The time constants at elevated and moderate frequencies are indicative of the responses exhibited by the deposition layer and the MAO outer layer, correspondingly. The electrochemical activity at the double layer between the MAO inner layer and the Mg substrate is reflected by the time constant at low frequencies. As shown in Figure 2, SEM observations have confirmed the presence of two protective layers on the magnesium surface. In accordance with the architecture of the coatings, the equivalent circuits employed for the purpose of fitting the impedance data are demonstrated in Figure 11c. In Figure 11, R_s represents the resistance of the electrolyte between the working and reference electrodes; in both circuits, Q_{2-n} and Q_{3-n} model the electrical properties of the outer and inner layers of the MAO coating, respectively; R_2 is the resistance of the outer layer saturated by the electrolyte; and R_3 represents the resistance of the inner layer of the MAO coating. For the MAO-LPD-1, MAO-LPD-2, and MAO-LPD-3 samples, R_1 and Q_{1-n} were added to the circuit model to describe the sealing treatment's effectiveness and obtain the best fit. The fitting results are provided in Table 2.

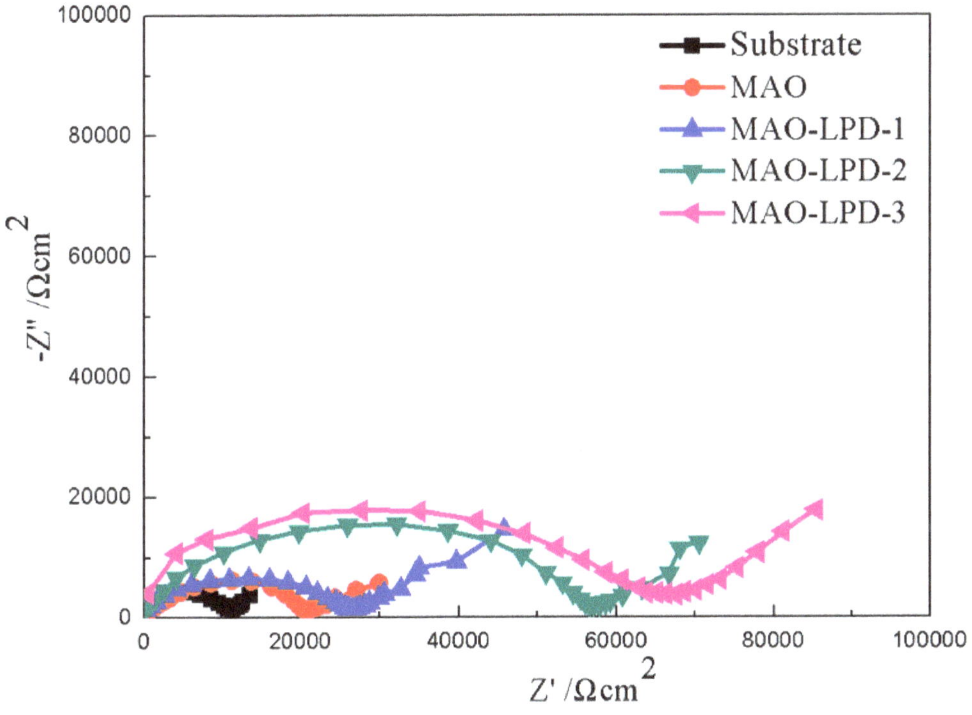

Figure 10. EIS spectra of the coatings.

To mitigate the scatter effect at the electrode/electrolyte interface, constant phase elements (CPEs), represented by the symbol Q, were employed as a replacement for capacitance. The formula for CPE impedance, in which Y_0 represents the CPE constant and n is the empirical exponent, is defined as follows: if $n = 1$, Q exhibits purely capacitive behavior, whilst if $n = 0$, Q exhibits purely resistive behavior. The behavior of CPE is typically attributed to factors such as distributed surface reactivity, surface inhomogeneity, roughness, or electrode porosity [37,38].

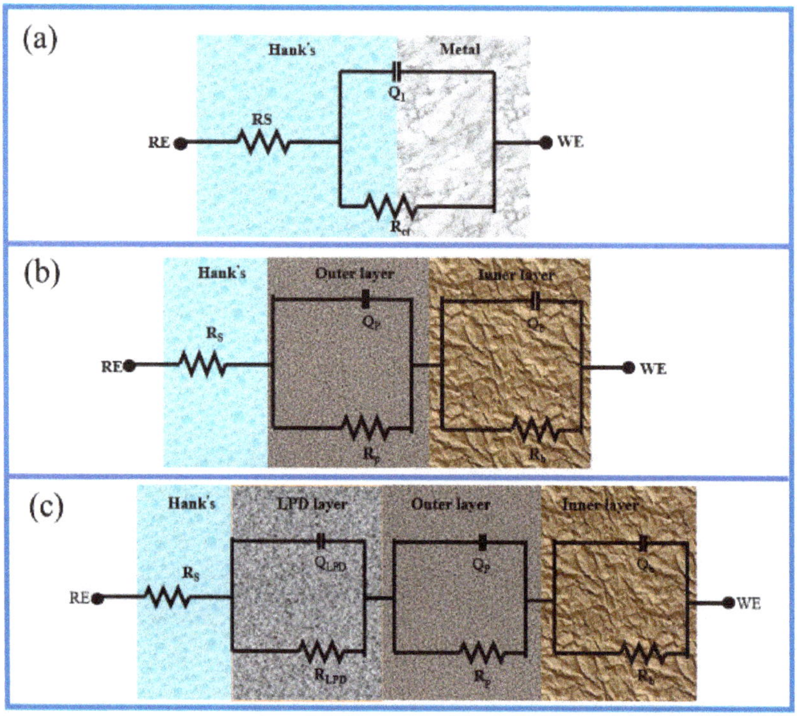

Figure 11. Simplified structure sketch of equivalent circuit: (**a**) bare magnesium alloy, (**b**) MAO, and (**c**) MAO-LPD coatings.

Table 2. Fitting results of EIS plots of the magnesium alloy AZ31 with MAO coatings and MAO-LPD coatings.

Samples	Rs(Ω·cm²)	$Q_1 - Y_0$ (Ω^{-1} cm^{-2} s^{-n})	Q_{1-n}	R_1(Ω·cm²)	$Q_2 - Y_0$ (Ω^{-1} cm^{-2} s^{-n})	Q_{2-n}	R_2(Ω·cm²)	$Q_3 - Y_0$ (Ω^{-1} cm^{-2} s^{-n})	Q_{3-n}	R_3(Ω·cm²)
Sub	7.3×10^{-2}							1.65×10^{-5}	0.87	5015
MAO	9.4×10^{-4}				7.88×10^{-7}	0.83	5712	3.76×10^{-4}	0.81	7132
MAO-LPD-1	1.7×10^{-6}	7.53×10^{-5}	0.81	972	1.65×10^{-6}	0.76	1.1×10^4	4.72×10^{-4}	0.79	8163
MAO-LPD-2	7.2×10^{-6}	1.87×10^{-7}	0.77	2873	9.55×10^{-6}	0.71	1.6×10^4	8.12×10^{-4}	0.74	9447
MAO-LPD-3	9.5×10^{-6}	2.61×10^{-7}	0.79	3223	9.78×10^{-6}	0.67	1.8×10^4	7.75×10^{-4}	0.71	9852

The R_s values of the samples tested in solutions are comparable due to the consistent composition of the solutions and the fixed positions of the working electrode and reference electrode during the tests. The MAO impedance fitting results from the circuit diagram clearly indicate an increase in the resistance of the MAO coating. Specifically, the resistance value increased by two orders of magnitude, thereby enhancing corrosion protection through the formation of a ceramic oxide coating on the Mg alloy surface. Although the MAO coating surface contains numerous micropores, these pores are not interconnected and the inner MAO layer consists of dense MgO, thereby providing effective corrosion protection for the Mg alloy. Additionally, in MAO coatings the Rb values were found to exceed their corresponding Rp values. This is attributed to the dense and flawless morphology of the inner layer of the coatings, while the outer layer displays a more porous structure.

The additional components, R_1 with resistance values of 972, 2873, and 3223 Ω·cm², were provided by the MAO-LPD-1, MAO-LPD-2, and MAO-LPD-3 samples. Furthermore, the R_2 values of MAO-LPD-2 and MAO-LPD-3 are approximately three and four orders of magnitude greater than those of the MAO-coated Mg, respectively. This suggests that sealing the pores in the MgO layer with HA effectively prevents the penetration of corrosive solutions into the coatings. Additionally, it is noteworthy that the fitted Q_{2-n} values of the coatings follow the order MAO > MAO-LPD-1 > MAO-LPD-2 > MAO-LPD-3. This indicates that MAO-LPD-3 coatings have a dielectric constant lower than that of the MAO-LPD-1 coating and significantly lower than that of the MAO coating. It is well-established that more porous and defective coatings exhibit higher dielectric constants. Therefore, the significantly lower Q_{3-n} of the MAO-LPD-3 coating is attributed to its much more compact structure in comparison to the MAO coating, making it more effective in preventing corrosion of the underlying Mg. Additionally, the lower Q_{3-n} value of MAO-LPD-3-coated Mg compared to MAO-coated Mg, also indicates superior anticorrosion performance, as suggested in other studies. The R_3 value increased with increasing NaH_2PO_4 concentration in the solution, whereas the Q_{3-n} value decreased with increasing NaH_2PO_4 concentration. This indicates that the performance of the inner layer of the coating is enhanced by solution treatment to a certain extent, providing improved corrosion resistance to the magnesium alloy. This conclusion is consistent with the results of polarization curve measurements and morphological analysis.

Mineralization is a critical factor in enhancing implant–tissue integration post-implantation for biomaterials. Previous research typically evaluates the bioactivity of biomaterials indirectly by examining apatite formation on their surfaces in simulated body fluid (SBF) [39]. Apatite formation has traditionally been considered a predictor of favorable bioactivity. Figure 12 presents SEM images of the surface morphologies of the MAO-LPD-1, MAO-LPD-2, and MAO-LPD-3 samples after immersion in SBF for 3 days. After 3 days of immersion, a substantial number of irregular spherical deposits appeared on the surface of the MAO-LPD-1 sample (Figure 12a), leading to the disappearance of its porous structure. The MAO-LPD-2 sample exhibited a flower-like morphology with a spherical structure after 3 days of immersion: the layers of these "flowers" covered the entire surface. For the MAO-LPD-3 sample, its entire surface was densely covered by cluster-like particles (Figure 12c), with significant deposition of irregular spherical matter. Notably, the ability to induce new deposition formation not only demonstrated improved surface bioactivity but also enhanced long-term corrosion resistance as a self-repair process could occur, wherein the previous HA layer degraded and a new HA layer simultaneously formed.

Figure 12. Surface morphologies of the coatings immersed in SBF solution for 3 days. (**a**) MAO-LPD-1, (**b**) MAO-LPD-2, and (**c**) MAO-LPD-3.

4. Conclusions

Hydroxyapatite layers have been successfully deposited on an MAO-coated AZ31B Mg alloy via solution treatment in an aqueous NaH_2PO_4 solution. The microstructures of composite coatings prepared at different NaH_2PO_4 concentrations have been studied. Wettability and bonding strength were investigated. The corrosion behavior and in vitro mineralization of the samples have also been analyzed.

The following conclusions were drawn:

1. The composite coatings comprise an inner MAO coating and an outer HA layer. The solution penetrates the porous structure of the MAO coating, deposits within it, and subsequently fuses with the porous layer.
2. Solution treatment enhances the hydrophobicity of the MAO coatings. A maximum contact angle of 161° is achieved at NaH_2PO_4 concentration of 0.1 mmol/L. This effect is attributed to the formation of dandelion-like clusters on the surface.
3. The bonding strength of the composite coatings is enhanced by the incorporation of the MAO coating. All composite coatings exhibit a tensile strength exceeding 20 MPa.
4. Electrochemical measurements revealed that the composite coatings exhibit significantly greater corrosion resistance than bare and MAO-coated magnesium, owing to the sealing effect and enhanced hydrophobicity.
5. The composite coatings demonstrate superior apatite-forming bioactivity. The formation of new hydroxyapatite on the surface of the composite coatings was observed after 3 days of immersion in SBF.

Therefore, the MAO-LPD composite coatings developed in this study possess excellent mechanical properties, corrosion resistance, and bioactivity, providing a promising strategy for the clinical application of biodegradable magnesium alloys.

Supplementary Materials: The following supporting information can be downloaded at: https://www.mdpi.com/article/10.3390/coatings15060675/s1, Figure S1. The variation of the Ca/P ratios of the coatings; Figure S2. Variation of thickness of the coatings; Figure S3. Variation of roughness of the coatings; Figure S4. Schematic diagram of structural superhydrophobic.

Author Contributions: Conceptualization, Y.H. and H.T.; methodology, Y.H.; validation, Y.H., X.L. and H.T.; investigation, Y.H., Y.Y. and F.J.; writing—original draft preparation, Y.H.; writing—review and editing, Y.H., X.L. and H.T. All authors have read and agreed to the published version of the manuscript.

Funding: This research received no external funding.

Institutional Review Board Statement: Not applicable.

Informed Consent Statement: Not applicable.

Data Availability Statement: Data are contained within the article and Supplementary Materials.

Conflicts of Interest: Authors Yanqing Hu and Xin Liang were employed by Chengdu Aircraft Industry Group Co., Ltd. The remaining authors declare that the research was conducted in the absence of any commercial or financial relationships that could be construed as a potential conflict of interest.

References

1. Kim, J.; Gilbert, J.L.; Lv, W.W.; Du, P.; Pan, H. Reduction reactions dominate the interactions between Mg alloys and cells: Understanding the mechanisms. *Bioact. Mater.* **2025**, *45*, 363–387. [CrossRef]
2. Staiger, M.P.; Pietak, A.M.; Huadmai, J.; Dias, G. Magnesium and its alloys as orthopedic biomaterials: A review. *Biomaterials* **2006**, *27*, 1728–1734. [CrossRef]

3. Zhang, Y.; Xu, J.; Ruan, Y.C.; Yu, M.K.; O'Laughlin, M.; Wise, H.; Chen, D.; Tian, L.; Shi, D.; Wang, J.; et al. Implant-derived magnesium induces local neuronal production of CGRP to improve bone-fracture healing in rats. *Nat. Med.* **2016**, *22*, 1160–1169. [CrossRef]
4. Yang, J.; Koons, G.L.; Cheng, G.; Zhao, L.; Mikos, A.G.; Cui, F.-Z. A review on the exploitation of biodegradable magnesium-based composites for medical applications. *Biomed. Mater.* **2018**, *13*, 022001. [CrossRef]
5. Zhao, D.; Witte, F.; Lu, F.; Wang, J.; Li, J.; Qin, L. Current status on clinical applications of magnesium-based orthopaedic implants: A review from clinical translational perspective. *Biomaterials.* **2017**, *112*, 287–302. [CrossRef]
6. Xu, L.; Pan, F.; Yu, G.; Yang, L.; Zhang, E.; Yang, K. In vitro and in vivo evaluation of the surface bioactivity of a calcium phosphate coated magnesium alloy. *Biomaterials.* **2009**, *30*, 1512–1523. [CrossRef]
7. Uppal, G.; Thakur, A.; Chauhan, A.; Bala, S. Magnesium based implants for functional bone tissue regeneration—A review. *J. Magnes. Alloy.* **2022**, *10*, 356–386. [CrossRef]
8. Zhang, A.; Fan, X.; Yang, Z.; Xie, Y.; Wu, T.; Zhang, M.; Xue, Y.; Wang, Y.; Zhao, Y.; Wu, X.; et al. Optimized design and biomechanical evaluation of biodegradable magnesium alloy vascular stents. *Acta Mech. Sin.* **2024**, *41*, 1–14. [CrossRef]
9. Tang, H.; Han, Y.; Wu, T.; Tao, W.; Jian, X.; Wu, Y.; Xu, F. Synthesis and properties of hydroxyapatite-containing coating on AZ31 magnesium alloy by micro-arc oxidation. *Appl. Surf. Sci.* **2017**, *400*, 391–404. [CrossRef]
10. Podgorbunsky, A.; Imshinetskiy, I.; Mashtalyar, D.; Sidorova, M.; Gnedenkov, A.; Sinebryukhov, S.; Gnedenkov, S. Bioresorbable composites based on magnesium and hydroxyapatite for use in bone tissue engineering: Focus on controlling and minimizing corrosion activity. *Ceram. Int.* **2024**, *51*, 423–436. [CrossRef]
11. Zhu, Y.; Zhou, M.; Zhao, W.; Geng, Y.; Chen, Y.; Tian, H.; Zhou, Y.; Chen, G.; Wu, R.; Zheng, Y.; et al. Insight the long-term biodegradable Mg-RE-Sr alloy for orthopaedics implant via friction stir processing. *Bioact. Mater.* **2024**, *41*, 293–311. [CrossRef]
12. Hashemi, T.S.; Jaiswal, S.; Celikin, M.; McCarthy, H.O.; Levingstone, T.J.; Dunne, N.J. Strategically designed bioactive dual-layer coating of octacalcium phosphate and dicalcium phosphate dihydrate for enhancement of the corrosion resistance of pure magnesium for orthopaedic applications. *Surf. Coatings Technol.* **2024**, *495*. [CrossRef]
13. Naftchali, N.K.; Aghdam, R.M.; Najjari, A.; Dehghanian, C. Investigating a nanocomposite coating of cerium ox-ide/merwinite via PEO/EPD for enhanced biocorrosion resistance, bioactivity and antibacterial activity of magnesium-based implants. *Ceram. Int.* **2024**, *50*, 42766–42779. [CrossRef]
14. Tang, H.; Gao, Y. Preparation and characterization of hydroxyapatite containing coating on AZ31 magnesium alloy by micro-arc oxidation. *J. Alloy Compd.* **2016**, *688*, 699–708. [CrossRef]
15. Tang, H.; Wu, T.; Wang, H.; Jian, X.; Wu, Y. Corrosion behavior of HA containing ceramic coated magnesium alloy in Hank's solution. *J. Alloy Compd.* **2017**, *698*, 643–653. [CrossRef]
16. Wang, J.; Dou, J.; Wang, Z.; Hu, C.; Liu, J.; Yu, H.; Chen, C. Corrosion resistance and biodegradability of micro-arc oxidation coatings with the variable sodium fluoride concentration on ZM21 magnesium alloys. *J. Alloy Compd.* **2023**, *962*. [CrossRef]
17. Tang, H.; Li, D.; Chen, X.; Wu, C.; Wang, F. FORMATION OF HA-CONTAINING COATING ON AZ31 MAGNESIUM ALLOY BY MICRO-ARC OXIDATION. *Surf. Rev. Lett.* **2013**, *20*. [CrossRef]
18. Tang, H.; Wang, F. Synthesis and properties of CaTiO3-containing coating on AZ31 magnesium alloy by micro-arc oxidation. *Mater. Lett.* **2013**, *93*, 427–430. [CrossRef]
19. Li, B.; Han, Y.; Qi, K. Formation Mechanism, Degradation Behavior, and Cytocompatibility of a Nanorod-Shaped HA and Pore-Sealed MgO Bilayer Coating on Magnesium. *ACS Appl. Mater. Interfaces.* **2014**, *6*, 18258–18274. [CrossRef]
20. Tang, H.; Tao, W.; Wang, C.; Yu, H. Fabrication of hydroxyapatite coatings on AZ31 Mg alloy by micro-arc oxidation coupled with sol–gel treatment. *RSC Adv.* **2018**, *8*, 12368–12375. [CrossRef]
21. Wang, C.; Jiang, B.; Liu, M.; Ge, Y. Corrosion characterization of micro-arc oxidization composite electrophoretic coating on AZ31B magnesium alloy. *J. Alloy Compd.* **2015**, *621*, 53–61. [CrossRef]
22. Li, J.; Song, Y.; Zhang, S.; Zhao, C.; Zhang, F.; Zhang, X.; Cao, L.; Fan, Q.; Tang, T. In vitro responses of human bone marrow stromal cells to a fluoridated hydroxyapatite coated biodegradable Mg–Zn alloy. *Biomaterials* **2010**, *31*, 5782–5788. [CrossRef]
23. Zhang, X.; Cui, S.-D.; Zhou, L.; Lian, J.-B.; He, J.; Li, X.-W. Preparation and characterization of calcium phosphate containing coating on plasma electrolytic oxidized magnesium and its corrosion behavior in simulated body fluids. *J. Alloy Compd.* **2022**, *896*. [CrossRef]
24. Farshid, S.; Kharaziha, M.; Salehi, H.; Hakemi, M.G. Morphology-Dependent Immunomodulatory Coating of Hydroxyapatite/PEO for Magnesium-Based Bone Implants. *ACS Appl. Mater. Interfaces* **2023**, *15*, 48996–49011. [CrossRef]
25. Farshid, S.; Kharaziha, M.; Atapour, M.; Di Franco, F.; Santamaria, M. Duplex plasma electrolytic oxidation/hydroxyapatite-polydopamine coating on WE43 alloy for bone implants: Long-term corrosion resistance and biological properties. *Surf. Coatings Technol.* **2024**, *493*. [CrossRef]
26. Tang, H.; Xin, T.; Sun, Q.; Yi, C.; Jiang, Z.; Wang, F. Influence of FeSO4 concentration on thermal emissivity of coatings formed on titanium alloy by micro-arc oxidation. *Appl. Surf. Sci.* **2011**, *257*, 10839–10844. [CrossRef]

27. Wang, Y.; Zhang, J.; Wang, W.; Yu, F.; Cao, W.; Hu, S. Effect of sodium fluoride additive on microstructure and corrosion performance of micro-arc oxidation coatings on EK30 magnesium alloy. *Surf. Coatings Technol.* **2024**, *496*. [CrossRef]
28. Taş, A.C. Molten Salt Synthesis of Calcium Hydroxyapatite Whiskers. *J. Am. Ceram. Soc.* **2001**, *84*, 295–300. [CrossRef]
29. Mujahid, K.; Iqbal, F.; Ali, A.; Butt, M.; Bukhari, N.; Nosheen, S.; Sharif, F.; Abbas, Z. Zinc-doped phosphate coatings for en-hanced corrosion resistance, antibacterial properties, and biocompatibility of AZ91D Mg alloy. *J. Alloys Com-Pounds* **2024**, 1005.
30. Guan, Q.; Hu, T.; Zhang, L.; Yu, M.; Niu, J.; Ding, Z.; Yu, P.; Yuan, G.; An, Z.; Pei, J. Concerting magnesium implant degradation facilitates local chemotherapy in tumor-associated bone defect. *Bioact. Mater.* **2024**, *40*, 445–459. [CrossRef]
31. Kong, J.; Kolooshani, A.; Kolahdouz, A.; Nejad, M.G.; Toghraie, D. Fabrication and characterization of magnesium implants coated with magnetic nanoparticles-wollastonite-hydroxyapatite for medical and sports injury applications: Finite element analysis. *Ceram. Int.* **2023**, *50*, 5755–5765. [CrossRef]
32. Drotarova, L.; Slamecka, K.; Balint, T.; Remesova, M.; Hudak, R.; Zivcak, J.; Schnitzer, M.; Celko, L.; Montufar, E.B. Biodegradable WE43 Mg alloy/hydroxyapatite interpenetrating phase composites with reduced hydrogen evolution. *Bioact. Mater.* **2024**, *42*, 519–530.
33. Ahmadkhaniha, D.; Fedel, M.; Sohi, M.H.; Hanzaki, A.Z.; Deflorian, F. Corrosion behavior of magnesium and magnesium–hydroxyapatite composite fabricated by friction stir processing in Dulbecco's phosphate buffered saline. *Corros. Sci.* **2016**, *104*, 319–329. [CrossRef]
34. Chen, Z.; Ji, H.; Geng, X.; Chen, X.; Yong, X.; Zhang, S. 3-D distribution characteristics of the micro-defects in the PEO coating on ZM6 mg-alloy during corrosion. *Corros. Sci.* **2020**, *174*. [CrossRef]
35. Cui, L.-Y.; Gao, S.-D.; Li, P.-P.; Zeng, R.-C.; Zhang, F.; Li, S.-Q.; Han, E.-H. Corrosion resistance of a self-healing micro-arc oxidation/polymethyltrimethoxysilane composite coating on magnesium alloy AZ31. *Corros. Sci.* **2017**, *118*, 84–95. [CrossRef]
36. Cui, X.-J.; Lin, X.-Z.; Liu, C.-H.; Yang, R.-S.; Zheng, X.-W.; Gong, M. Fabrication and corrosion resistance of a hydrophobic micro-arc oxidation coating on AZ31 Mg alloy. *Corros. Sci.* **2015**, *90*, 402–412. [CrossRef]
37. Dai, J.; Yang, J.; Zhang, X.; Zhang, L.; Sun, B.; Li, X.; Bai, J.; Xue, F.; Chu, C. Synergistic effects of BSA adsorption and shear stress on corrosion behaviors of WE43 alloy under simulated physiological flow field. *Corros. Sci.* **2024**, *237*. [CrossRef]
38. Zhang, J.; Dai, C.; Wei, J.; Wen, Z.; Zhang, S.; Lin, L. Calcium phosphate/chitosan composite coating: Effect of different concentrations of Mg2+ in the m-SBF on its bioactivity. *Appl. Surf. Sci.* **2013**, *280*, 256–262. [CrossRef]
39. Bohner, M.; Lemaitre, J. Can bioactivity be tested in vitro with SBF solution? *Biomaterials* **2009**, *30*, 2175–2179. [CrossRef]

coatings

MDPI

Article

Comparative Effects of Fine and Conventional Shot Peening on Surface Morphology, Topography, Wettability, and Antibacterial Activity of Biomedical Ti6Al4V Alloy

Egemen Avcu [1,2,3,*], Mert Guney [4,*], Yasemin Yıldıran Avcu [1,3], Mine Sulak [5], Hüseyin Uzuner [6], Meltem İlçe Bahadır [1], Eray Abakay [7], Mustafa Armağan [8], Rıdvan Yamanoğlu [9], Cagatay Elibol [3] and Martin F.-X. Wagner [3]

1 Department of Mechanical Engineering, Kocaeli University, Kocaeli 41001, Türkiye
2 Ford Otosan Ihsaniye Automotive Vocational School, Kocaeli University, Kocaeli 41650, Türkiye
3 Institute of Materials Science and Engineering, Chemnitz University of Technology, Erfenschlager Str. 73, 09125 Chemnitz, Germany; martin.wagner@mb.tu-chemnitz.de (M.F.-X.W.)
4 Department of Civil and Environmental Engineering, School of Engineering and Digital Sciences, The Environment & Resource Efficiency Cluster (EREC), Nazarbayev University, Astana 010000, Kazakhstan
5 Department of Medical Pharmacology, Faculty of Medicine, Atatürk University, Erzurum 25100, Türkiye
6 Department of Medical Services and Techniques, Kocaeli Vocational School of Health Services, Kocaeli University, Kocaeli 41001, Türkiye
7 Department of Metallurgy and Materials Engineering, Sakarya University, Sakarya 54050, Türkiye
8 Department of Mechanical Engineering, Istanbul Medeniyet University, Istanbul 34700, Türkiye
9 Department of Metallurgy and Materials Engineering, Kocaeli University, Kocaeli 41001, Türkiye
* Correspondence: avcuegemen@gmail.com (E.A.); mert.guney@nu.edu.kz (M.G.); Tel.: +90-262-4356167 (E.A.); +7-7172-704553 (M.G.)

Abstract: Interest in textured surfaces for biomaterials and implants is increasing, with shot peening emerging as a promising method for surface modification. This study investigates the influence of conventional and fine shot peening on the surface morphology, topography, wettability, and antibacterial properties of biomedical-grade Ti6Al4V alloy. Peening was conducted using a custom-built, fully automated system, employing fine (100–200 μm) and coarse (700–1000 μm) shots using well-controlled sets of parameters. Both treatments introduced severe plastic deformation on the surface, resulting in increased roughness. Conventionally shot-peened samples exhibited deeper and wider dimples compared to finely peened ones. Surface wettability shifted from hydrophilic (contact angle: ~4°, untreated) to hydrophobic, reaching contact angles of ~91° and ~100° for fine and conventional shot peening, respectively. Antibacterial assays against *Staphylococcus aureus* (*S. aureus*) and *Escherichia coli* (*E. coli*) were evaluated by normalizing colony counts to the untreated Ti6Al4V surface. The inherent antibacterial response of Ti6Al4V against *E. coli* was preserved after both shot peening treatments, showing no notable increase in bacterial proliferation. In contrast, adhesion of *S. aureus* increased, more notably on fine shot-peened surfaces, indicating a strain-specific response influenced by surface roughness and wettability. In summary, both fine and conventional shot peening altered the surface morphology, topography, and wettability of Ti6Al4V. At the same time, their antibacterial influence was strain-dependent, underscoring the need for careful parameter selection in biomedical applications.

Keywords: antibacterial properties; biomaterials; implants; shot peening; surface metrology; surface roughness; surface wettability; titanium alloys

1. Introduction

Titanium alloys have been preferred materials in various biomedical applications for a long time due to their good mechanical properties, including biocompatibility, strong acid and alkali resistance, toughness, and fatigue and yield strength [1–6]. However, these alloys are rarely employed in engineering disciplines, particularly associated with tribology due to certain disadvantages that include low surface hardness, an elevated coefficient of friction, and inadequate wear resistance [3,4]. Their tribological, corrosion, and biological characteristics, including wear and corrosion resistance, antibacterial activity, and osteointegration, are directly influenced by the surface properties of these materials [3]. Thus, there is a substantial demand for enhancements in their surface properties (such as roughness, wettability, and morphology), mechanical properties (including hardness, wear resistance, corrosion resistance, fatigue behavior, and toughness), and antibacterial properties of titanium alloys. Roughened surfaces featuring patterns are generally more useful than smooth, patternless surfaces in numerous respects, particularly when considered for biomedical properties [7]. Patterned surfaces are typically achieved through the deposition of coatings or by processing the material's surface at the sub-micrometer or micrometer scale [8,9].

Methods employed for surface modifications of titanium alloys in biomedical applications include: laser surface texturing (LST) [10,11], laser peening (LSP) [12], electro-spark surface texturing (ESST) [13], electro-erosion (EDM), chemical reactive ion etching (RIE) [14], lithography and anisotropic etching (LIAE) [15], abrasive jet machining (AJM) [16], and physical vapor deposition (PVD) [17–19], among others [3,6]. The predominant technique for surface patterning seems to be LST. Liu et al. [4] found that laser processing of a titanium alloy to create cell-like patterns enhanced surface microhardness while decreasing the friction coefficient and wear volume. Milles et al. [20] generated patterns at varying depths by fabricating hexagonal honeycomb structures on the aluminum surface using a laser processing technique, concluding that these patterns enhanced surface wettability. Shivakoti et al. [1] developed pit and pillar surface patterns on titanium and zirconia alloys employed in the biomedical sector through laser processing, concluding that the corrosion and wear resistance, and cell adhesion, increased after the patterning process. Erdoğan et al. [21] employed a laser method to process dot and straight-line patterns on Ti6Al4V, concluding that the quality of cell adhesion improved following the patterning procedure. Numerous studies have demonstrated that laser treatment can enhance the microhardness of titanium surfaces, refine grain size, improve micro morphology, and enhance corrosion resistance [6]. Nevertheless, the development of cracks on both the surface and subsurface of the material, with crack size directly proportional to the processing duration, has also been documented. Even though laser treatment has proven to be an effective method for producing surface patterns on titanium alloys and improving the surface properties, it also often leads to undesirable surface and subsurface defects. Therefore, to present a relatively more robust method, researchers have recently focused on using mechanical surface treatment methods (e.g., shot peening) to modify the surface properties of titanium alloys and to obtain specific surface patterns for biomedical applications [6,22–24]. Among these, there has been a particular interest in utilizing shot peening methods to tailor the surface properties of titanium alloys for biomedical applications.

In the shot peening process, peened material in the form of spheres impacts the surface at high velocities, resulting in plastic deformation and grain refinement in both the surface and subsurface regions [25]. Grain refinement enhances the mechanical properties of the material, including its hardness and toughness. Furthermore, shot peening notably modifies the surface characteristics of the treated alloy [26], such as topography, morphology, and wettability, particularly in relation to the applied shot peening parameters. Shot peening offers several potential advantages over other surface processing techniques, including reduced equipment requirements, adjustable surface and subsurface characteristics based on adjustable process parameters, lower energy consumption, shorter processing durations, higher efficiency, cost-effectiveness, and the capability to treat more complex-shaped components with ease [27]. Nevertheless, suboptimal shot peening conditions may also result in detrimental surface effects, including excessive roughening, microcrack initiation, and particle embedding [28], which can negatively influence mechanical integrity and in vivo performance. A rigorous optimization and comprehensive evaluation of shot peening parameters are imperative to maximize its functional benefits while mitigating potential surface damage. There are some well-documented mechanical benefits of shot peening; however, research is also required to elucidate the specific effects of both conventional and fine shot peening on surface roughness, wettability, and biological responses of titanium alloys. These surface-related attributes are critical, particularly for biomedical applications where cell adhesion, protein adsorption, and antibacterial performance are highly sensitive to micro- and nano-scale surface features.

Recent studies addressing the influence of shot peening parameters, especially in relation to surface morphology, topography, and biofunctionality, are critically discussed as follows to provide a clearer understanding of current research trends and remaining knowledge gaps: Wang et al. [29] applied fine particle shot peening (FPSP) to Ti6Al4V, generating half-dimpled surfaces that improved tribological performance; however, assessment of wettability or antibacterial effects was not conducted. Zhu et al. [30] showed that ultrasonic shot peening (USP) on commercially pure Ti largely enhanced hydrophilicity, directly linking surface roughness modifications to greater antibacterial behavior. Hsiao et al. [31] combined conventional shot peening (CSP) with HF/HNO_3 etching to achieve Ra $\approx$ 0.9 μm on Ti6Al4V, promoting osteoblast adhesion and spreading. Uanlee & Khantachawana [32] applied FPSP to Ti6Al4V, resulting in higher surface hardness and improved fibroblast viability, though an assessment of wettability and antibacterial effect was not provided. Khandaker et al. [33] used laser peening to create microgrooves on Ti surfaces that notably boost osteoblast adhesion, proliferation, and differentiation. Shen et al. [34] combined shot peening with Si/Cu-doped micro-arc oxidation (MAO) coatings to achieve >95% antibacterial efficacy against *S. aureus* and stimulate osteogenic marker expression. Li et al. [35] implemented copper-powder ultrasonic shot peening (C-USSP) on cp-Ti, demonstrating important antibacterial effects directly due to peening-induced surface changes. Finally, Zheng et al. [36] applied shot peening and gas nitriding to TC4 and found that, although the mechanical properties were substantially improved, antibacterial effectiveness rates did not change much, highlighting the complexity of peening vs. antimicrobial surface behavior. Altogether, these studies confirm that shot peening, especially when combined with additional treatments or optimized as USP/C-USSP, has the potential to substantially modulate antibacterial responses and cell behavior through control of surface topography and wettability. At the same time, not all peening variants largely enhance antibacterial performance, indicating the need for careful parameter optimization and comprehensive biological testing to identify protocols that consistently benefit implant integration and microbial resistance.

To date, a range of surface treatment techniques has been applied to Ti6Al4V alloys for biomedical applications, each having distinct advantages and limitations. Laser-based methods (e.g., laser surface texturing and laser shock peening) can generate well-defined micro/nano-patterns, enhance hardness, and improve wear and corrosion resistance; however, they often induce surface/subsurface cracks and require high-cost equipment and long processing times [4,6,12,21]. Chemical and electrochemical treatments such as reactive ion etching, anodization, and micro-arc oxidation are effective in tailoring surface chemistry and bioactivity, but they typically produce brittle oxide layers and may lack long-term stability in physiological environments [14,34]. Physical vapor deposition (PVD) coatings provide excellent control over surface chemistry and tribological behavior; and yet, challenges remain in achieving strong adhesion, avoiding delamination, and maintaining antibacterial efficacy under cyclic loading [17–19]. Abrasive and jet-based methods (e.g., abrasive jet machining, electro-spark texturing) offer rapid texturing capabilities but can introduce uncontrolled roughness and surface defects [13,16]. By contrast, mechanical surface treatments such as shot peening present a cost-effective, scalable, and equipment-flexible alternative that can refine surface and subsurface properties through severe plastic deformation. Shot peening avoids many of the brittle surface defects associated with laser or chemical treatments and allows parameter tuning to balance roughness, wettability, and functional biological responses [22–24,27].

The present study provides (to the authors' knowledge) the first systematic comparison between fine and conventional shot peening on biomedical-grade Ti6Al4V, focusing on their influences on surface morphology, topography, wettability, and antibacterial activity. This direct comparison aims to clarify the potential of fine shot peening as an optimized surface modification route for biomedical implants. Using a fully automated, custom-designed shot peening system, samples were treated under controlled parameters by using coarse and fine media to generate distinct surface textures. Through a comparative approach, the study mainly explores how shot size influences biologically relevant surface features and bacterial adhesion behavior.

2. Materials and Methods

Conventional and fine shot peening were conducted on biomedical Ti6Al4V alloy samples utilizing fine (100–200 μm) and coarse shots (700–1000 μm) under regulated conditions with a custom-designed fully automated peening system. Subsequently, optical profilometry, scanning electron microscopy (SEM), and contact angle measurements were performed to characterize the surface properties of fine-peened and shot-peened samples. The antibacterial efficacy of the samples against *S. aureus* and *E. coli* was assessed.

2.1. Materials

Ti6Al4V alloy (grade 5) samples with equiaxed microstructure were commercially supplied in rod form (Ø20 mm × 1000 mm) by TIMET (Kocaeli, Türkiye). These were cut into slices that were 10 mm thick using a semi-automatic band saw. Before shot peening, the samples were ground using sandpapers of 360, 600, 1000, and 2000 mesh and polished sequentially with diamond solutions of 9 μm, 3 μm, and 1 μm. The chemical composition of the alloy (Table 1) was determined using Optical Emission Spectroscopy (OES) (Hitachi High-Tech WAS Foundry-Master, Westford, MA, USA).

Table 1. Chemical composition of the investigated Ti6Al4V alloy (wt.%).

Al	V	Fe	Si	Sn	Mn	Cr	Mo	Cu	Zr	Nb	Ti
5.75	3.93	0.054	0.011	0.050	0.018	0.010	0.010	0.010	0.014	0.045	Balance

2.2. Shot Peening of Ti6Al4V Alloy

Shot peening treatments were carried out using a CNC-controlled, custom-designed shot peening system with the configuration and operating principles previously described in detail and illustrated schematically in our earlier work [24]. The samples were subjected to shot peening at an applied pressure of 5 bar for 2.5 min from a distance of 40 mm, with the peening parameters detailed in Table 2. Two distinct sizes of stainless-steel shots were utilized in the shot peening processes. The choice of shot sizes (100–200 μm and

700–1000 μm) was based on their well-established use, i.e., these ranges are the most widely reported in both industrial and biomedical applications of shot peening. They may also help provide a representative comparison between fine and conventional peening modes under realistic processing conditions. This setup ensures novelty by directly contrasting the two most practically relevant size ranges within a single systematic study.

Table 2. Shot peening parameters used for peening of the Ti6Al4V alloy.

Parameter	Value
Shot type	Stainless steel shots
Shot hardness	460 HV
Shot size	100–200 μm (Fine shot peening)
	700–1000 μm (Conventional shot peening)
Acceleration pressure	5 bar
Duration	2.5 min
Standoff distance	40 mm
Nozzle diameter	4.76 mm

The morphology and chemical composition of the stainless-steel shots were examined by SEM coupled with energy-dispersive X-ray spectroscopy (EDS). As shown in Figure 1a,b, the coarse (700–1000 μm) and fine (100–200 μm) shots exhibited spherical morphology with surface texture features. EDS spectra revealed that both shot types primarily consisted of Fe, Cr, Ni, Mn, and Si, in agreement with their stainless-steel composition (Figure 1, Table 3).

Table 3. EDS-derived chemical composition of stainless-steel shots used for shot peening (wt.%).

Shot Size	C	O	Si	Cr	Mn	Fe	Ni
Fine (100–200 μm)	0.04	10.08	5.28	18.52	2.78	56.95	6.36
Coarse (700–1000 μm)	0.07	10.92	5.93	17.79	3.44	55.69	6.16

2.3. *Surface Characterization of Shot-Peened Alloy*

Figure 2 illustrates the methodology employed for characterizing the surface roughness of the peened samples. First, the three-dimensional topographies of the sample surfaces were acquired utilizing an optical profilometer (Huvitz, Gyeonggi, Republic of Korea). Then, the surface roughness characteristics (i.e., surface topography, areal roughness parameters, roughness profiles) of the samples were analyzed by Mountains® 9 software (Digital Surf, Besançon, France).

After the characterization of surface roughness, the surface morphologies of the shot-peened samples were analyzed utilizing an SEM (Jeol JSM-6060, Tokyo, Japan) coupled with an energy dispersive spectroscopy (EDS) detector (Oxford Instruments, Oxford, UK). The wettability of the unpeened, fine-peened, and shot-peened samples was assessed using an optical contact angle measurement system (Kruess GmbH, Hamburg, Germany). Contact angle tests were conducted using distilled water, maintaining a constant droplet volume of 3 mL, with each specimen tested five times.

2.4. *Antibacterial Activity of Shot-Peened Surface*

Antibacterial tests of samples before and after shot peening were conducted using *Staphylococcus aureus* strain ATCC 29213 and *Escherichia coli* strain ATCC 25922 (Mueller-Hinton Agar, Merck, Germany) as model microorganisms. All tests were performed in triplicate (n = 3). Nutrient broth was prepared by dissolving 34 g of Mueller-Hinton Agar (MHA) in 1000 mL of distilled water, sterilized at 121 °C for 15 min, and poured into Petri dishes under sterile conditions. *S. aureus* and *E. coli* were grown in the broth at 37 °C to a concentration of 10^8 CFU·mL^{-1}. The broth was then diluted 10-fold with phosphate-buffered saline (PBS) to obtain a bacterial suspension of 10^4 CFU·mL^{-1}. For each test, 0.4 mL of bacterial suspension was pipetted onto the sample surfaces, and the Petri dishes were incubated at 37 °C for 18–24 h under 90% humidity. After incubation, the suspension was collected in sterile tubes containing 3.6 mL of PBS. The samples were also transferred into the same tubes, and 0.02 mL of the solution was subsequently spread on MHA plates using sterile glass rods. Plates were incubated at 37 °C for 18–24 h, after which the number of viable bacteria was quantified by manually counting colony-forming units (CFUs) using a standard colony counter.

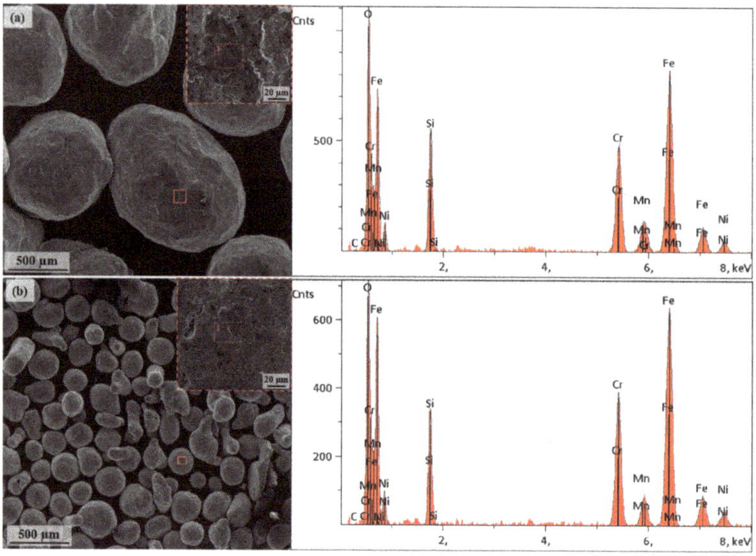

Figure 1. SEM images and EDS analysis of stainless-steel shots used in the present study: (**a**) coarse shots (700–100 µm); (**b**) fine shots (100–200 µm).

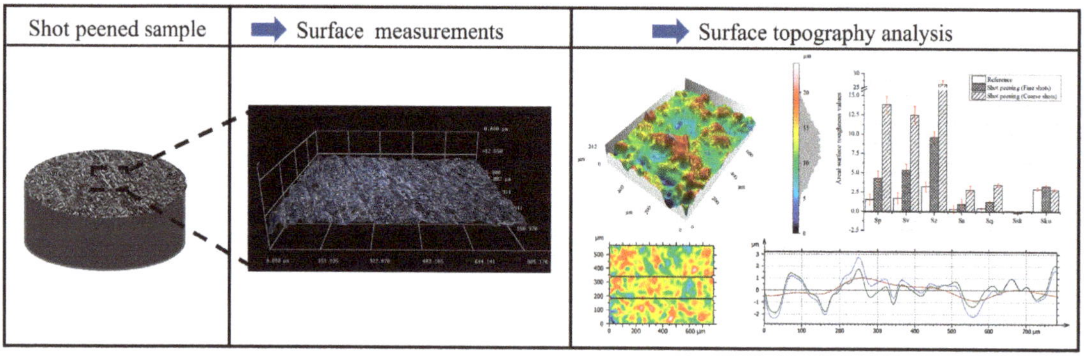

Figure 2. Representation of surface topography analysis of shot-peened Ti6Al4V alloy.

The untreated Ti-6Al-4V surface was used as the internal control for all antibacterial comparisons. Agar-only plates were included solely as viability controls to confirm bacterial growth and exclude contamination, i.e., not used for calculations or for discussion as Ti-6Al-4V is bio-inert and thus cannot be directly compared to a nutrient-rich agar plate. Finally, results were normalized to the untreated alloy to disclose the effect of shot peening on bacterial adhesion and proliferation. Data analysis was carried out in two complementary formats:

- Relative bacterial count ($N/N_{untreated}$): For each replicate, the CFU count on a treated sample was divided by the mean CFU count of the untreated reference. Results were expressed as mean ± standard deviation (SD), with untreated set to 1.0. A reduction percentage could be derived as $R(\%) = (1 - N/N_{untreated}) \times 100$, where negative values indicate increased adhesion relative to untreated.
- Log_{10}-reduction (vs. untreated): The difference in bacterial load between untreated and treated surfaces was expressed as $\log_{10}(N_{untreated}) - \log_{10}(N_{sample})$, where untreated set to 0. Positive values indicate reduced bacterial adhesion, while negative values indicate increased adhesion.

Both metrics were reported as mean ± SD (n = 3). Graphical presentation was made in two panels: (a) relative bacterial counts normalized to untreated and (b) $\log_{10}$-reduction with respect to untreated.

3. Results and Discussion

3.1. Surface Morphology After Shot Peening

Figures 3 and 4 display low and high-magnification SEM micrographs of the Ti6Al4V alloy surfaces subjected to conventional and fine shot peening, respectively. Both treatments induced severe plastic deformation, with characteristic surface features directly influenced by the size and kinetic energy of the impacting media. The surfaces exhibit dense impact craters and material pileups, typical of shot peening-induced topographical modifications [24,37,38]. In the conventionally shot-peened sample (Figure 3), the morphology is dominated by larger and deeper dimples, associated with high-energy impacts from coarse shots (700–1000 μm). The surface shows pronounced ridges and valleys with irregular crater geometry and evidence of micro-cutting or plowing, suggesting intense localized plastic deformation. Similar morphological features have been reported in Ti-based substrates processed by high-intensity peening, where the severity of plastic flow correlates with surface work hardening and increased roughness [24,39]. In contrast, the surface treated with fine shot peening (Figure 4) reveals more uniform and shallower dimples, attributed to the lower kinetic energy of fine shots (100–200 μm). The resulting surface layer thus exhibits a more homogeneous texture, with finer-scale peak-valley features and less severe surface distortion. This microtopography reflects a more controlled peening action, capable of refining the surface without extensive surface damage or excessive roughening, a behavior that is consistent with prior observations on fine shot or ultrasonic peening of titanium alloys [32,40–43]. No significant cracks, microcracks, fractures, or particle dislodgement were seen on either surface, demonstrating the absence of harmful surface degradation. This observation corroborates the efficacy of the employed process parameters in attaining advantageous surface modification while preserving material integrity.

From a mechanistic standpoint, the microstructural refinement and surface texture variation stem from shot-induced strain localization. The higher dimple density and smoother contours in the fine shot-peened sample suggest a more uniform energy transfer, whereas coarse shot impacts lead to heterogeneous strain zones and larger deformation footprints. These distinct morphological features are known to critically influence surface properties such as protein adsorption, wettability, and cellular interaction, particularly in biomedical applications [32–35]. Moreover, the differences in morphology align with surface roughness measurements (Figures 5 and 6), wherein conventionally peened surfaces exhibit greater roughness amplitudes. The fine peening technique offers a desirable balance between surface modification and preservation of topological integrity, supporting its suitability for biofunctional surface engineering of Ti6Al4V implants.

3.2. Surface Topography After Shot Peening

Surface topography plays a critical role in governing the mechanical and biological performance of titanium-based biomaterials. The results presented in Figures 5–7 reveal the profound influence of shot peening parameters on the three-dimensional surface features of Ti6Al4V alloy. Conventional shot peening, characterized by larger shot sizes and higher kinetic energy, produced an irregular surface profile with pronounced peaks (Sp) and valleys (Sv), as indicated by increased areal roughness values—maximum height (Sz) $\approx$ 24.20 μm and arithmetic mean height (Sa) $\approx$ 2.78 μm. In contrast, fine shot peening generated a much smoother surface ($Sz \approx$ 11.20 μm, $Sa \approx$ 1.69 μm), while still inducing beneficial topographic features necessary for mechanical interlocking and cellular interaction [44–47]. These findings agree with the previously discussed SEM-based morphology results (Figures 3 and 4), where larger and deeper impact craters were observed on the conventionally shot-peened surface, whereas the fine shot-peened surface exhibited more uniform and shallower surface features. These morphological observations support the interpretation that shot size plays a dominant role in determining local deformation mechanisms and surface reconstruction.

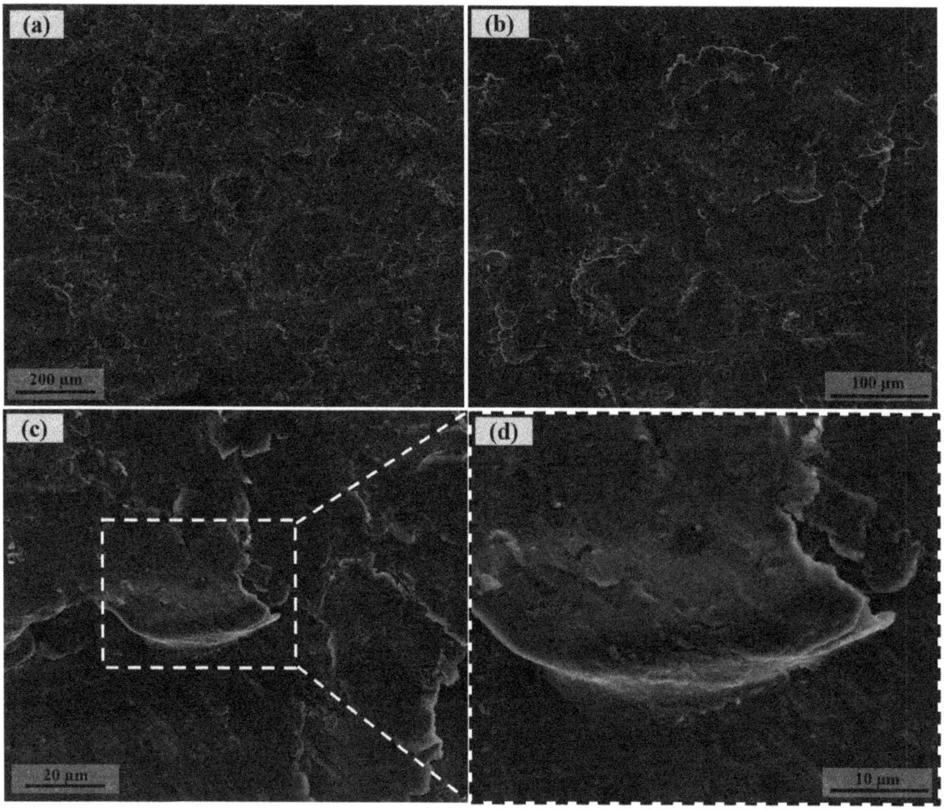

Figure 3. SEM micrographs showing the surface morphology of conventionally shot-peened Ti6Al4V alloy at different magnifications: (**a**) low magnification overview; (**b**) higher magnification revealing plastically deformed and roughened regions; (**c**) localized surface damage and flattened areas induced by repeated shot impacts; and (**d**) enlarged view of a plastically extruded region indicating severe plastic deformation and material flow.

Although smaller shot sizes (<100 μm) could potentially generate finer topographies, such cases remain uncommon in biomedical surface treatments, and their antibacterial efficacy has not yet been systematically explored. For this reason, the present study focused on the practically relevant ranges of 100–200 μm and 700–1000 μm. Finally, while the use of shots <100 μm in biomedical surface treatments is relatively limited, it may also be interesting to see its performance for fine shot peening in future work.

The data extracted from ISO 25178 [48] parameters (Figure 6c) and comparative bar plots (Figure 7a) highlight a clear correlation between shot size and roughness amplitude. *Sq* (root mean square height) increased with shot size, confirming higher roughness amplitude for conventional shot peening. Both shot-peened surfaces showed slightly negative *Ssk* (skewness), indicating valley-dominated morphologies favorable for fluid retention and potential cell adhesion. *Sku* (kurtosis) values > 3 reflected sharper peak–valley features, enhancing potential mechanical interlocking, which are critical for evaluating functional surfaces in biomedical implants [49]. In addition, roughness profiles (Figure 7b) demonstrate the amplitude modulation and wavelength distribution across the treated surfaces, with fine shot peening yielding a more periodic and less jagged profile than the conventional counterpart.

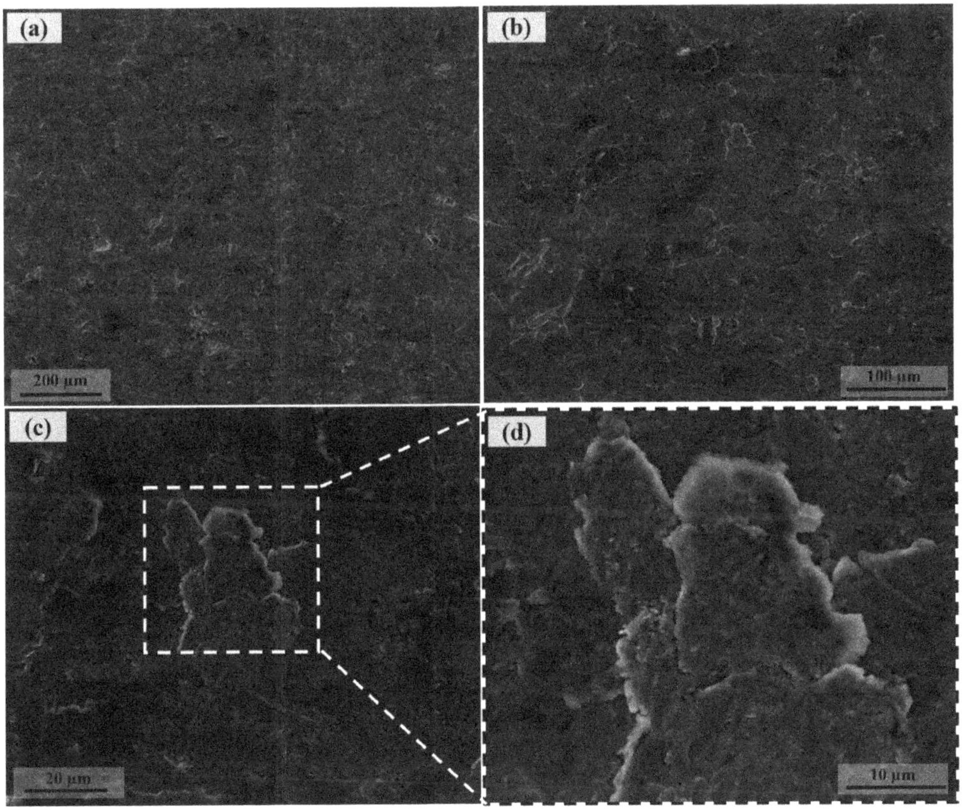

Figure 4. SEM micrographs showing the surface morphology of fine shot-peened Ti6Al4V alloy at different magnifications: (**a**) low magnification overview; (**b**) higher magnification highlighting plastically deformed surface with finer grooves and asperities; (**c**) localized surface feature associated with repeated fine shot impacts; and (**d**) enlarged view revealing fragmented and plastically extruded regions indicating severe surface deformation at the microscale.

From a fundamental standpoint, the observed differences in roughness and topography arise from the interplay between shot diameter, kinetic energy per impact, and impact frequency. Larger shots used in conventional peening impart higher momentum and localized energy, leading to deeper plastic deformation zones, irregular asperity formation, and increased surface roughness. Fine shot peening, on the other hand, involves smaller media with lower energy per impact but a higher number of impacts per unit area. This results in more homogeneous plastic flow, reduced micro-cutting, and smoother surface reconstruction due to cumulative overlapping deformation events [50,51]. Such differences in surface characteristics directly impact the implant's performance in vivo. Excessively rough surfaces may enhance mechanical anchorage, but at the same time risk promoting bacterial colonization and debris entrapment [52]. Conversely, ultra-smooth surfaces may reduce biological fixation. Therefore, the moderately rough and homogeneously peened surfaces achieved via fine shot peening represent an optimized compromise, ensuring both structural integration and, potentially, a reduced infection risk. Therefore, the moderately rough and homogeneously peened surfaces achieved via fine shot peening may represent an optimized compromise, ensuring both structural integration and, potentially, a reduced infection risk [47,49].

The present findings align with the literature on shot peening–induced surface texturing of Ti6Al4V alloys, where controlled roughness plays a pivotal role in modulating osteoblastic activity, corrosion behavior, and long-term implant performance. These morphological and topographical improvements serve as a mechanistic foundation for understanding the antibacterial responses discussed below.

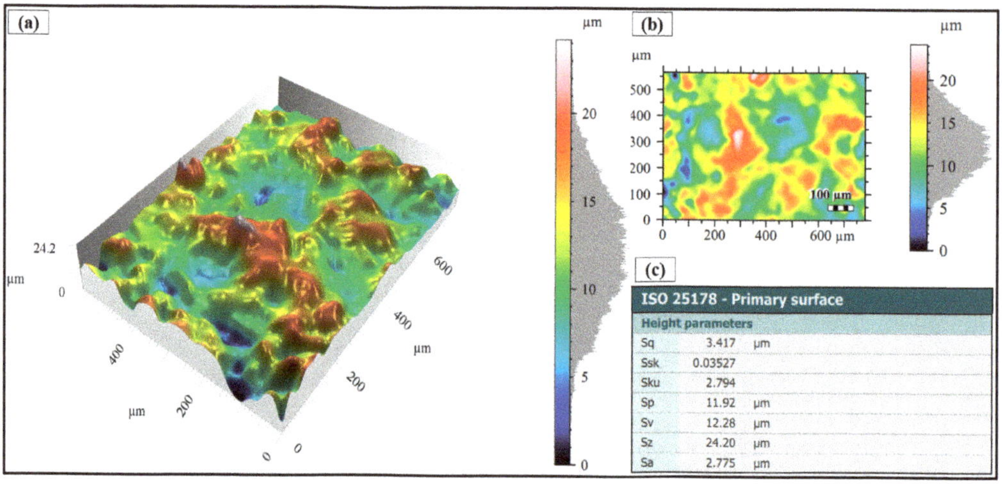

Figure 5. (**a**) Three-dimensional surface topography, (**b**) 2D roughness map, and (**c**) areal surface roughness values of conventionally shot-peened Ti6Al4V.

3.3. Wettability After Shot Peening

Surface wettability plays a pivotal role in the biological performance of metallic implants, influencing protein adsorption, initial cell attachment, and bacterial adhesion. The results of the contact angle measurements presented in Figure 8 reveal a notable change in wettability following both fine and conventional shot peening treatments compared to the reference condition. The reference sample, which possessed an ultrafine and mirror-polished surface, exhibited a low contact angle (~4°), confirming its hydrophilic nature. This behavior is consistent with its smooth topography and minimal surface roughness, as quantified in Figures 5–7. After shot peening, a clear transition toward hydrophobic behavior was observed. Specifically, fine shot peening increased the contact angle to ~90°, while conventional shot peening further elevated it to ~100°, indicating strong water repellency. These wettability shifts are closely linked to the morphological and topographical alterations induced by the peening treatments. As shown in Figures 4 and 6, the generation of micro-craters and overlapping deformation zones increased the surface roughness and introduced hierarchical surface features. These surface modifications reduce the real contact area and alter the solid–liquid interaction, resulting in an apparent increase in contact angle [53–55]. These observations are consistent with Cassie–Baxter wetting model, in which air entrapment within surface asperities reduces the real solid–liquid contact area and increases the apparent contact angle [56]. Such pronounced changes in wettability are expected to have direct implications for the biological response of the alloy surface. In particular, the transition from a hydrophilic to a hydrophobic state may influence subsequent protein adsorption, cell–material interactions, and bacterial adhesion, as will be discussed in the following sections.

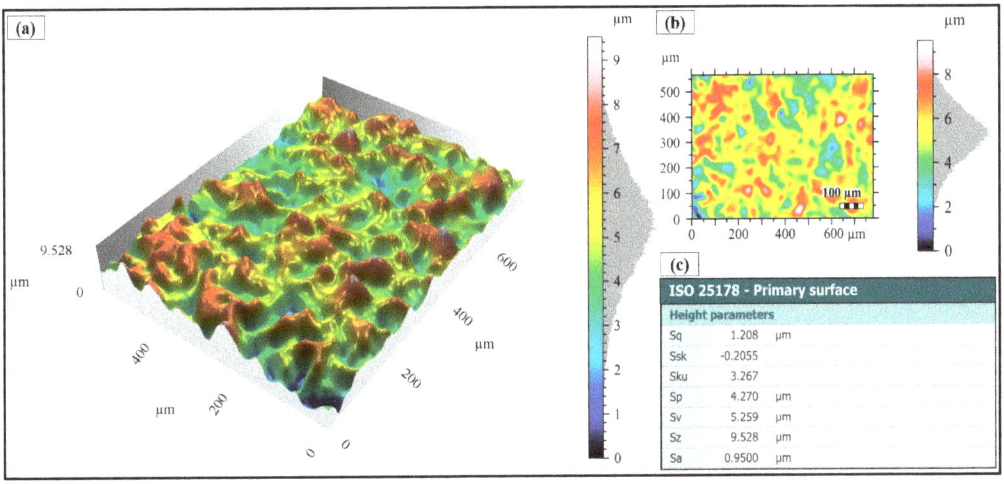

Figure 6. (**a**) Three-dimensional surface topography, (**b**) 2D roughness map, and (**c**) areal surface roughness values of fine shot-peened Ti6Al4V.

Fine shot peening applied in the present study appears to provide an optimized surface condition by tuning wettability and surface texture simultaneously. The finer and more uniformly distributed dimples generated by fine shot peening produced a moderately rough but topographically ordered surface, supporting a transition to near-hydrophobic conditions while potentially maintaining favorable biological responses. In contrast, the irregular and sharper asperities caused by conventional shot peening likely reduce droplet spreading and enhance water repellency. From a functional perspective, the altered wettability is expected to influence subsequent biological interactions. Moderately hydrophobic surfaces can inhibit bacterial adhesion and delay biofilm formation, while still supporting protein anchorage and cell adhesion when combined with bioactive chemistry [44].

3.4. Antibacterial Activity After Shot Peening

Antibacterial activity was assessed against *S. aureus* and *E. coli* using the untreated Ti6Al4V surface as the baseline reference (Figures 9–11). Agar-only plates were included solely as viability controls, i.e., not used for calculations or discussion. The results presented thus reflect the influence of surface modification relative to the bio-inert alloy. As it is a bio-inert material, bacterial colony formation on Ti6Al4V was notably lower compared to the open agar plate (Figures 9 and 10). The differences in bacterial growth between the nutrient-rich agar and the bio-inert alloy surface are as expected, since Ti6Al4V has no intrinsic antibacterial effect [57,58].

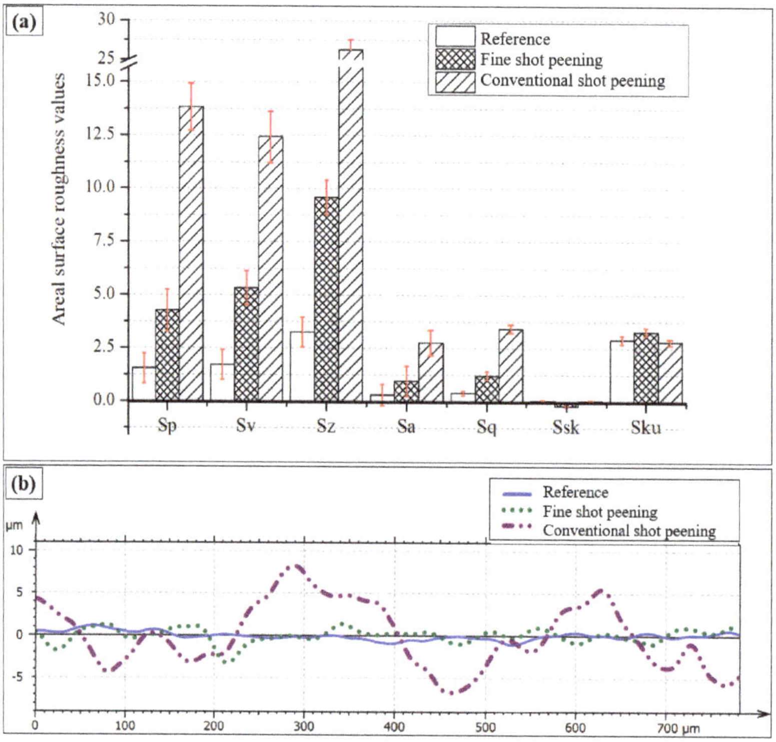

Figure 7. (**a**) Areal surface roughness results and (**b**) roughness profiles of conventionally and fine shot-peened Ti6Al4V.

Shot peening increased surface roughness, which in turn promoted bacterial growth relative to the very smooth reference surface. Quantitative data revealed that *E. coli* adhesion remained largely unchanged compared to untreated surfaces, indicating preserved antibacterial response, whereas *S. aureus* adhesion increased, particularly on fine shot-peened surfaces (Figure 11). We observed a higher proliferation of *S. aureus* on fine shot-peened surfaces ($S_Z \approx 9.528$ µm), as seen in Figure 9. This result aligns with prior studies by Cunha et al. [57] and Truong et al. [58], which reported *S. aureus* adhesion to microgrooves (10–20 µm) created by femtosecond laser treatment on titanium. For Ti6Al4V, which lacks intrinsic antibacterial activity [59–61], overall, shot peening did not compromise the antibacterial response against *E. coli* while promoting a strain-specific increase in *S. aureus* adhesion.

Among the peening techniques employed, conventionally shot-peened samples exhibited a notably higher surface roughness compared to fine shot-peened ones. It is well established that surface topography and roughness critically affect bacterial adhesion and proliferation. Surfaces with higher roughness may serve as a scaffold for bacterial colonization, especially when the average roughness (*Sa*) exceeds 0.8 µm, facilitating bacterial attachment by providing protective niches [49]. In our study, both treatment methods yielded $Sa > 0.8$ µm, while conventionally shot-peened surfaces exhibited the higher roughness value ($Sa > 2.775$ µm). Additionally, the spacing of the surface grooves plays an important role in bacterial colonization. When the groove spacing is smaller than the bacterial cell size, adhesion becomes less favorable. In our case, the spacing formed by coarse shots ($S_Z \approx 24.20$ µm) was much larger than that of fine shots ($S_Z \approx 9.528$ µm), indicating higher potential bacterial adhesion on conventionally shot-peened surfaces. Priyanka et al. [49] reported greater bacterial adhesion on rougher TiN-Ag coated surfaces, which supports this claim.

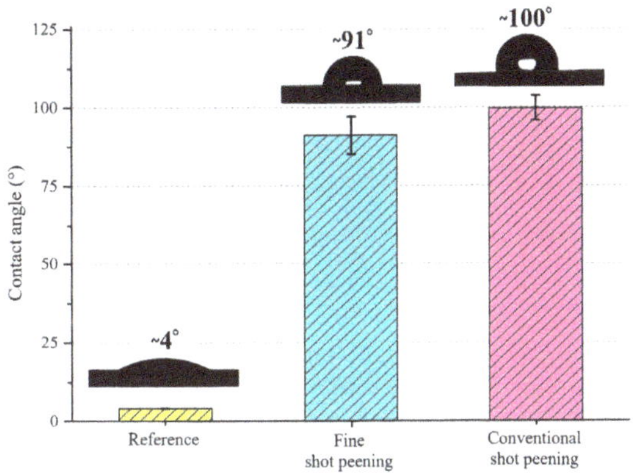

Figure 8. Contact angle measurements of Ti6Al4V surfaces: reference (mirror-polished), fine shot-peened, and conventional shot-peened samples. The results demonstrate a notable increase in hydrophobicity after shot peening, with the reference surface being highly hydrophilic (~4°), while fine and conventional shot peening produced contact angles of ~91° and ~100°, respectively.

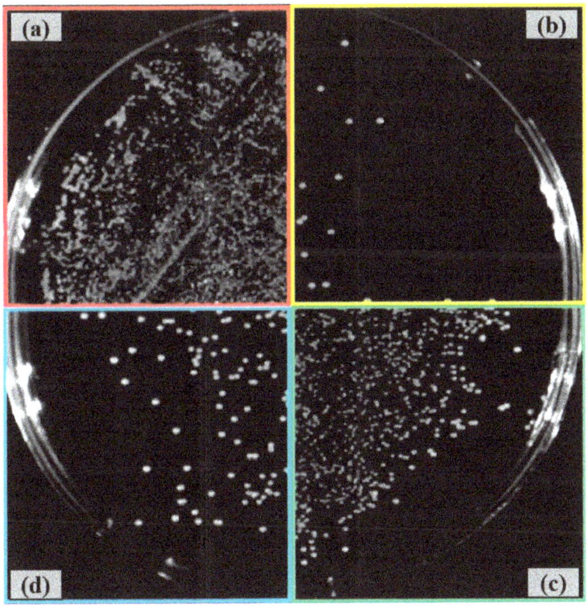

Figure 9. Antibacterial performance of samples against *S. aureus* after 6 h of incubation. Representative images of bacterial colony formation on (**a**) control agar plate (no material), (**b**) untreated Ti6Al4V reference sample, (**c**) fine shot-peened Ti6Al4V surface, and (**d**) conventionally shot-peened Ti6Al4V surface.

Nishitani et al. [42] demonstrated that at lower surface roughness via fine particle bombardment, antibacterial activity improved, particularly against both *E. coli* and *S. aureus*. Our results partly agree with these findings: While *E. coli* adhesion seems unchanged or slightly reduced on smoother, fine shot-peened surfaces, *S. aureus* adhesion increased (Figures 10 and 11). The ~200 nm convex features formed during fine peening may have interacted with *E. coli* cell membranes, potentially limiting adhesion; however, further live/dead assays are required to confirm this mechanism.

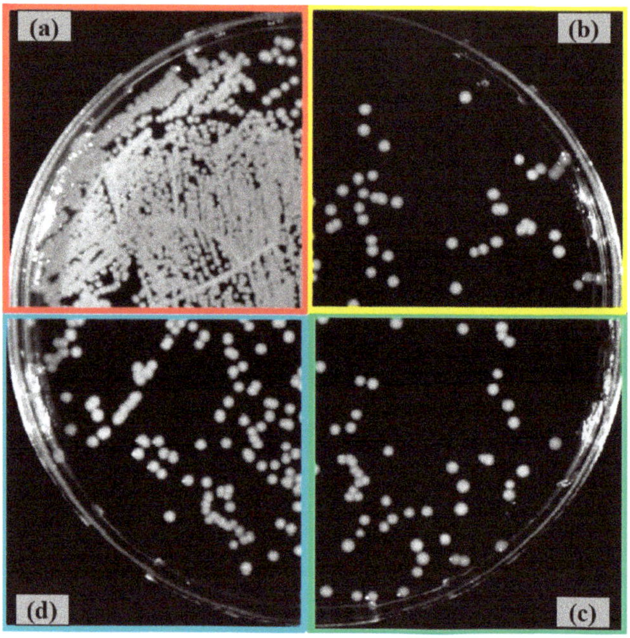

Figure 10. Antibacterial performance of samples against *E. Coli* after 6 h of incubation. Representative images of bacterial colony formation on (**a**) control agar plate (no material), (**b**) untreated Ti6Al4V reference sample, (**c**) fine shot-peened Ti6Al4V surface, and (**d**) conventional shot-peened Ti6Al4V surface.

The wettability of the surface is determined by its roughness, topography, and chemical composition. Untreated Ti6Al4V surfaces are typically hydrophilic (Figure 8), and both peening methods increased hydrophobicity. Trapped air within the hydrophobic interface can act as a barrier, preventing bacterial attachment and motility, thereby enhancing antibacterial resistance [56]. Bright et al. [62] also observed that *S. aureus* preferred proliferation on hydrophilic surfaces. Thus, the increased proliferation of *S. aureus* on the less hydrophobic, fine shot-peened surfaces in our study (Figure 9) is consistent with the literature.

Biofilm formation typically begins with the weak, reversible attachment of bacterial cells to the surface. Environmental factors such as surface roughness, hydrophobicity, and surface charge, as well as bacterial traits and surrounding pH or flow conditions, strongly influence adhesion and biofilm development. Once adhered, microorganisms produce extracellular polymeric substances (EPS), forming a protective matrix that increases resistance to antibiotics and environmental stressors [63]. Biofilms are therefore a major concern in the context of chronic polymicrobial infections and medical device-related complications [56]. Wu et al. [64] reported that micro- and nanoscale surface roughness notably affects bacterial adhesion and early biofilm development.

The observed proliferation of *S. aureus*, particularly on fine shot-peened surfaces, may indicate a higher potential for biofilm formation. In contrast, *E. coli* adhesion remained limited across all conditions, suggesting preserved antibacterial response. However, to confirm these strain-specific trends, time-dependent antibacterial testing at various incubation durations (e.g., 12, 18, and 24 h) is needed.

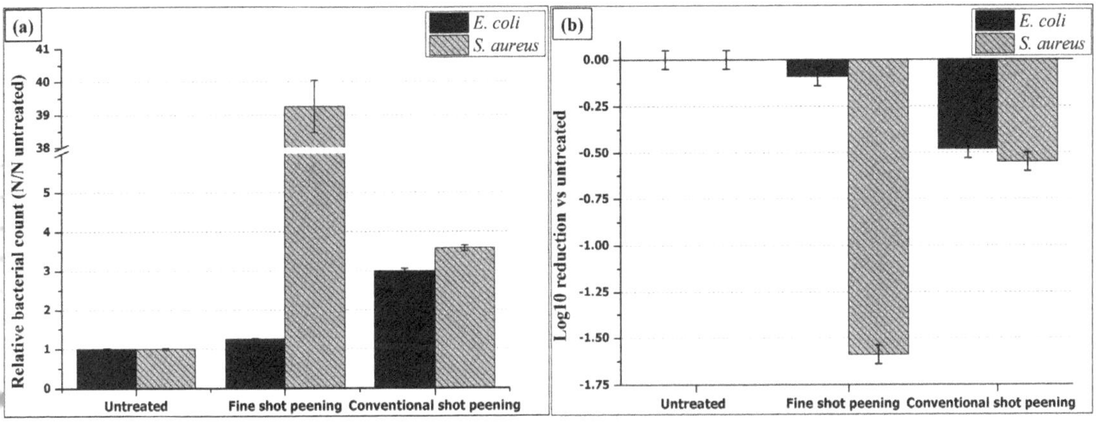

Figure 11. Antibacterial response of Ti6Al4V surfaces after 6 h of incubation. (**a**) Relative bacterial counts normalized to untreated ($N/N_{untreated}$). (**b**) Log_{10}-reduction values with respect to the untreated. Untreated samples form a baseline, set to 1.0 for (**a**), and to 0 for (**b**). Bars represent mean ± SD (n = 3). The results indicate that shot peening did not improve antibacterial performance against *E. coli*, and that *S. aureus* adhesion increased on fine shot-peened surfaces.

Antibacterial testing was performed at 6 h to evaluate the response of Ti6Al4V surfaces, as rapid antibacterial activity is relevant for biomedical applications. The literature reports that antibacterial properties can be assessed within 2–24 h, depending on the material and bacterial strain [56,62]. In our study, the 6 h time point was selected as the endpoint, since early-stage bacterial adhesion is particularly critical for implant surfaces. It should be noted that, as the antibacterial assessment in this study was limited to a single incubation period (6 h), future work should include additional time points (e.g., 12–24 h) to provide a more comprehensive understanding of strain-specific and time-dependent bacterial responses.

4. Conclusions

In this study, a comparative investigation was conducted to evaluate the effects of conventional and fine shot peening techniques on the surface morphology, roughness, wettability, and antibacterial response of biomedical-grade Ti6Al4V alloy. Using a custom-designed automated shot peening system, surface modification was achieved through the use of either coarse or fine stainless-steel media to explore their impact on biofunctional surface design. Critical findings of the study include the following:

- Surface characteristics were strongly influenced by shot size and energy. Conventional shot peening (700–1000 μm) induced rougher, irregular surfaces with higher peak-valley features ($Sa \approx 2.78$ μm), while fine shot peening (100–200 μm) produced more homogeneous textures with smoother topographies ($Sa \approx 1.69$ μm), minimizing potential surface damage.
- Wettability shifted from hydrophilic to hydrophobic after peening, with fine shot peening achieving a contact angle of ~91° and conventional peening exceeding 99°. These changes are attributed to the formation of micro-dimples and hierarchical

textures, which reduce solid–liquid interaction and influence subsequent protein adsorption, bacterial adhesion, and cell–material interaction.

- Antibacterial assays revealed strain-specific effects: While the inherent antibacterial response of Ti6Al4V against *E. coli* was preserved after both treatments, *S. aureus* adhesion increased particularly on fine shot-peened surfaces. This highlights the importance of surface roughness and wettability in modulating bacterial behavior and indicates that careful parameter selection is required to balance antibacterial performance with biofunctional integration

From an application-oriented perspective, both treatments provide viable routes to tailor surface morphology, topography, and wettability of Ti6Al4V. However, the strain-specific response observed (particularly the increased adhesion of *S. aureus* on fine shot-peened surfaces) highlights the need for careful parameter optimization to balance antibacterial performance with desirable implant–cell interactions. Tailoring shot-peening parameters can offer a practical, low-cost, and scalable post-processing strategy for Ti6Al4V implants, providing a feasible pathway for clinical translation and industrial implementation.

Author Contributions: Conceptualization, E.A. (Egemen Avcu), Y.Y.A., R.Y. and M.G.; methodology, E.A. (Egemen Avcu), Y.Y.A., H.U., M.S. and M.G.; software, E.A. (Egemen Avcu) and M.G.; validation, E.A. (Egemen Avcu), Y.Y.A., H.U. and R.Y.; formal analysis, E.A. (Egemen Avcu), H.U., M.S. and M.G.; investigation, E.A. (Egemen Avcu), Y.Y.A., H.U., M.İ.B. and E.A. (Eray Abakay); resources, E.A. (Egemen Avcu), H.U. and M.G.; data curation, E.A. (Egemen Avcu), M.İ.B. and E.A. (Eray Abakay); writing—original draft preparation, E.A. (Egemen Avcu), M.A., M.S. and M.G.; writing—review and editing, E.A. (Egemen Avcu), Y.Y.A., R.Y., C.E., M.F.-X.W. and M.G.; visualization, E.A. (Egemen Avcu), E.A. (Eray Abakay), M.A. and M.G.; supervision, E.A. (Egemen Avcu) and M.G.; project administration, E.A. (Egemen Avcu); funding acquisition, E.A. (Egemen Avcu) and M.G. All authors have read and agreed to the published version of the manuscript.

Funding: The authors acknowledge the financial support provided by the Kocaeli University Scientific Research Projects Coordination Unit under project numbers FKA-2024-3749, FKA-2024-3934, and FBA-2025-4619.

Data Availability Statement: The original contributions presented in this study are included in the article. Further inquiries can be directed to the corresponding author.

Conflicts of Interest: The authors declare no conflicts of interest. The funders had no role in the design of the study; in the collection, analyses, or interpretation of data; in the writing of the manuscript; or in the decision to publish the results.

References

1. Shivakoti, I.; Kibria, G.; Cep, R.; Pradhan, B.B.; Sharma, A. Laser Surface Texturing for Biomedical Applications: A Review. *Coatings* **2021**, *11*, 124. [CrossRef]
2. Sandoval-Robles, J.A.; Rodríguez, C.A.; García-López, E. Laser Surface Texturing and Electropolishing of CoCr and Ti6Al4V-ELI Alloys for Biomedical Applications. *Materials* **2020**, *13*, 5203. [CrossRef]
3. Lin, N.; Li, D.; Zou, J.; Xie, R.; Wang, Z.; Tang, B. Surface Texture-Based Surface Treatments on Ti6Al4V Titanium Alloys for Tribological and Biological Applications: A Mini Review. *Materials* **2018**, *11*, 487. [CrossRef]
4. Liu, Q.; Liu, Y.; Li, X.; Dong, G. Pulse Laser-İnduced Cell-Like Texture On Surface Of Titanium Alloy For Tribological Properties İmprovement. *Wear* **2021**, *477*, 203784. [CrossRef]
5. Shen, X.; Shukla, P.; Nath, S.; Lawrence, J. Improvement in Mechanical Properties of Titanium Alloy (Ti-6Al-7Nb) Subject to Multiple Laser Shock Peening. *Surf. Coat.* **2017**, *327*, 101–109. [CrossRef]
6. Abakay, E.; Armağan, M.; Yıldıran Avcu, Y.; Guney, M.; Yousif, B.F.; Avcu, E. Advances in improving tribological performance of titanium alloys and titanium matrix composites for biomedical applications: A critical review. *Front. Mater.* **2024**, *11*, 1452288. [CrossRef]
7. Gedikoğlu, M.; Kolsal, A.; Tutuş, H.; Toker, S.M. Current Approaches in Surface Processing of Biomedical Alloys; Laser Processes. *GU J. Sci. Part C* **2021**, *9*, 413–431.

8. Günther, D.; Scharnweber, D.; Hess, R.; Wolf-Brandstetter, C.; Holthaus, M.G.; Lasagni, A.F. High Precision Patterning of Biomaterials Using the Direct Laser İnterference Patterning Technology. In *Laser Surface Modification of Biomaterials*; Woodhead Publishing: Cambridge, UK, 2016.
9. Niemann, R.; Hahn, S.; Diestel, A.; Backen, A.; Schultz, L.; Nielsch, K.; Wagner, M.F.X.; Fähler, S. Reducing the nucleation barrier in magnetocaloric Heusler alloys by nanoindentation. *APL Mater.* **2016**, *4*, 064101. [CrossRef]
10. Wu, S.; Wang, Y.; Jiang, X.; Hu, S.; Yao, C.; Dong, B.; Ma, C.; Lu, L.; Chen, J. Bioinspired femtosecond laser surface texturing for enhancing tribological properties of Ti6Al4V titanium alloy. *Tribol. Int.* **2026**, *213*, 110987. [CrossRef]
11. Zhao, W.; Hu, J.; Yu, Z.; Pan, J. Effect of pulsed laser biomimetic patterns on surface properties and cell migration of titanium alloys. *Surf. Interfaces* **2025**, *68*, 106621. [CrossRef]
12. Song, B.; Wang, X.; Xie, L.; Xiang, J.; Zou, X.; Zou, S. Effect of laser shock peening on the surface integrity and fretting fatigue properties of high-strength titanium alloy TC21. *J. Mater. Res. Technol.* **2024**, *33*, 4533–4547. [CrossRef]
13. Puoza, J.C. Research progress of surface texturing and thermal diffusion technology to improve tribological properties of materials. *Prog. Eng. Sci.* **2025**, *2*, 100034. [CrossRef]
14. Vlcak, P.; Fojt, J.; Koller, J.; Drahokoupil, J.; Smola, V. Surface pre-treatments of Ti-Nb-Zr-Ta beta titanium alloy: The effect of chemical, electrochemical and ion sputter etching on morphology, residual stress, corrosion stability and the MG-63 cell response. *Results Phys.* **2021**, *28*, 104613. [CrossRef]
15. Löffler, R.; Fleischer, M.; Kern, D.P. An anisotropic dry etch process with fluorine chemistry to create well-defined titanium surfaces for biomedical studies. *Microelectron. Eng.* **2012**, *97*, 361–364. [CrossRef]
16. Zhang, P.; Gao, Y.; Yu, Y.; Sun, Y.; Zhou, H.; Zhang, J. Effects of water-guided laser surface strengthening on surface properties and fatigue life of TC4 titanium alloy in tension-tension fatigue tests. *Vacuum* **2025**, *232*, 113863. [CrossRef]
17. Ben Aissa, C.; Khlifi, K. CAE-PVD synthesis and characterization of titanium-based biocompatible coatings deposited on titanium alloy for biomedical application. *Mater. Today Proc.* **2021**, *42*, A10–A17. [CrossRef]
18. Bilen, M. Enhancing biomedical applications of Ti–6Al–4V alloys: The role of boron nitride and titanium diboride coatings in mechanical and radiation shielding performance. *Radiat. Phys. Chem.* **2025**, *230*, 112597. [CrossRef]
19. Noori, M.; Atapour, M.; Ashrafizadeh, F.; Elmkhah, H.; di Confiengo, G.G.; Ferraris, S.; Perero, S.; Cardu, M.; Spriano, S. Nanostructured multilayer CAE-PVD coatings based on transition metal nitrides on Ti6Al4V alloy for biomedical applications. *Ceram. Int.* **2023**, *49*, 23367–23382. [CrossRef]
20. Milles, S.; Dahms, J.; Soldera, M.; Lasagni, A.F. Stable Superhydrophobic Aluminum Surfaces Based on Laser-Fabricated Hierarchical Textures. *Materials* **2021**, *14*, 184. [CrossRef]
21. Erdoğan, M.; Öktem, B.; Kalaycıoğlu, H.; Yavaş, S.; Mukhopadhyay, P.K.; Eken, K.; Özgören, K.; Aykaç, Y.; Tazebay, U.H.; Ilday, F.Ö. Texturing of titanium (Ti6Al4V) medical implant surfaces with MHz-repetition-rate femtosecond and picosecond Yb-doped fiber lasers. *Opt. Express* **2011**, *19*, 10986–10996. [CrossRef] [PubMed]
22. Avcu, Y.Y.; Iakovakis, E.; Guney, M.; Çalım, E.; Özkılınç, A.; Abakay, E.; Sönmez, F.; Koç, F.G.; Yamanoğlu, R.; Cengiz, A.; et al. Surface and Tribological Properties of Powder Metallurgical Cp-Ti Titanium Alloy Modified by Shot Peening. *Coatings* **2023**, *13*, 89. [CrossRef]
23. Avcu, E.; Abakay, E.; Yıldıran Avcu, Y.; Çalım, E.; Gökalp, İ.; Iakovakis, E.; Koç, F.G.; Yamanoglu, R.; Akıncı, A.; Guney, M. Corrosion Behavior of Shot-Peened Ti6Al4V Alloy Produced via Pressure-Assisted Sintering. *Coatings* **2023**, *13*, 2036. [CrossRef]
24. Yildiran Avcu, Y.; Yetik, O.; Guney, M.; Iakovakis, E.; Sinmazcelik, T.; Avcu, E. Surface, Subsurface and Tribological Properties of Ti6Al4V Alloy Shot Peened under Different Parameters. *Materials* **2020**, *13*, 4363. [CrossRef]
25. Žagar, S.; Grum, J. Surface Modification Analysis after Shot Peening of AA 7075 in Different States. *Mater. Sci. Forum* **2013**, *768–769*, 519–525. [CrossRef]
26. Tini, S.C.; Zeren, A.; Avcu, Y.Y.; Abakay, E.; Iakovakis, E.; Sınmazçelik, T.; Guney, M.; Avcu, E. Effect of shot peening on the cavitation erosion of manganese aluminium bronze alloy: Microstructural and morphological insights. *Mater. Lett.* **2025**, *394*, 138659. [CrossRef]
27. Lin, Q.; Liu, H.; Zhu, C.; Chen, D.; Zhou, S. Effects of Different Shot Peening Parameters on Residual Stress, Surface Roughness and Cell Size. *Surf. Coat. Technol.* **2020**, *398*, 126054. [CrossRef]
28. Avcu, Y.Y.; Gonul, B.; Yetik, O.; Sonmez, F.; Cengiz, A.; Guney, M.; Avcu, E. Modification of Surface and Subsurface Properties of AA1050 Alloy by Shot Peening. *Materials* **2021**, *14*, 6575. [CrossRef] [PubMed]
29. Wang, Y.M.; Guo, J.W.; Zhuang, J.P.; Jing, Y.B.; Shao, Z.K.; Jin, M.S.; Zhang, J.; Wei, D.Q.; Zhou, Y. Development and characterization of MAO bioactive ceramic coating grown on micro-patterned Ti6Al4V alloy surface. *Appl. Surf. Sci.* **2014**, *299*, 58–65. [CrossRef]
30. Zhu, L.; Guan, Y.; Wang, Y.; Xie, Z.; Lin, J.; Zhai, J. Influence of process parameters of ultrasonic shot peening on surface roughness and hydrophilicity of pure titanium. *Surf. Coat. Technol.* **2017**, *317*, 38–53. [CrossRef]
31. Hsiao, V.K.S.; Shih, M.-H.; Wu, H.-C., Wu, T.-I. Comparative Study of Surface Modification Techniques for Enhancing Biocompatibility of Ti-6Al-4V Alloy in Dental Implants. *Appl. Sci.* **2024**, *14*, 10904. [CrossRef]

32. Uanlee, T.; Khantachawana, A. Effect of fine shot peening on physical and biological properties of SUS316L and Ti-6Al-4V for medical applications. In Proceedings of the 2017 10th Biomedical Engineering International Conference (BMEiCON), Hokkaido, Japan, 31 August–2 September 2017; pp. 1–4.
33. Khandaker, M.; Riahinezhad, S.; Sultana, F.; Vaughan, M.B.; Knight, J.; Morris, T.L. Peen treatment on a titanium implant: Effect of roughness, osteoblast cell functions, and bonding with bone cement. *Int. J. Nanomed.* **2016**, *11*, 585–594. [CrossRef]
34. Shen, X.; Hu, W.; Ping, L.; Liu, C.; Yao, L.; Deng, Z.; Wu, G. Antibacterial and Osteogenic Functionalization of Titanium with Silicon/Copper-Doped High-Energy Shot Peening-Assisted Micro-Arc Oxidation Technique. *Front. Bioeng. Biotechnol.* **2020**, *8*, 573464. [CrossRef] [PubMed]
35. Li, J.; Chen, J.; Chen, Y.; Ren, J.; Li, J.; Wu, L.; Liu, H.; Cao, F.; Sun, Q. Copper powder ultrasonic shot peening and its application in antifouling of Ti alloy. *Surf. Coat. Technol.* **2025**, *505*, 132089. [CrossRef]
36. Zheng, H.; You, X.; Zhang, J.; Liu, F.; Lou, C. Enhanced mechanical properties and antimicrobial performance of TC4 titanium alloy via surface nanocrystallization and gas nitriding. *Dig. J. Nanomater. Biostruct.* **2025**, *20*, 545–558. [CrossRef]
37. Linjee, S.; Bumrungpon, M.; Newyawong, P.; Wutikhun, T.; Songkuea, S.; Muengto, S.; Tange, M.; Suwanpreecha, C.; Manonukul, A. On the severe shot peening effect to generate nanocrystalline surface towards enhancing fatigue life of injection-moulded Ti-6Al-4V alloy. *J. Mater. Sci.* **2023**, *58*, 15513–15528. [CrossRef]
38. Bagherifard, S.; Fernandez-Pariente, I.; Ghelichi, R.; Guagliano, M. Effect of severe shot peening on microstructure and fatigue strength of cast iron. *Int. J. Fatigue* **2014**, *65*, 64–70. [CrossRef]
39. Yang, Q.; Cheng, J.-H.; Guan, H.-J.; Tan, W.-J.; Zhang, Y.-H. Investigation of wet shot peening on microstructural evolution and tensile-tensile fatigue properties of Ti-6Al-4V alloy. *Mater. Chem. Phys.* **2022**, *291*, 126635. [CrossRef]
40. Zhang, Q.; Duan, B.; Zhang, Z.; Wang, J.; Si, C. Effect of ultrasonic shot peening on microstructure evolution and corrosion resistance of selective laser melted Ti–6Al–4V alloy. *J. Mater. Res. Technol.* **2021**, *11*, 1090–1099. [CrossRef]
41. Nishitani, T.; Masuda, K.; Mimura, S.; Hirokawa, T.; Ishiguro, H.; Kumagai, M.; Ito, T. Antibacterial effect on microscale rough surface formed by fine particle bombarding. *AMB Express* **2022**, *12*, 9. [CrossRef]
42. Nishitani, T.; Hirokawa, T.; Ishiguro, H.; Ito, T. Mechanism of antibacterial property of micro scale rough surface formed by fine-particle bombarding. *Sci. Technol. Adv. Mater.* **2024**, *25*, 2376522. [CrossRef]
43. Agrawal, R.K.; Pandey, V.; Barhanpurkar-Naik, A.; Wani, M.R.; Chattopadhyay, K.; Singh, V. Effect of ultrasonic shot peening duration on microstructure, corrosion behavior and cell response of cp-Ti. *Ultrasonics* **2020**, *104*, 106110. [CrossRef]
44. Avcu, E.; Yıldıran Avcu, Y.; Baştan, F.E.; Rehman, M.A.U.; Üstel, F.; Boccaccini, A.R. Tailoring the surface characteristics of electrophoretically deposited chitosan-based bioactive glass composite coatings on titanium implants via grit blasting. *Prog. Org. Coat.* **2018**, *123*, 362–373. [CrossRef]
45. Vishnu, J.; Ansheed, A.R.; Hameed, P.; Praveenkumar, K.; Pilz, S.; Andrea Alberta, L.; Swaroop, S.; Calin, M.; Gebert, A.; Manivasagam, G. Insights into the surface and biocompatibility aspects of laser shock peened Ti-22Nb alloy for orthopedic implant applications. *Appl. Surf. Sci.* **2022**, *586*, 152816. [CrossRef]
46. Chauhan, P.; Shadangi, Y.; Bhatnagar, A.; Singh, V.; Chattopadhyay, K. Influence of Surface Nano-Structuring on Microstructure, Corrosion Behavior and Osteoblast Response of Commercially Pure Titanium Treated Through Ultrasonic Shot Peening. *JOM* **2022**, *74*, 584–595. [CrossRef]
47. Bagherifard, S.; Hickey, D.J.; de Luca, A.C.; Malheiro, V.N.; Markaki, A.E.; Guagliano, M.; Webster, T.J. The influence of nanostructured features on bacterial adhesion and bone cell functions on severely shot peened 316L stainless steel. *Biomaterials* **2015**, *73*, 185–197. [CrossRef] [PubMed]
48. *ISO 25178*; Geometrical Product Specifications (GPS)—Surface Texture: Areal. International Organization for Standardization: Geneva, Switzerland, 2016.
49. Priyanka, C.P.; Keerthi Krishnan, K.; Sudeep, U.; Ramachandran, K.K. Osteogenic and antibacterial properties of TiN-Ag coated Ti-6Al-4V bioimplants with polished and laser textured surface topography. *Surf. Coat. Technol.* **2023**, *474*, 130058. [CrossRef]
50. Kikuchi, S.; Noguchi, S.; Mitake, K.; Kurosaka, S.; Doi, K.; Harada, H. Surface modification of low-alloy steels by fine particle peening using combination of iron sulfide and spring steel particles. *Mater. Lett.* **2025**, *391*, 138517. [CrossRef]
51. Takesue, S.; Morita, T.; Misaka, Y.; Komotori, J. Rapid formation of titanium nitride coating by atmospheric-controlled induction-heating fine particle peening and investigation of its wear and corrosion resistance. *Results Surf. Interfaces* **2023**, *11*, 100121. [CrossRef]
52. Huang, L.Z.Y.; Truong, V.K.; Murdoch, B.J.; Elbourne, A.; Caruso, R.A. Inherent variation in surface roughness of Selective Laser Melting (SLM) printed titanium caused by build angle changes the mechanomicrobiocidal effectiveness of nanostructures. *J. Colloid Interface Sci.* **2025**, *696*, 137866. [CrossRef]
53. Chen, M.; Deng, W.; Li, J.; Jiang, C.; Ji, V. Study on surface characteristics and wear resistance of PBF-LB Ti-6Al-4V alloy treated by shot peening. *Mater. Today Commun.* **2025**, *46*, 112901. [CrossRef]
54. Sivasubramanian, J.; Upadhyay, P.; Mallik, A.; Basu, A. Electrochemical corrosion behaviour of electrodeposited nickel coating subjected to ultrasonic shot peening treatment. *Surf. Coat. Technol.* **2025**, *503*, 132005. [CrossRef]

55. Lopez-Ruiz, P.; Garcia-Blanco, M.B.; Vara, G.; Fernández-Pariente, I.; Guagliano, M.; Bagherifard, S. Obtaining tailored surface characteristics by combining shot peening and electropolishing on 316L stainless steel. *Appl. Surf. Sci.* **2019**, *492*, 1–7. [CrossRef]
56. Yuan, Y.; Hays, M.P.; Hardwidge, P.R.; Kim, J. Surface characteristics influencing bacterial adhesion to polymeric substrates. *RSC Adv.* **2017**, *7*, 14254–14261. [CrossRef]
57. Cunha, A.; Elie, A.-M.; Plawinski, L.; Serro, A.P.; Botelho do Rego, A.M.; Almeida, A.; Urdaci, M.C.; Durrieu, M.-C.; Vilar, R. Femtosecond laser surface texturing of titanium as a method to reduce the adhesion of Staphylococcus aureus and biofilm formation. *Appl. Surf. Sci.* **2016**, *360*, 485–493. [CrossRef]
58. Truong, V.K.; Lapovok, R.; Estrin, Y.S.; Rundell, S.; Wang, J.Y.; Fluke, C.J.; Crawford, R.J.; Ivanova, E.P. The influence of nano-scale surface roughness on bacterial adhesion to ultrafine-grained titanium. *Biomaterials* **2010**, *31*, 3674–3683. [CrossRef]
59. Ferraris, S.; Spriano, S. Antibacterial titanium surfaces for medical implants. *Mater. Sci. Eng. C Mater. Biol. Appl.* **2016**, *61*, 965–978. [CrossRef] [PubMed]
60. Ferraris, S.; Venturello, A.; Miola, M.; Cochis, A.; Rimondini, L.; Spriano, S. Antibacterial and bioactive nanostructured titanium surfaces for bone integration. *Appl. Surf. Sci.* **2014**, *311*, 279–291. [CrossRef]
61. Xu, J.; Wu, X.; Lian, Z.; Yu, H.; Hou, Y.; Nie, X. Micro-nano manufacturing of Ti6Al4V antibacterial surface. *Colloids Surf. A Physicochem. Eng. Asp.* **2021**, *625*, 126929. [CrossRef]
62. Bright, R.; Fernandes, D.; Wood, J.; Palms, D.; Burzava, A.; Ninan, N.; Brown, T.; Barker, D.; Vasilev, K. Long-term antibacterial properties of a nanostructured titanium alloy surface: An in vitro study. *Mater. Today Bio* **2022**, *13*, 100176. [CrossRef]
63. Burgueño-Barris, G.; Camps-Font, O.; Figueiredo, R.; Valmaseda-Castellón, E. The Influence of Implantoplasty on Surface Roughness, Biofilm Formation, and Biocompatibility of Titanium Implants: A Systematic Review. *Int. J. Oral Maxillofac. Implant.* **2021**, *36*, e111–e119. [CrossRef]
64. Wu, S.; Altenried, S.; Zogg, A.; Zuber, F.; Maniura-Weber, K.; Ren, Q. Role of the Surface Nanoscale Roughness of Stainless Steel on Bacterial Adhesion and Microcolony Formation. *ACS Omega* **2018**, *3*, 6456–6464. [CrossRef] [PubMed]

coatings

MDPI

Article

Influence of TiO_2 Nanoparticle Concentration on Micro-Arc Oxidized Calcium–Phosphate Coatings: Corrosion Resistance and Biological Response

Ainur Zhassulan [1,2], Bauyrzhan Rakhadilov [1,3], Daryn Baizhan [1,2], Aidar Kengesbekov [1,3,4,*], Dauir Kakimzhanov [1,3] and Nazira Musataeva [2]

1 Research Center "Protective and Functional Coatings", D. Serikbayev East Kazakhstan Technical University, Ust-Kamenogorsk 070000, Kazakhstan; a.zhassulan@shakarim.kz (A.Z.); rakhadilovb@mail.ru (B.R.); daryn.baizhan@mail.ru (D.B.); dauir_97@mail.ru (D.K.)

2 Engineering Center, Shakarim University NJSC, Glinka Street, 20A, Semey 071412, Kazakhstan; naziramusataeva51@gmail.com

3 PlasmaScience LLP, Ust-Kamenogorsk 070000, Kazakhstan

4 Institute of Composite Materials, Ust-Kamenogorsk 070000, Kazakhstan

* Correspondence: kenesbekovaidar@gmail.com

Abstract: Titanium and its alloys are widely used in biomedical implants due to their favorable mechanical properties and corrosion resistance; however, their natural surface lacks sufficient bioactivity and antibacterial performance. Micro-arc oxidation is a promising approach to producing bioactive coatings, and the incorporation of nanoparticles such as TiO_2 may further improve their functionality. This study aimed to determine the optimal TiO_2 nanoparticle concentration in the micro-arc oxidation electrolyte that ensures coating stability and biological safety. Calcium–phosphate coatings were fabricated on commercially pure titanium using micro-arc oxidation with two TiO_2 concentrations: 0.5 wt.% (MAO 1) and 1 wt.% (MAO 2). Surface morphology, porosity, and phase composition were analyzed by scanning electron microscopy, energy-dispersive spectroscopy, and X-ray diffraction. Corrosion resistance was evaluated via potentiodynamic polarization in NaCl and Ringer's solutions, while biocompatibility was assessed in vitro using HOS human osteosarcoma cells and MTT assays. Increasing the TiO_2 content to 1% decreased coating porosity (13.7% vs. 26.3% for MAO 1), enhanced corrosion protection, and reduced the friction coefficient compared to bare titanium. However, MAO 2 exhibited high cytotoxicity (81% cell death) and partial structural degradation in the biological medium. MAO 1 maintained integrity and showed no toxic effects (3% cell death). These results suggest that 0.5% TiO_2 is the optimal concentration, providing a balance between corrosion resistance, mechanical stability, and biocompatibility, supporting the development of safer implant coatings.

Keywords: micro-arc oxidation; titanium; TiO_2 nanoparticles; corrosion; biocompatibility; cytotoxicity

1. Introduction

Titanium and its alloys are among the most widely used materials in modern medicine due to their high corrosion resistance, low density, favorable mechanical properties, and good bioinertness under physiological conditions [1,2]. However, the natural surface of titanium has several limitations, such as low wear resistance, limited osseointegration, and lack of pronounced antibacterial activity [3,4]. These factors drive the search for effective surface modification methods capable of improving the functional characteristics of titanium implants [5].

Various technologies are employed for the surface modification of titanium implants, including anodic oxidation, plasma spraying, sol-gel methods, laser texturing, ion implantation, and others [6–9]. These approaches allow the formation of oxide or bioactive layers that enhance cell adhesion and improve the corrosion resistance of the material. However, they also have several limitations, including poor coating adhesion to the substrate, insufficient porosity that slows down the osseointegration process, and challenges in the controlled incorporation of functional elements into the layer composition [10,11]. Micro-arc oxidation (MAO) is one of the most promising methods for producing bioactive coatings on titanium [12,13]. During MAO, a multilayered oxide film with a well-developed porous structure forms on the surface, increasing the contact area with biological tissues and enhancing osseointegration [14]. The additional incorporation of nanoparticles of various elements (Ag, Zn, Cu, TiO_2, CeO_2, etc.) into the coating can impart antibacterial, osteogenic, and photocatalytic properties [15–17], making MAO a versatile tool for the development of medical implants.

Particular attention has been given to TiO_2 nanoparticles, which are widely used due to their bioinertness, photocatalytic activity, and ability to enhance the corrosion and mechanical resistance of coatings [18–20]. However, growing evidence indicates potential toxicological risks associated with the migration of TiO_2 nanoparticles from coatings into biological environments. In vivo and in vitro studies have demonstrated that nanoparticles can penetrate tissues and organs, accumulate in the liver, kidneys, and brain, and cause oxidative stress, cell membrane damage, and apoptosis [21–24]. These effects depend on the concentration, size, and degree of nanoparticle aggregation, as well as the stability of the coating in physiological media.

Recent review and experimental studies have confirmed the promising potential of MAO coatings while highlighting the need for precise control over the composition and dosage of incorporated ions. For instance, electrochemical modification of titanium with Zn, Cu, Ag, Sr, and Ce ions via MAO has been shown to enhance osteogenic, angiogenic, and antibacterial properties. However, systematic reviews emphasize the risk of uncontrolled ion release and coating instability during long-term exposure to physiological environments [25–29].

In vitro studies on cell cultures further underline the dose-dependent nature of TiO_2 toxicity: high nanoparticle concentrations can induce oxidative stress, DNA damage, and cell cycle disruptions, particularly in three-dimensional models that mimic in vivo conditions. This highlights the increased cellular sensitivity to nanoparticle dosage and morphology and underscores the need for a comprehensive analysis of both the physicochemical and biological properties of such coatings [30,31].

Despite the widespread use of TiO_2-modified coatings in implantology, the mechanisms underlying their potential cytotoxicity at different nanoparticle concentrations remain unclear. A comprehensive study, combining physicochemical characterization and biological testing, is required to assess coating stability and safety for living cells.

The aim of this work is to provide a systematic evaluation of the influence of TiO_2 nanoparticle concentration in the MAO electrolyte on the morphology, phase composition, corrosion resistance, and cytotoxicity of calcium–phosphate coatings formed on titanium. The findings of this study are expected to help define optimal MAO processing conditions that ensure a balance between mechanical and corrosion stability and biological safety, which is essential for the development of next-generation implantable medical devices.

2. Materials and Methods

Commercially pure titanium (grade VT1-0) samples with dimensions of 10 × 10 × 3 mm were used in this study. Prior to treatment, the sample surfaces were subjected to mechanical grinding, ultrasonic cleaning in ethanol, rinsing with deionized water, and sandblasting to remove the native oxide layer and level the surface. Calcium–phosphate coatings were produced by micro-arc oxidation (MAO) in an aqueous electrolyte containing 30% phosphoric acid (H_3PO_4), 60 g L^{-1} hydroxyapatite (HA), and 90 g L^{-1} calcium carbonate ($CaCO_3$). To investigate the effect of TiO_2 nanoparticle concentration, two electrolyte compositions were prepared, containing 0.5% and 1% TiO_2 nanoparticles, respectively. Samples obtained under these conditions are hereafter designated as MAO 1 (0.5% TiO_2) and MAO 2 (1% TiO_2). In addition to the TiO_2-containing coatings, a control PEO coating was prepared in the same calcium–phosphate electrolyte without TiO_2 nanoparticles (denoted MAO 0) to provide a reference for comparison of the morphological and physicochemical properties. We used commercial titanium (IV) oxide nanopowder (anatase phase) purchased from AldrichSt. Louis, MI, USA (catalogue No. 637254-50G, Lot No. MKCN0838). According to the manufacturer's specifications, the particles have an average size of <25 nm, a purity of 99.7%, and the crystalline phase is anatase. The base electrolyte for micro-arc oxidation was a calcium–phosphate solution consisting of 30% H_3PO_4, 60 g L^{-1} hydroxyapatite, and 90 g L^{-1} $CaCO_3$. For the MAO 1 and MAO 2 samples, this same calcium–phosphate electrolyte was used, with the only difference being the addition of 0.5 wt.% or 1 wt.% of the TiO_2 nanopowder described above. The MAO process was performed using a KP-HI-F-40A600V pulsed power supply under the following parameters: voltage—300 V, processing time—600 s, frequency—200 Hz, duty cycle—15%, and current density—0.23–0.35 A/cm^2. After treatment, the samples were rinsed with deionized water and dried at room temperature.

The surface morphology and elemental composition of the coatings were analyzed using a Tescan scanning electron microscope equipped with an energy-dispersive spectroscopy (EDS) detector. Phase composition was determined by X-ray diffraction (X'Pert PRO, PANalytical, Almelo, The Netherlands). Porosity and average pore size were quantified based on SEM images using the ImageJ softwareversion:1.54d.

Corrosion resistance was evaluated by potentiodynamic polarization using a CS350M potentiostat-galvanostat (Corrtest Instruments) with a flat-type corrosion cell (model CS936) in a standard three-electrode configuration. The working electrode was a coated or uncoated sample with an exposed area of 1 cm^2. A saturated Ag/AgCl electrode served as the reference, and a platinum mesh was used as the counter electrode. Tests were conducted in two media: 3.5 wt.% NaCl and Ringer's solution at 25 °C. Prior to polarization, the open-circuit potential (OCP) was stabilized for 30 min. The potential sweep was carried out from −0.25 V to 0.00 V versus Ag/AgCl at a scan rate of 0.5 mV/s. The open surface porosity and pore size were quantified from plan-view SEM micrographs in ImageJ. After scale calibration images were segmented (Otsu threshold) with watershed separation; edge-touching features and objects < 2 μm were excluded. For each condition, 5 random fields of view were analyzed; the results are reported as mean ± SD. Static water contact angles were measured using a goniometer SDA-100 contact goniometer (Qchaida, Dongguan, China, sessile-drop method (3 μL), at 23 ± 1 °C.

Tribological tests were performed using an Anton Paar TRB3 tribometer in a ball-on-disk configuration in Ringer's solution. As the counterbody, we used a 6 mm silicon–nitride (Si_3N_4) ceramic ball, supplied with the instrument, with an applied load of 2 N and a sliding distance of 100 m. The friction coefficient was recorded in real time. Si_3N_4 was chosen because it is chemically inert and does not corrode in Ringer's solution, allowing the tribological response of the coating itself to be evaluated without interference from the counterbody.

For cytotoxicity evaluation, 10 samples were prepared (5 from each coating group). The study was carried out using the HOS human osteosarcoma cell line in accordance with ISO 10993-5 [32]. The cell concentration was 10^5 cells/mL in culture medium (90% DMEM, 10% FBS, antibiotics, and L-glutamine). Samples were incubated with the cell culture for 24 h at 37 °C, 5% CO_2, and 100% humidity. A culture without samples served as the control. After incubation, samples were removed, and 0.5% MTT solution (NeoFroxx, Einhausen Germany) was added and incubated for an additional 4 h, followed by dissolution of the formazan crystals in 10% dodecyl sulfate in 0.01 M HCl. Optical density was measured using a Multiskan FC spectrophotometer (Thermo Fisher Scientific, Shanghai China) at 540 nm with a 630 nm reference. The cytotoxicity index (*CI*, %) was calculated using the following formula:

$$CI(\%) = \frac{K - O}{K} \times 100\% \quad (1)$$

where K is the absorbance of the control and O is the absorbance of the sample. Samples were considered non-cytotoxic when $CI \leq 30\%$. Macro- and microscopic images of the surfaces before and after the experiments were taken using a Canon Power Shot A630 digital camera and an ADF I350 inverted metallographic microscope after staining fixed cells with a mixture of Azure II and eosin. Statistical analysis was performed in RStudio (R) software version:6.3.1128.1 using the Shapiro–Wilk test (with Royston's correction), Welch's ANOVA, the Brunner–Munzel test, and Games–Howell multiple comparisons, with a significance level of $p < 0.05$. Prior to cell contact, specimens were rinsed 3× with deionized water (and PBS), then air-dried; after MAO, all samples were rinsed and dried as described.

3. Results

3.1. Surface Morphology of MAO Coatings

Figure 1 presents the surface morphology of the coatings produced in electrolytes with different TiO_2 nanoparticle contents. The MAO 0 surface, obtained without nanoparticle addition, exhibits the typical micro-arc oxidation (MAO) morphology: a porous layer with rounded discharge craters of various sizes and relatively smooth, dense pore walls. When 0.5 wt.% TiO_2 nanoparticles are introduced into the electrolyte (MAO 1, Figure 1b), the coating retains the characteristic MAO structure but shows a slightly denser arrangement of pores and more uniform pore distribution. Increasing the nanoparticle concentration to 1 wt.% produces a surface with reduced total open porosity and partial closure of some discharge channels; the large craters remain but their number is lower, indicating a more compact oxide layer. XRD analysis confirms the presence of TiO_2 signals in both MAO 1 and MAO 2 coatings, demonstrating that the TiO_2 nanoparticles become incorporated into the MAO layer during plasma-electrolytic oxidation. The amount of incorporated TiO_2 increases with the nanoparticle content in the electrolyte, but no qualitative difference in the type of incorporated phase is observed. Thus, the main effect of increasing TiO_2 concentration is a denser, less porous microstructure with a higher level of TiO_2 incorporation, which can influence the subsequent corrosion resistance and biological response.

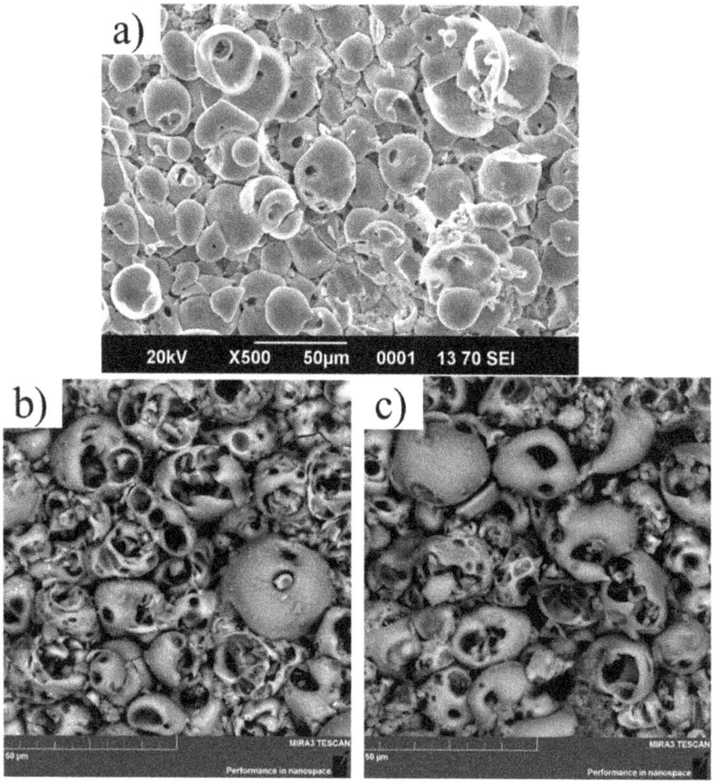

Figure 1. Surface SEM images of the coatings: (**a**) MAO 0—coating formed in the calcium–phosphate electrolyte without TiO_2 nanoparticles, (**b**) MAO 1—coating formed with 0.5 wt.% TiO_2, (**c**) MAO 2—coating formed with 1 wt.% TiO_2.

3.2. *Phase Composition*

Figure 2 presents the X-ray diffraction (XRD) patterns of the untreated titanium surface and the MAO coatings produced with different TiO_2 nanoparticle concentrations. For the uncoated titanium substrate, characteristic diffraction peaks of α-Ti are observed at $2\theta \approx 35°$, 38°, 40°, 53°, and 63°, corresponding to the hexagonal close-packed crystal structure of metallic titanium. After MAO treatment, additional diffraction peaks appear for both coating types (MAO 1 and MAO 2), which can be attributed to titanium dioxide phases. Distinct peaks of anatase TiO_2 are detected at approximately $2\theta \approx 25°$, 37°, 48°, and 55°, while rutile peaks are observed to a lesser extent near $2\theta \approx 27°$. This indicates the formation of an oxide film with a mixed polymorphic composition. Increasing the TiO_2 nanoparticle concentration in the electrolyte (MAO 2) results in a higher intensity of both anatase and rutile peaks compared to MAO 1, suggesting a greater amount of crystalline TiO_2 phases in the coating [33,34]. The metallic titanium peaks remain visible, which is attributed to partial penetration of the X-ray beam through the porous coating layer to the underlying substrate. Thus, a higher TiO_2 nanoparticle content promotes the formation of a coating with enhanced crystallinity and enrichment in the anatase and rutile phases.

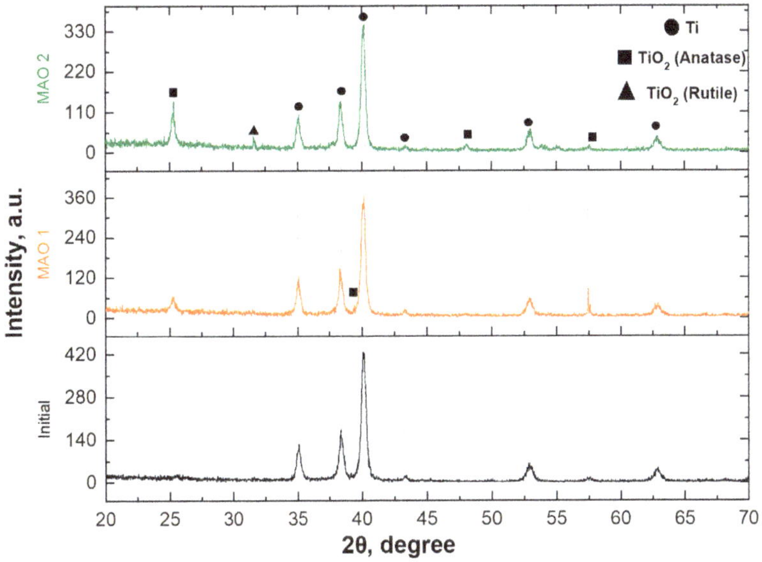

Figure 2. X-ray diffraction patterns of initial titanium and coatings containing TiO_2.

3.3. Porosity and Pore Size

In aqueous electrolytes, such particles can form loose agglomerates of several hundred nanometers, which is typical for unmodified TiO_2 suspensions in near-neutral or weakly alkaline solutions [35]. During micro-arc oxidation, the local plasma micro-discharges reach transient temperatures of 2000–10,000 K, but the residence time of individual nanoparticles in the discharge channel is extremely short (micro- to milliseconds) and rapid quenching occurs [36]. Numerous studies have shown that TiO_2 nanoparticles do not melt completely; instead, they become partially sintered and are incorporated into the growing calcium–phosphate/TiO_2 oxide layer as crystalline anatase/rutile phases [37]. Our own XRD and EDS analyses (Figure 2) confirm the presence of anatase and rutile peaks in the final coating, demonstrating that TiO_2 nanoparticles survive the MAO process and are embedded within the oxide matrix rather than being dissolved or lost.

Figure 3 presents the porosity and average pore-size values of the coatings. For the MAO 1 sample, the total open surface porosity is 26.3%, with an average pore size of 22.4 μm. Increasing the TiO_2 nanoparticle concentration to 1% (MAO 2) reduces the open porosity to 13.7%, while the mean pore size decreases only slightly to 19.7 μm. SEM observations (Figure 1) show that a few large discharge craters remain visible on MAO 2, but their areal density is lower, so that the overall mean pore diameter is marginally smaller. The reduction in open porosity is attributed to a more intense micro-arc oxidation process and the consequent densification of the oxide layer, which leads to partial closure of discharge channels [38–40]. Although high porosity can facilitate osseointegration by increasing the surface area available for cell attachment, excessive porosity may compromise corrosion resistance and mechanical strength. Thus, a TiO_2 nanoparticle concentration of 1% promotes the formation of a denser coating with potentially improved structural stability while maintaining the characteristic MAO surface morphology.

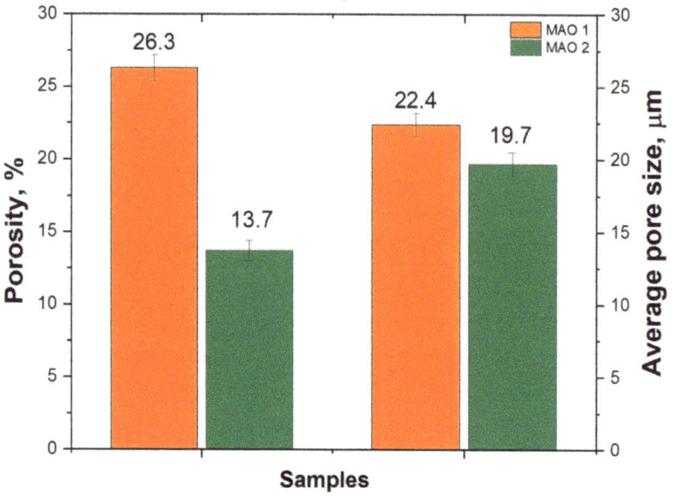

Figure 3. Porosity and average pore size of MAO coatings formed with different TiO_2 nanoparticle concentrations.

3.4. Coating Thickness (Cross-Sectional SEM)

Cross-sectional SEM micrographs (Figure 4) reveal the typical duplex MAO structure with a porous outer layer and a dense inner barrier layer. The coating thickness, determined from three independent measurements on each sample, is about 72 ± 8 μm for MAO 1 and about 123 ± 5 μm for MAO 2. Although no statistical analysis was performed, these values clearly indicate that the coating obtained with 1 wt.% TiO_2 is markedly thicker than the one produced with 0.5 wt.% TiO_2. The increase in thickness can be attributed to the higher density and energy of micro-discharges in the electrolyte with a larger amount of TiO_2 nanoparticles, which accelerates oxide growth and promotes the formation of a thicker porous layer.

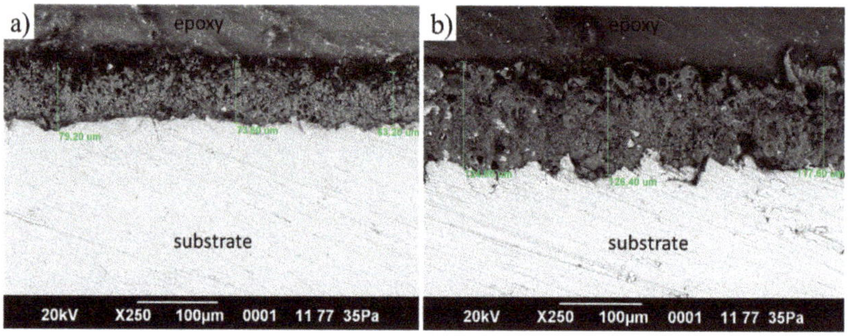

Figure 4. SEM cross-section images of MAO coatings: (**a**) MAO 1, (**b**) MAO 2.

3.5. Surface Roughness

The surface roughness (Ra) values obtained from contact profilometry are summarized in Table 1 and representative profiles are shown in Figure 5. The untreated titanium surface exhibits an average roughness of 3.55 ± 1.5 μm, which reflects the initial sand-blasted preparation. The TiO_2-free MAO coating (MAO 0) shows a comparable Ra of 3.96 ± 1.2 μm, indicating that the micro-arc oxidation process without nanoparticle additives does not substantially change the macro-scale roughness. In contrast, incorporation of TiO_2 nanoparticles leads to a slight but measurable decrease: MAO 1—3.05 ± 0.4 μm and MAO 2—2.98 ± 0.7 μm. This trend is consistent with the SEM observations (Figure 1), where the coatings containing nanoparticles display partial closure of large discharge channels and a denser surface. A lower Ra can improve the uniformity of cell attachment while maintaining the characteristic porous microstructure required for osseointegration.

Table 1. Average surface roughness (Ra) of untreated titanium and MAO coatings with different TiO_2 nanoparticle contents.

Group	Roughness Ra, μm
Initial	3.55 ± 1.5
MAO 0	3.96 ± 1.2
MAO 1	3.05 ± 0.4
MAO 2	2.98 ± 0.7

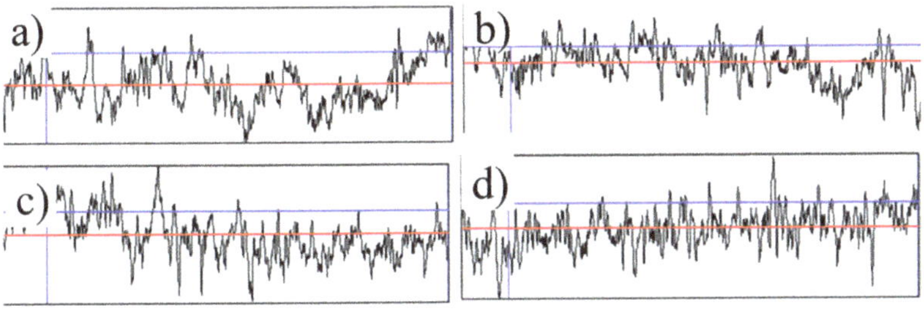

Figure 5. Surface roughness profiles of the tested samples: (**a**) initial, (**b**) MAO 0, (**c**) MAO 1, (**d**) MAO 2.

3.6. Corrosion Behavior

The anodic polarization curves (Figure 6) demonstrate clear differences in the corrosion behavior of bare titanium and MAO-coated samples. In 3.5 wt.% NaCl solution (Figure 6a), the corrosion current density (Icorr) of untreated titanium is 1.48×10^{-5} A cm^{-2}, whereas MAO 1 and MAO 2 show markedly lower values of 3.50×10^{-6} A cm^{-2} and 8.89×10^{-6} A cm^{-2}, respectively. Both coatings also exhibit a positive shift of the corrosion potential, indicating a slowdown of anodic dissolution processes. Among them, the MAO 2 coating provides the most pronounced improvement in the chloride medium. In Ringer's solution (Figure 6b), the absolute Icorr values for all samples are lower, reflecting the formation of a protective passive TiO_2 film promoted by the calcium- and phosphate-containing ions of the medium: 2.27×10^{-6} A cm^{-2} for bare titanium, 9.68×10^{-7} A cm^{-2} for MAO 1, and 1.13×10^{-5} A cm^{-2} for MAO 2. Owing to this strong spontaneous passivation, the untreated titanium shows a slightly lower Icorr than the MAO coatings. This effect is well known for titanium in simulated body fluids and represents a temporary passivation phenomenon rather than superior long-term protection. Importantly, in the more aggressive chloride environment (NaCl), the MAO coatings, particularly MAO 2, still provide markedly better protection than bare titanium. The superior corrosion performance of MAO 2 compared with MAO 1 can be attributed to its lower open porosity (13.7% vs. 26.3%), which limits the penetration of aggressive ions, and to the higher crystallinity of the oxide layer with an increased content of stable TiO_2 phases [41–43].

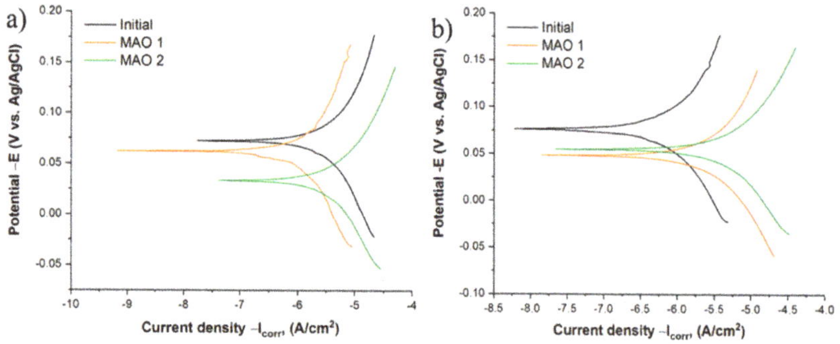

Figure 6. Polarization curves of the samples: (**a**) in 3.5 wt.% NaCl solution; (**b**) in Ringer's solution.

3.7. Tribological Performance

Figure 7 shows the evolution of the friction coefficient (μ) during ball-on-disk sliding in Ringer's solution. All samples display an initial running-in stage with a rapid rise in μ, followed by a quasi-steady regime. The untreated titanium and the MAO 0 coating (formed without TiO_2 nanoparticles) exhibit the highest steady-state friction coefficients, stabilizing at about 0.55–0.60 and 0.50–0.55, respectively. For the TiO_2-containing coatings the friction levels are lower overall: MAO 1 reaches approximately 0.46–0.50, while MAO 2 remains in the range 0.38–0.42 throughout the test. It should be noted that, at longer sliding distances, the μ value of MAO 1 slightly exceeds that of the untreated reference, which we have now stated explicitly; this reflects the combined effect of its relatively high open porosity and the formation of a stable tribolayer rather than any measurement artefact. The progressive reduction of μ from MAO 0 to MAO 2 correlates with the decrease in open surface porosity and the more compact microstructure (Figure 3), which favor better retention of the lubricating medium and reduce adhesive interactions with the Si_3N_4 counterbody. Thus, the incorporation of TiO_2 nanoparticles, particularly at 1 wt.%, results in coatings with improved tribological behavior compared to both untreated titanium and MAO 0.

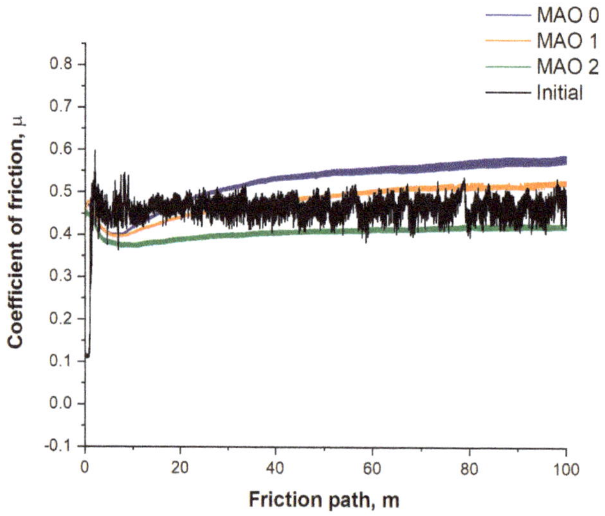

Figure 7. Tribological test results.

3.8. Wettability (Water Contact Angle)

Figure 8 shows the static water contact angles of the investigated surfaces. The untreated titanium exhibits a contact angle of 68.0°, corresponding to a weakly hydrophilic surface. After micro-arc oxidation without TiO_2 (MAO 0) the contact angle decreases to 52.0°, and further to 47.2° for MAO 1 (0.5 wt.% TiO_2), indicating a more pronounced hydrophilic character. This trend correlates with the higher surface porosity and the moderately rough topography revealed by SEM and profilometry (Table 1), which favor the capillary penetration of water and thus enhance wetting. For MAO 2 (1 wt.% TiO_2), the contact angle slightly increases to 60.3°, remaining within the hydrophilic range (<90°). The small rise compared with MAO 1 can be related to the lower open porosity (13.7% vs. 26.3%) and the denser outer layer, which reduce the capillary effect despite similar Ra values. Overall, all coatings remain hydrophilic, a surface property known to be beneficial for early cell adhesion and subsequent osseointegration of implant materials [44]. All coatings remain hydrophilic (<90°), which is favorable for early cell adhesion.

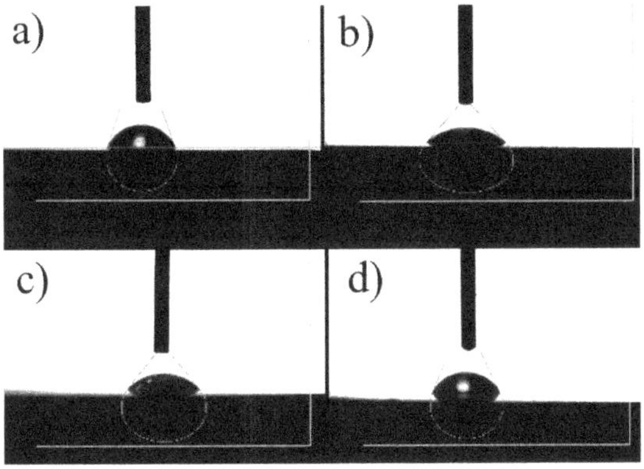

Figure 8. Water contact angle images of the tested surfaces: (**a**) initial, (**b**) MAO 0, (**c**) MAO 1, (**d**) MAO 2.

3.9. Cytotoxicity and Surface Stability in Cell Culture

In this study, dark-field light microscopy was employed to assess the microrelief of the coating surfaces before and after thermal sterilization, as well as to analyze the uniformity and structural changes induced by technological treatments. Reflective dark-field microscopy of the samples (Figure 9) revealed a pronounced microrelief typical of calcium–phosphate coatings formed by the micro-arc oxidation process, consisting of bright spherulitic structures and darker surface depressions.

After sterilization at 160 °C for 60 min (prior to contact with the liquid cell culture), no additional surface integrity changes (such as cracks or delamination of structural elements) were observed. However, when the sterile samples were placed in the cell suspension, yellowing of the culture medium was noticeable in the MAO 2 group (Figure 10). Only sample No. 25 displayed a medium color similar to that of the control group (cells without samples, K1 for comparison) and the MAO 1 test group (sample No. 35 for comparison). In addition, a white amorphous, flake-like precipitate formed at the bottom of the test tubes containing samples No. 21–24 (MAO 2 group), which was absent in the control tubes and in tubes containing samples No. 25 and No. 35. Synthetic cell culture media have their own buffering systems that protect against pH fluctuations. The yellowing observed in group 2 indicates acidification of the medium caused by chemical products released from the samples (e.g., detachment of structural coating elements or residual products of the micro-arc oxidation process that were not fully removed during sample preparation). pH fluctuations of ± 1 unit from the neutral range (7.0–7.4) indicate that the sample did not pass sanitary-chemical suitability tests for potential implant use. Therefore, samples from the MAO 2 group may subsequently demonstrate excessive cytotoxicity in the MTT assay.

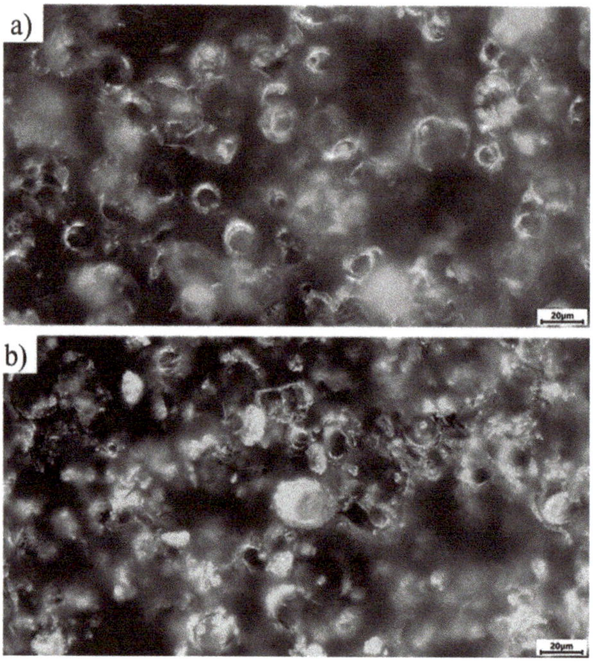

Figure 9. Microscopic appearance of the sample surfaces before dry-heat sterilization (dark-field light microscopy), magnification $\times 500$; (**a**) sample No. 31 from the MAO 1 group; (**b**) sample No. 21 from the MAO 2 group.

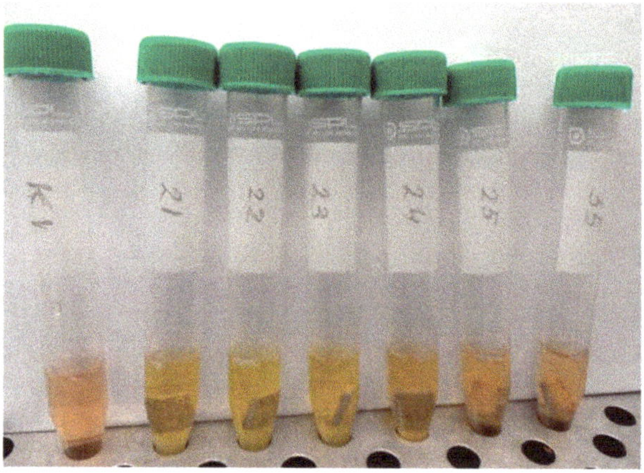

Figure 10. Change in the color of the culture medium immediately after adding samples from the MAO 2 group (Nos. 21–24).

In this context, it should be noted that after contact with the cell culture, coating defects were observed on both sides of the MAO 2 samples, particularly along the edges, which may indicate partial dissolution of the coating and precipitation of its components into the medium (Figure 10). In the MAO 1 group, this effect was much less pronounced. Edge defects of MAO coatings have been reported in the literature in cases of excessive coating dissolution and are presumably associated with "stress concentrators." The condition of a normal distribution of sample characteristics was not met ; therefore, a nonparametric test was used to compare the results between the MAO 1 and MAO 2 groups. Verification according to of the PI guidelines (Table 2) showed that the median cytotoxicity index (CI) in the MAO 2 group significantly exceeded the 30% in vitro cytotoxicity threshold recommended by ISO 10993-5, reaching 81% cell death compared to the control. In contrast, the MAO 1 group had a median CI of 3% (Table 3), indicating the non-cytotoxic nature of the samples in this group. Statistically significant differences ($p < 0.001$) in the median CI values were identified between the two groups (Table 3 and Figure 11).

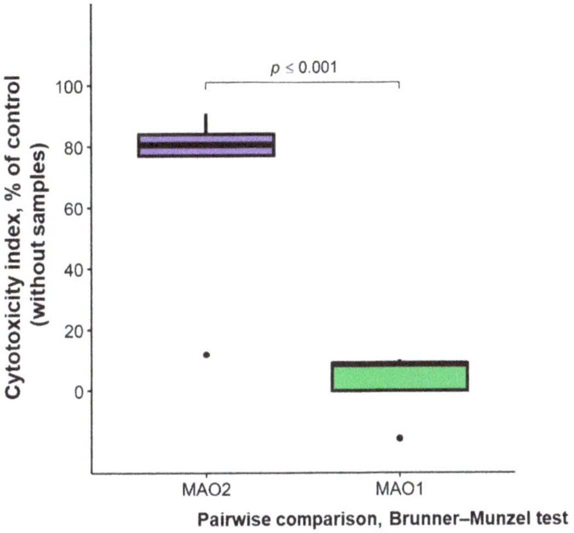

Figure 11. Graphical representation of MTT test results for in vitro cytotoxicity of different sample groups (five samples per group).

Table 2. Verification of MTT test results for the normal distribution of variables in the studied samples.

Group	Shapiro–Wilk Test with Royston's Correction	Normal Distribution Law of the Variable
MAO 1	SW = 0.94, $p = 0.13$	yes
MAO 2	SW = 0.66, $p < 0.001$	no

Table 3. Intergroup comparison of MTT test results in the studied samples.

Group	Cytotoxicity Index Values, % of Control, Me (Q1; Q3)	Pairwise Comparison, Brunner–Munzel Test
MAO 1	3 (−2; 10)	Brunner–Munzel Test Statistic = 17.63 $p < 0.001$
MAO 2	81 (76; 84)	

Thus, the MAO 2 group samples exhibit a significant cytotoxic effect (>30% relative to control) on the in vitro culture of HOS human osteosarcoma cells under direct short-term contact for 24 h. The results presented in Tables 4 and 5 indicate that the cytotoxic effect in the MAO 1 group varies around zero. Notably, sample No. 32 demonstrates statistically significant differences from the other samples in this group, showing a cytoprotective effect (enhanced cell viability).

Table 4. Intragroup comparison of MTT test results among samples within the MAO 1 group.

Group	n	Cytotoxicity Index Values, % of Control (X ± SD)	Multiple Comparison, Welch's ANOVA	Pairwise Comparison, Games–Howell Test
Sample 31	5	10 ± 7	F = 7.21 p = 0.0054	p_{31-32} = 0.0067 p_{31-33} = 0.99 p_{31-34} = 0.28 p_{31-35} = 0.99 p_{32-33} = 0.0091 p_{32-34} = 0.099 p_{32-35} = 0.0095 p_{33-34} = 0.39 p_{33-35} = 0.99 p_{34-35} = 0.41
Sample 32	5	−15.80 ± 8.82		
Sample 33	5	8.80 ± 7.19		
Sample 34	5	−0.2 ± 7.82		
Sample 35	5	8.6 ± 7.16		

n—sample size (number of measurement replicates); the intragroup distribution of the variable was consistent with normality according to the Shapiro–Wilk test with Royston's correction.

Table 5. Intragroup comparison of MTT test results among samples within the MAO 2 group.

Group	n	Cytotoxicity Index Values, % of Control (X ± SD)	Multiple Comparison, Welch's ANOVA	Pairwise Comparison, Games–Howell Test
Sample 21	5	80.60 ± 1.67	F = 201.47 p < 0.001	p_{21-22} < 0.001 p_{21-23} = 0.042 p_{21-24} = 0.071 p_{21-25} < 0.001 p_{22-23} < 0.001 p_{22-24} < 0.001 p_{22-25} < 0.001 p_{23-24} = 0.0014 p_{23-25} < 0.001 p_{24-25} < 0.001
Sample 22	5	90.80 ± 0.84		
Sample 23	5	84 ± 1.23		
Sample 24	5	77 ± 1.87		
Sample 25	5	11.80 ± 6.61		

n—sample size (number of measurement replicates); the intragroup distribution of the variable was consistent with normality according to the Shapiro–Wilk test with Royston's correction.

In turn, with the exception of sample No. 25 (CI = 11.8%), the remaining four samples showed CI values in the range of 77.0%–90.8% (Table 5), which is 2.5–3 times higher than the threshold level recommended by ISO 10993-5 (not exceeding 30%).

The significant intragroup differences in CI values observed among the MAO 2 samples may indicate variations in the structure and composition of the coatings, which are likely related to the visual non-uniformity of the coatings across the sample surfaces, as revealed by macroscopic imaging before and after contact with the cell culture, as well as partial detachment of the coatings into the cell suspension. For reader clarity, Figure 12 shows front and back views of both sample groups before and after cell contact and staining (MAO 1—Nos. 31-35; MAO 2—Nos. 21-25). Prior to exposure (Figure 12a,b), both groups exhibit uniform, intact surfaces. After 24 h in the HOS suspension and subsequent Azure II/eosin staining (Figure 12c,d), the MAO 1 samples retain a more uniform appearance with scattered stained spots corresponding to adhered cells, whereas the MAO 2 samples display more pronounced color heterogeneity and local surface changes, consistent with the higher cytotoxicity and partial coating instability reported below.

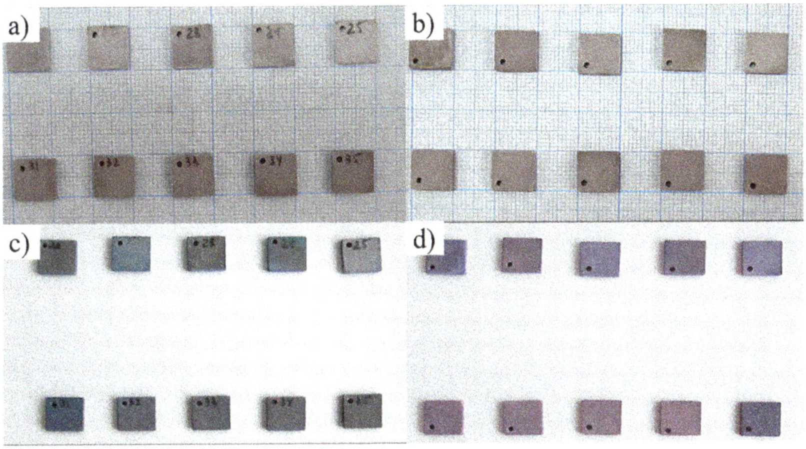

Figure 12. Macroscopic appearance of the samples before and after 24 h contact with the HOS cell suspension followed by Azure II/eosin staining. (**a**) Front view, before exposure; (**b**) back view, before exposure; (**c**) front view, after exposure and staining; (**d**) back view, after exposure and staining. Sample groups: MAO 1—Nos. 31–35; MAO 2—Nos. 21–25 (engraved IDs are visible on each tile). Images in (**a**,**b**) were taken on graph paper for alignment; images in (**c**,**d**) were taken on a white background to visualize staining contrast.

Microscopic imaging of the sample surfaces after Azure II/eosin staining was used to visualize HOS cells adhered to the coatings (Figure 13). Because of the irregular micro-relief and variable depth of the MAO surface it was not possible to obtain sharply focused micrographs at all sites, even at low magnification (×200). Nevertheless, representative low-magnification images of the MAO 1 group (Figure 13a) show multiple scattered stained areas corresponding in size to individual cells. Examination of several fields of view at higher magnification (×500) confirmed the presence of cell nuclei-like structures in these stained regions (Figure 13b). For the MAO 2 group (Figure 13c), stained elements were also found, but in repeated fields the overall number of attached cells was lower than on MAO 1, even though the single representative image shows two stained spots. The lower apparent count in the selected micrographs should therefore be interpreted qualitatively; the conclusion of higher cell adhesion on MAO 1 is based on observation of multiple areas rather than only the frames shown.

Figure 14 presents the SEM results of the samples after cytotoxicity testing. For the MAO 1 group samples (Figure 14a,b), the surface largely retains its original coating structure with minimal signs of degradation. The morphology remains relatively stable, with individual areas showing adhered cellular elements, confirming the low cytotoxicity of the coating and its ability to support cell adhesion. No significant dissolution or delamination of the coating was observed. In contrast, the MAO 2 group samples (Figure 14c,d) exhibit more pronounced coating degradation, including localized dissolution, the appearance of microcracks, and areas where structural elements have delaminated. Some regions display residual decomposition products of the coating, likely released into the culture medium, correlating with the color change observed at the early stages of the experiment (Figure 10). Such surface alterations indicate coating instability and potential release of particles or ions that may cause toxic effects on cells [45–48].

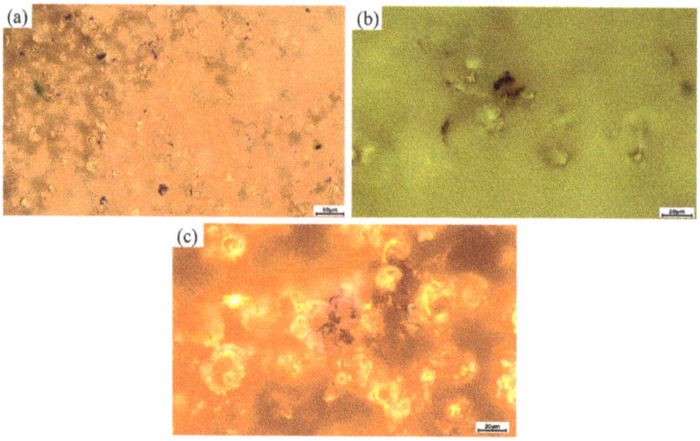

Figure 13. Microscopic appearance of sample surfaces after 24 h contact with HOS cells and Azure II/eosin staining. (**a**) MAO 1 No. 33, dark-field, ×200: numerous stained regions corresponding in size to individual cells are visible across the surface; (**b**) MAO 1, bright-field, ×500: single stained element showing a cell-nucleus-like structure; (**c**) MAO 2 No. 22, dark-field, ×500: representative view with two stained elements.

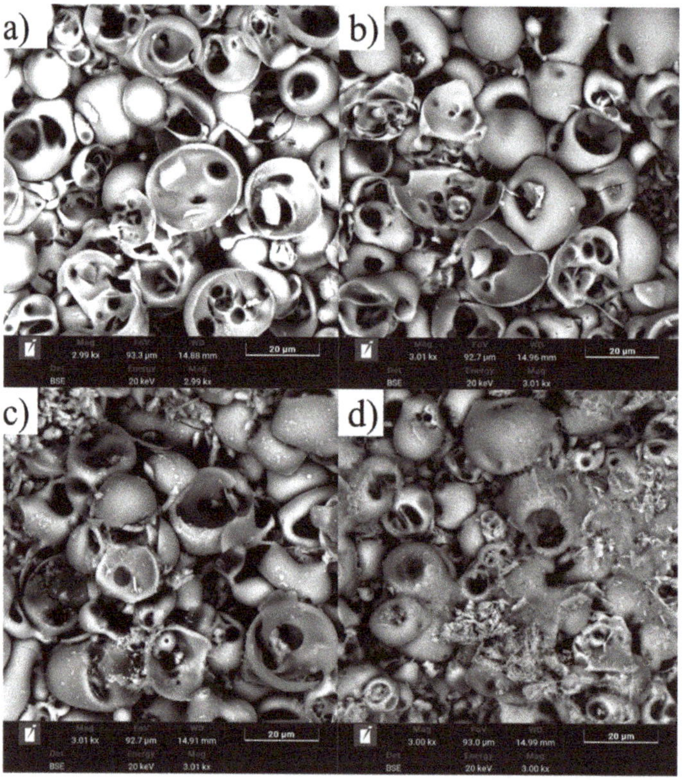

Figure 14. SEM micrographs of MAO coatings after cytotoxicity testing: (**a**,**b**) MAO 1 sample No. 32, (**c**,**d**) MAO 2 sample No. 22.

EDS elemental mapping and spectra (Figure 15) confirm the presence of titanium, calcium, phosphorus, oxygen, and trace amounts of carbon in both coatings. In the MAO 1 sample (non-cytotoxic coating), the distribution of calcium and phosphorus remains relatively uniform across the surface, and the overall composition shows minimal changes after contact with the cell culture. This indicates good chemical stability of the coating and low release of degradation products into the surrounding medium, correlating with the absence of cytotoxic effects. In contrast, the MAO 2 sample exhibits decreased phosphorus and calcium signals accompanied by an increased titanium and oxygen contribution, suggesting partial dissolution of the calcium–phosphate layer and exposure of the underlying titanium oxide phase. These compositional changes are consistent with the previously observed yellowing of the culture medium and the higher cytotoxicity index (>30%). Thus, coatings formed with 0.5% TiO_2 nanoparticles (MAO 1) demonstrate better chemical stability and biocompatibility compared to coatings formed with 1% TiO_2 (MAO 2).

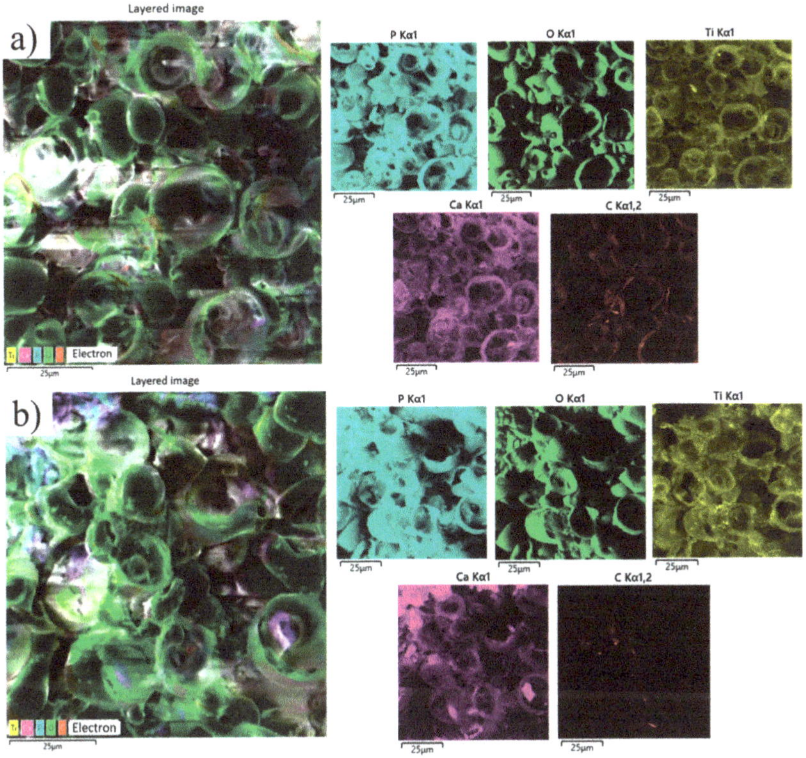

Figure 15. EDS analysis of MAO coatings after cytotoxicity testing: (**a**) MAO 1; (**b**) MAO 2.

Table 6 summarizes the elemental composition of the coatings obtained from EDS mapping. Both coatings consist primarily of oxygen, phosphorus, and calcium, confirming the formation of a calcium–phosphate–titania layer. The titanium content is about 10.17 wt.% in MAO 1 and 7.24 wt.% in MAO 2, indicating that TiO_2 nanoparticles are successfully incorporated into the oxide layer. The slightly lower Ti content at 1 wt.% TiO_2 addition is consistent with the thicker Ca–P rich outer layer and agrees with the XRD identification of anatase and rutile phases. These EDS results, together with the XRD and SEM analyses, confirm the presence and stable incorporation of TiO_2 nanoparticles in the MAO coatings and address the reviewers' request for quantitative elemental data.

Table 6. Elemental composition (EDS, weight %) of the MAO coatings.

Element MAO 1	Weight, %	Element MAO 2	Weight, %
O	38.24	O	41.72
P	30.12	P	23.68
Ti	10.17	Ti	7.24
Ca	15.31	Ca	20.18
C	6.16	C	7.18

The observed differences between coatings produced with 0.5 and 1 wt.% TiO_2 nanoparticles can be explained by the influence of nanoparticle concentration on the micro-arc discharge behavior and the kinetics of oxide growth. TiO_2 nanoparticles in the electrolyte act as additional charge carriers and heterogeneous nucleation sites, which modify the breakdown voltage and increase the density and energy of micro-discharges [49]. At a higher nanoparticle content (1 wt.%), the higher discharge energy promotes more intense local melting and re-solidification of the oxide, leading to partial closure of discharge channels and formation of a denser, more crystalline TiO_2-containing layer. Conversely, at 0.5 wt.%, the lower concentration of particles results in a more open, highly porous structure with larger areal density of discharge craters. These microstructural changes directly affect the corrosion behavior, tribological performance, and cytotoxicity of the coatings.

4. Conclusions

In this study, a comprehensive evaluation was conducted on the effect of TiO_2 nanoparticle concentration in the micro-arc oxidation (MAO) electrolyte on the morphology, phase composition, corrosion resistance, and biological safety of calcium–phosphate coatings on titanium. The results demonstrated that increasing the TiO_2 content from 0.5% to 1% led to coating densification, reduced porosity, and improved corrosion resistance; however, it was also associated with decreased biocompatibility and increased in vitro cytotoxicity. SEM and EDS analyses after the cytotoxicity test revealed partial degradation and leaching of the coating in the MAO 2 group, which correlated with higher cell death rates and pH changes in the culture medium. In contrast, MAO 1 coatings maintained stable morphology, preserved their elemental composition, and exhibited no toxic effects on cells. Therefore, to enhance the biological safety of implantable devices, MAO coatings with an optimized TiO_2 nanoparticle concentration ($\leq$0.5%) are preferable, providing a balance between corrosion protection and biocompatibility. These findings may serve as a basis for further optimization of electrolyte composition and MAO processing parameters in the development of bioinert and bioactive implant materials. A limitation of the present study is that cytotoxicity was not assessed for the TiO_2-free MAO 0 coating, since the biological experiments were focused on elucidating the specific influence of TiO_2 nanoparticles.

Author Contributions: Conceptualization, A.Z. and B.R.; methodology, D.B.; software, D.K.; validation, D.B., N.M., and B.R.; formal analysis, D.K.; investigation, A.Z.; resources, D.B.; data curation, N.M.; writing—original draft preparation, A.Z.; writing—review and editing, A.K.; visualization, D.B.; supervision, A.K.; project administration, A.K.; funding acquisition, B.R. All authors have read and agreed to the published version of the manuscript.

Funding: This research has been funded by the Committee of Science of the Ministry of Science and Higher Education of the Republic of Kazakhstan (Grant No. BR24992862).

Institutional Review Board Statement: Not applicable.

Informed Consent Statement: Informed consent was obtained from all subjects involved in this study.

Data Availability Statement: Data are available from the authors upon reasonable request.

Conflicts of Interest: Author Bauyrzhan Rakhadilov, Aidar Kengesbekov and Dauir Kakimzhanov were employed by the company PlasmaScience LLP. The remaining authors declare that the research was conducted in the absence of any commercial or financial relationships that could be construed as a potential conflict of interest.

References

1. Akshaya, S.; Rowlo, P.K.; Dukle, A.; Nathanael, A.J. Antibacterial coatings for titanium implants: Recent trends and future perspectives. *Antibiotics* **2022**, *11*, 1719. [CrossRef] [PubMed]
2. Besisa, N.H.A.; Yajima, T. Titanium-Based Alloys: Classification and Diverse. *Titan.-Based Alloys-Charact. Appl.* **2024**, *125*.
3. Spriano, S.; Yamaguchi, S.; Baino, F.; Ferraris, S. A critical review of multifunctional titanium surfaces: New frontiers for improving osseointegration and host response, avoiding bacteria contamination. *Acta Biomater.* **2018**, *79*, 1–22. [CrossRef]
4. Liu, Z.; Liu, X.; Ramakrishna, S. Surface engineering of biomaterials in orthopedic and dental implants: Strategies to improve osteointegration, bacteriostatic and bactericidal activities. *Biotechnol. J.* **2021**, *16*, 2000116. [CrossRef]
5. Kengesbekov, A.B.; Serikbaykyzy, A.; Bayandinova, M.B.; Batanov, E.E.; Bazarov, N.E.; Askhatov, A.N. Development of biocompatible coatings for orthopedic joint implants. *Phys. Sci. Technol.* **2025**, *12*, 84–94. [CrossRef]
6. Ben Arbia, M.; Helal, H.; Comini, E. Recent advances in low-dimensional metal oxides via sol-gel method for gas detection. *Nanomaterials* **2024**, *14*, 359. [CrossRef]
7. Prasad, K.N.; Syed, I.; Subbu, S.K. Laser dimple texturing–applications, process, challenges, and recent developments: A review. *Aust. J. Mech. Eng.* **2022**, *20*, 316–331. [CrossRef]
8. Roccaforte, F.; Giannazzo, F.; Greco, G. Ion implantation doping in silicon carbide and gallium nitride electronic devices. *Micro* **2022**, *2*, 23–53. [CrossRef]
9. Xie, J.; Zhang, C.; Waite, T.D. Hydroxyl radicals in anodic oxidation systems: Generation, identification and quantification. *Water Res.* **2022**, *217*, 118425. [CrossRef] [PubMed]
10. Xue, T.; Attarilar, S.; Liu, S.; Liu, J.; Song, X.; Li, L.; Tang, Y. Surface modification techniques of titanium and its alloys to functionally optimize their biomedical properties: Thematic review. *Front. Bioeng. Biotechnol.* **2020**, *8*, 603072. [CrossRef] [PubMed]
11. Alontseva, D.; Safarova, Y.; Voinarovych, S.; Obrosov, A.; Yamanoglu, R.; Khoshnaw, F.; Weiß, S. Biocompatibility and corrosion of microplasma-sprayed titanium and tantalum coatings versus titanium alloy. *Coatings* **2024**, *14*, 206. [CrossRef]
12. Zhang, Z.Y.; Huang, T.Y.; Zhai, D.J.; Wang, H.B.; Feng, K.Q.; Xiang, L. Study on strontium doped bioactive coatings on titanium alloys surfaces by micro-arc oxidation. *Surf. Coat. Technol.* **2022**, *451*, 129045. [CrossRef]
13. Wen, X.; Liu, Y.; Xi, F.; Zhang, X.; Kang, Y. Micro-arc oxidation (MAO) and its potential for improving the performance of titanium implants in biomedical applications. *Front. Bioeng. Biotechnol.* **2023**, *11*, 1282590. [CrossRef]
14. Wang, R.; Ni, S.; Ma, L.; Li, M. Porous construction and surface modification of titanium-based materials for osteogenesis: A review. *Front. Bioeng. Biotechnol.* **2022**, *10*, 973297. [CrossRef]
15. Manisekaran, R.; Chettiar, A.D.R.; Marasamy, L.; Arthikala, M.K.; Kandasamy, G.; Ibarra, V.C.; Rathore, H.S. Copper, Zinc, and Titanium-Based Semiconductor Nanomaterials for Antimicrobial Coatings and Their Mechanisms. *Nano Select* **2025**, *6*, e202400155. [CrossRef]
16. Rezić, I.; Meštrović, E. Characterization of Nanoparticles in Antimicrobial Coatings for Medical Applications—A Review. *Coatings* **2023**, *13*, 1830. [CrossRef]
17. Parham, S.; Wicaksono, D.H.; Bagherbaigi, S.; Lee, S.L.; Nur, H. Antimicrobial treatment of different metal oxide nanoparticles: A critical review. *J. Chin. Chem. Soc.* **2016**, *63*, 385–393. [CrossRef]
18. Vera, M.L.; Traid, H.D.; Dwojak, A.N.; Rosenberger, M.R.; Schvezov, C.E. Advances in nanostructured TiO_2 coatings for industrial applications. In *Industrial Applications of Nanoparticles*; CRC Press: Boca Raton, FL, USA, 2023; pp. 228–255. [CrossRef]
19. Rakhadilov, B.; Zhassulan, A.; Ormanbekov, K.; Shynarbek, A.; Baizhan, D.; Aldabergenova, T. Surface Modification and Tribological Performance of Calcium Phosphate Coatings with TiO2 Nanoparticles on VT1-0 Titanium by Micro-Arc Oxidation. *Crystals* **2024**, *14*, 945. [CrossRef]
20. Kengesbekov, A.; Sagdoldina, Z.; Torebek, K.; Baizhan, D.; Kambarov, Y.; Yermolenko, M.; Maulet, M. Synthesis and Formation Mechanism of Metal Oxide Compounds. *Coatings* **2022**, *12*, 1511. [CrossRef]
21. Abbasi-Oshaghi, E.; Mirzaei, F.; Pourjafar, M. NLRP3 inflammasome, oxidative stress, and apoptosis induced in the intestine and liver of rats treated with titanium dioxide nanoparticles: In vivo and in vitro study. *Int. J. Nanomed.* **2019**, *14*, 1919–1936. [CrossRef] [PubMed]
22. Kumar, V.; Sharma, N.; Maitra, S.S. In vitro and in vivo toxicity assessment of nanoparticles. *Int. Nano Lett.* **2017**, *7*, 243–256 [CrossRef]
23. Ajdary, M.; Moosavi, M.A.; Rahmati, M.; Falahati, M.; Mahboubi, M.; Mandegary, A.; Varma, R.S. Health concerns of various nanoparticles: A review of their in vitro and in vivo toxicity. *Nanomaterials* **2018**, *8*, 634. [CrossRef]

24. Salama, B.; Alzahrani, K.J.; Alghamdi, K.S.; Al-Amer, O.; Hassan, K.E.; Elhefny, M.A.; Fathalla, A.S. Silver nanoparticles enhance oxidative stress, inflammation, and apoptosis in liver and kidney tissues: Potential protective role of thymoquinone. *Biol. Trace Elem. Res.* **2023**, *201*, 2942–2954. [CrossRef]
25. Zhang, X.; Zhou, W.; Xi, W. Advancements in incorporating metal ions onto the surface of biomedical titanium and its alloys via micro-arc oxidation: A research review. *Front. Chem.* **2024**, *12*, 1353950. [CrossRef] [PubMed]
26. Xu, N.; Fu, J.; Zhao, L.; Chu, P.K.; Huo, K. Biofunctional elements incorporated nano/microstructured coatings on titanium implants with enhanced osteogenic and antibacterial performance. *Adv. Healthc. Mater.* **2020**, *9*, 2000681. [CrossRef] [PubMed]
27. Ming, X.; Wu, Y.; Zhang, Z.; Li, Y. Micro-arc oxidation in titanium and its alloys: Development and potential of implants. *Coatings* **2023**, *13*, 2064. [CrossRef]
28. Rakhadilov, B.; Zhassulan, A.; Baizhan, D.; Shynarbek, A.; Ormanbekov, K.; Aldabergenova, T. The effect of the electrolyte composition on the microstructure and properties of coatings formed on a titanium substrate by microarc oxidation. *AIMS Mater. Sci.* **2024**, *11*, 547–564. [CrossRef]
29. Bayatanova, L.B.; Zhassulankyzy, A.Z.; Magazov, N.M.; Rakhadilov, B.K.; Muktanova, N.; Uazyrkhanova, G.K. Effect of plasma-electrolytic oxidation on mechanical properties of titanium coatings. *Bull. Karaganda Univ. Phys. Ser.* **2023**, *111*, 65–74. [CrossRef]
30. Kashin, A.D.; Sedelnikova, M.B.; Uvarkin, P.V.; Ugodchikova, A.V.; Luginin, N.A.; Sharkeev, Y.P.; Bakina, O.V. Functionalizing diatomite-based micro-arc coatings for orthopedic implants: Influence of TiO_2 addition. *Biomimetics* **2023**, *8*, 280. [CrossRef]
31. Makurat-Kasprolewicz, B.; Wekwejt, M.; Ronowska, A.; Gajowiec, G.; Grodzicka, M.; Dzionk, S.; Ossowska, A. Influence of Ultrasound on the characteristics of CaP Coatings Generated Via the micro-arc oxidation process in relation to Biomedical Engineering. *ACS Biomater. Sci. Eng.* **2024**, *10*, 2100–2115. [CrossRef]
32. *ISO 10993-5*; Biological Evaluation of Medical Devices—Part 5: Tests for In Vitro Cytotoxicity. International Organization for Standardization: Geneva, Switzerland, 2009.
33. Wang, P.; Wei, X.W.; Cao, W.J.; Tang, Y.T.; Wang, Y.; Gong, Z.Y.; Zu, X.T. Effect of TiO_2 nanoparticles on the characteristics of MAO coatings. *Int. J. Electrochem. Sci.* **2019**, *14*, 9311–9325. [CrossRef]
34. Woo, S.R.; Sung, Y.M. Enhanced photoelectrochemical water splitting of micro-arc oxidized TiO_2 via anatase/rutile phase control and nitrogen doping. *J. Electrochem. Soc.* **2016**, *163*, H278. [CrossRef]
35. Yerokhin, A.L.; Nie, X.; Leyland, A.; Matthews, A. Characterisation of oxide films produced by plasma electrolytic oxidation of a Ti–6Al–4V alloy. *Surf. Coat. Technol.* **2000**, *130*, 195–206. [CrossRef]
36. Ceriani, F.; Casanova, L.; Massimini, L.; Brenna, A.; Ormellese, M. TiO_2 microparticles incorporation in coatings produced by plasma electrolytic oxidation (PEO) on Titanium. *Coatings* **2023**, *13*, 1718. [CrossRef]
37. Erfanifar, E.; Aliofkhazraei, M.; Nabavi, H.F.; Rouhaghdam, A.S. Growth kinetics and morphology of microarc oxidation coating on titanium. *Surf. Coat. Technol.* **2017**, *315*, 567–576. [CrossRef]
38. Zhang, P.; Zuo, Y.; Nie, G. The pore structure and properties of microarc oxidation films on 2024 aluminum alloy prepared in electrolytes with oxide nanoparticles. *J. Alloys Compd.* **2020**, *816*, 152520. [CrossRef]
39. Jin, F.; Chu, P.K.; Tong, H.; Zhao, J. Improvement of surface porosity and properties of alumina films by incorporation of Fe micrograins in micro-arc oxidation. *Appl. Surf. Sci.* **2006**, *253*, 863–868. [CrossRef]
40. Li, Z.; Cai, Z.B.; Cui, X.J.; Liu, R.R.; Yang, Z.B.; Zhu, M.H. Influence of nanoparticle additions on structure and fretting corrosion behavior of micro-arc oxidation coatings on zirconium alloy. *Surf. Coat. Technol.* **2021**, *410*, 126949. [CrossRef]
41. Kumar, K.; Bhadauria, S.S.; Singh, A.P. Effect of post-coating heat treatment on corrosion and stress corrosion behaviors of NiCr/TiO_2-coated 316L stainless steel. *Arab. J. Sci. Eng.* **2023**, *48*, 3893–3908. [CrossRef]
42. Chen, X.W.; Song, H.; Pu, H.; Zheng, R.; Zhang, M.; Zhang, D. Study of wear and corrosion resistance of ATO nanoparticle–doped micro-arc oxide film layers. *Int. J. Appl. Ceram. Technol.* **2024**, *21*, 1078–1093. [CrossRef]
43. Steel, J.P. The Treatment of Certain Mental Diseases by Ringer-Locke Solution. *Br. Med. J.* **1927**, *2*, 1177. [CrossRef]
44. Zhu, G.; Wang, G.; Li, J.J. Advances in implant surface modifications to improve osseointegration. *Mater. Adv.* **2021**, *2*, 6901–6927. [CrossRef]
45. Suthar, J.K.; Rakesh, B.; Vaidya, A.; Ravindran, S. Comprehensive Analysis of Titanium Oxide Nanoparticle Size and Surface Properties on Neuronal PC-12 Cells: Unraveling Cytotoxicity, Dopaminergic Gene Expression, and Acetylcholinesterase Inhibition. *J. Xenobiotics* **2023**, *13*, 662–684. [CrossRef]
46. Line, F.C. Titanium Dioxide Nanoparticle (TiO2 NP) Induces Toxic Effects on LA-9 Mouse Fibroblast Cell Line. *Cell. Physiol. Biochem.* **2023**, *57*, 63–81. [CrossRef]
47. Mohammadalipour, Z.; Rahmati, M.; Khataee, A.; Moosavi, M.A. Different concentrations of titanium dioxide nanoparticles induce autophagy followed by growth inhibition or cell death in A375 melanoma cells. *J. Skin Stem Cell* **2017**, *4*, e63994. [CrossRef]

48. Jin, C.Y.; Zhu, B.S.; Wang, X.F.; Lu, Q.H. Cytotoxicity of titanium dioxide nanoparticles in mouse fibroblast cells. *Chem. Res. Toxicol.* **2008**, *21*, 1871–1877. [CrossRef]
49. Mamaeva, A.; Kenzhegulov, A.; Panichkin, A.; Abdulvaliyev, R.; Fischer, D.; Bakhytuly, N.; Toiynbaeva, N. Influence of Current Duty Cycle and Voltage of Micro-Arc Oxidation on the Microstructure and Composition of Calcium Phosphate Coating. *Coatings* **2024**, *14*, 766. [CrossRef]

MDPI AG
Grosspeteranlage 5
4052 Basel
Switzerland
Tel.: +41 61 683 77 34

Coatings Editorial Office
E-mail: coatings@mdpi.com
www.mdpi.com/journal/coatings

www.ingramcontent.com/pod-product-compliance
Lightning Source LLC
LaVergne TN
LVHW070540071126
851858LV00011BA/230
* 9 7 8 3 7 2 5 8 8 2 8 7 8 *